ON THE ORIGINS
OF CHLOROPLASTS

ON THE ORIGINS
OF CHLOROPLASTS

Edited by

JEROME A. SCHIFF

with the assistance of

HARVARD LYMAN

ELSEVIER / NORTH-HOLLAND
New York • Amsterdam • Oxford

QK
725
.O5

Elsevier North Holland, Inc.
52 Vanderbilt Avenue, New York, New York 10017

Sole distributors outside the USA and Canada:
Elsevier Science Publishers B.V.
P.O. Box 211, 1000 AE Amsterdam, The Netherlands   # 7837651

Library of Congress Cataloging in Publication Data

Main entry under title:

On the origins of chloroplasts.
    "Proceedings of the Conference on the Origins of Chloroplasts held at the Fogarty
International Center, National Institutes of Health . . . October third to fifth, 1979,
under the sponsorship of the Science and Education Administration, USDA, the
National Institute of Allergy and Infectious Diseases, NIH, and the Fogarty
International Center, NIH"—Acknowledgments
    Includes bibliographical references and index.
    1. Chloroplasts—Origin—Congresses. I. Schiff, Jerome A. II. Lyman,
Harvard. III. Conference on the Origins of Chloroplasts (1979 : Fogarty
International Center, National Institutes of Health) IV. United States. Science and
Education Administration. V. National Institute of Allergy and Infectious Diseases
(U.S.) VI. John E. Fogarty International Center for Advanced Study in the Health
Sciences. [DNLM: 1. Chloroplasts—Congresses. QK 898.C5 O58 1979]
QK725.O5   581.87'33   81-15217
ISBN 0-444-00669-9   AACR2

Manufactured in the United States of America

On the Origins of Chloroplasts
was organized and inspired by
Roger Y. Stanier
and is dedicated to him by his colleagues

"The restriction of major variations on the theme of photosynthesis to prokaryotes suggests that this metabolic process first arose and evolved in the context of the prokaryotic cell, and that its ultimate version was subsequently transferred to certain eukaryotic cell lines."

Roger Y. Stanier

For Roger Yate Stanier, who has been to all of his colleagues everything a great teacher should be: wise, understanding, skeptical, brilliant, inspiring, and original; who, by taking us along with him through the various stages of his understanding, has led us through exciting discoveries to reach beautiful insights of wide and general application.

Contents

Acknowledgments

The contents of this volume are taken from the proceedings of the Conference on the Origins of Chloroplasts held at the Fogarty International Center, National Institutes of Health, in Bethesda, Maryland, from October third to fifth, 1979, under the sponsorship of the Science and Education Administration, USDA, the National Institute of Allergy and Infectious Diseases, NIH, and the Fogarty International Center, NIH. Our thanks are due to the various people at the Fogarty International Center whose efforts resulted in a highly productive and enjoyable conference. These people include: Dr. Earl C. Chamberlayne, Chief of the Conference and Seminar Program Branch; Mrs. Michiko M. Cooper, Conference Coordinator; Mrs. Toby Levin, Conference Management Assistant; and Mrs. Janice Wahlmann, Conference Assistant. Thanks are also due to Dr. Edwin D. Becker, Acting Director, whose welcoming remarks were greatly appreciated. Without the interest and help of Dr. Peter G. Condliffe, Chief, Scholars and Fellowships Program Branch, this meeting would not have taken place.

List of Participants
Conference on the Origins of Chloroplasts

Dr. Fred Abeles
Associate Program Manager, USDA, Competitive Research Grants Office, 1300 Wilson Boulevard, Arlington, Virginia 22209

Dr. Essica Barnabas
Professor in Biology, Building 31, Room 4B03, National Institutes of Health, Bethesda, Maryland 20205

Dr. W. Edgar Barnett
Director, Oak Ridge Graduate School of Biomedical Science, Biology Division, Oak Ridge National Laboratory, Oak Ridge, Tennessee 37830

Dr. James A. Bassham
Senior Staff Scientist, Building 3—Chemical Biodynamics, Lawrence Berkeley Laboratory, 1 Cyclotron Road, Berkeley, California 94720

Dr. Pierre Bennoun
Charge de Recherche au C.N.R.S., Institut de Biologie Physico-Chimique, 13, rue Pierre et Marie Curie, 75005 Paris, France

Dr. Lawrence Bogorad
Professor of Biology, Harvard University, Biological Laboratories, 16 Divinity Avenue, Cambridge, Massachusetts 02138

Dr. Ora Canaani
Postdoctoral Fellow, Smithsonian Institute, Radiation Biology Laboratory, Parklawn Drive, Rockville, Maryland 20852

Dr. Paul A. Castelfranco
Professor of Botany, Department of Botany, University of California, Davis, California 95616

Dr. Earl C. Chamberlayne
Chief, Conference and Seminar Program Branch, Fogarty International Center, National Institutes of Health, Building 31, Room 2C15, Bethesda, Maryland 20205

Dr. Robert Chasson
Executive Secretary, Review Branch, National Heart, Lung, and Blood Institute, National Institutes of Health, Westwood Building, Room 550, Bethesda, Maryland 20205

Dr. Mary E. Clutter
Program Director, Developmental Biology, National Science Foundation, 940 25th Street, NW/A 615-S, Washington D.C. 20037

Dr. Seymour S. Cohen
Distinguished Professor, Department of Pharmacological Sciences, State University of New York, Stony Brook, Stony Brook, New York 11794

Dr. Peter Condliffe
Scholarships and Fellowship Program, Fogarty International Center, National Institutes of Health, Bethesda, Maryland 20014

Dr. Anne Datko
Botanist, National Institute of Mental Health, National Institutes of Health, Building 32A, Room 101, Bethesda, Maryland 20205

Dr. C. F. D'Elia
Assistant Professor, Ches. Biological Laboratory, University of Maryland, Solomons, Maryland 20088

Dr. Burt Endo
Plant Pathologist, Plant Protection Institute, Beltsville Agricultural Research Center, Beltsville, Maryland 20705

Dr. Estella K. Engel
Associate Program Director, Biochemistry Program, National Science Foundation, 1800 G Street, Northwest, Washington, D.C. 20550

Dr. David Filer
National Institute of Arthritis, Metabolism and Digestive Diseases, National Institutes of Health, Building 4, Room 105, Bethesda, Maryland 20205

Dr. Anthony Furano
National Institute of Arthritis, Metabolism and Digestive Diseases, National Institutes of Health, Building 4, Room 104, Bethesda, Maryland 20205

Dr. Elisabeth Gantt
Senior Research Biologist, Radiation Biology Laboratory, Smithsonian Institution, 12441 Parklawn Drive, Rockville, Maryland 20852

Dr. John Giovanelli
National Institute of Mental Health, National Institutes of Health, Building 32A, Room 101, Bethesda, Maryland 20205

Dr. Alexander N. Glazer
Professor of Bacteriology, Department of Bacteriology and Immunology, University of California, Berkeley, California 94720

A. Guranowski
National Institute of Mental Health, National Institutes of Health, Building 36, Room 3A19,
Bethesda, Maryland 20205

Dr. Richard B. Hallick
Associate Professor, Department of Chemistry, University of Colorado, Boulder, Colorado 80309

Dr. Robert W. Hartley
National Institute of Arthritis, Metabolism and Digestive Diseases, National Institutes of Health,
Building 6, Room B1-11, Bethesda, Maryland 20205

Dr. R. Wayne Hendren
Senior Biochemist, Chemistry and Life Sciences Division, Research Triangle Institute,
P.O. Box 12194, Research Triangle Park, North Carolina 27709

Dr. Pierre Joliot
Directeur de Recherche au C.N.R.S., Institut de Biologie Physico-Chimique, 13, rue Pierre et
Marie Curie, 75005 Paris, France

Dr. Martin D. Kamen
Professor Emeritus, Department of Chemistry, Box A-002, University of California, San Diego,
La Jolla, California 92093

Dr. Ellis Kempner
National Institute of Arthritis, Metabolism and Digestive Diseases, National Institutes of Health,
Building 6, Room 118, Bethesda, Maryland 20205

Dr. Shimon Klein
Professor of Botany, Department of Botany, The Hebrew University, Jerusalem, Israel

Dr. Paul Kolenbrander
Senior Staff Fellow, National Institute of Dental Research, National Institutes of Health,
Building 30, Room 310, Bethesda, Maryland 20205

Dr. David W. Krogmann
Professor, Department of Biochemistry, Purdue University, Lafayette, Indiana 47907

Dr. Peter John Lea
Principal Scientific Officer, Department of Biochemistry, Rothamsted Experimental Station,
Harpenden, Herts., United Kingdom

Claudia Lipschultz
Smithsonian Radiation Biology Laboratory, 12441 Parklawn Drive, Rockville, Maryland 20852

Dr. Jack London
Research Microbiologist, National Institute of Dental Research, National Institutes of Health,
Building 30, Room 312, Bethesda, Maryland 20205

Dr. Harvard Lyman
Associate Professor of Biology, Biology Department, State University of New York, Stony Brook,
Stony Brook, New York 11794

Dr. Ronald Magnusson
Postdoctoral Fellow, National Institutes of Health, Building 4, Room B1-31, Bethesda,
Maryland 20205

Dr. Maurice M. Margulies
Chemist, Smithsonian Institution, Radiation Biology Laboratory, 12441 Parklawn Drive, Rockville, Maryland 20852

Dr. J. M. Meyer
Visiting Fellow, National Institutes of Health, Building 3, Room 122, Bethesda, Maryland 20205

Dr. Harvey Mudd
National Institute of Mental Health, National Institutes of Health, Building 32A, Bethesda, Maryland 20205

Dr. J. B. Mudd
Professor, Department of Chemistry, University of California, Riverside, California 92521

Dr. Kathleen Mullinix
National Cancer Institute, National Institutes of Health, Building 37, Room 4A17, Bethesda, Maryland 20205

Dr. Robert G. E. Murray
Department of Bacteriology and Immunology, University of Western Ontario, London, Ontario, Canada NGA 3K7 01

Dr. Leonard Muscatine
Professor of Biology, Department of Biology, University of California, Los Angeles, California 90024

Dr. Itzhak Ohad
Professor of Biochemistry, Department of Biological Chemistry, Institute of Life Sciences, The Hebrew University of Jerusalem, Jerusalem, Israel

Dr. Daniel Orion
Nematologist, c/o BARC West, Nematology Laboratory, Beltsville, Maryland 20705

Dr. Arnold W. Ravin
Department of Biology, University of Chicago, 1103 East 57th Street, Chicago, Illinois 60637

Dr. T. E. Redlinger
Research Associate, Smithsonian Institution, Radiation Biology Laboratory, Rockville, Maryland 20852

Robert Ruether
Research Assistant, BARC-East, Building 177B, Beltsville, Maryland 20705

Carol Salzberg
Senior Consultant, Ra Associates, 9834 Marcliff Court, Vienna, Virginia 22180

Dr. Jerome A. Schiff
Professor of Biology and Director, Institute for Photobiology of Cells and Organelles, Brandeis University, Waltham, Massachusetts 02254

Dr. Ahlert Schmidt
Professor, Botanisches Institut, Universität München, Menzinger Strasse 67, D-8000 München 19, West Germany

A. M. Schumacher
Biologist, National Cancer Institute, National Institutes of Health, Blair Building, Room 528, Bethesda, Maryland 20205

Dr. Moshe Shilo
Professor, Division of Microbial and Molecular Ecology, Life Sciences Institute, Hebrew University, Jerusalem, Israel

Dr. Piotr Slonimski
Professor of Genetics, University of Paris, Director, Centre de Génétique Moleculaire du C.N.R.S., 91190 Gif sur Yvette, France

Dr. Germaine Cohen-Bazire Stanier
Directeur de Recherche au C.N.R.S., Institut Pasteur, Unité de Physiologie Microbienne, 28, rue du Docteur Roux, 75724 Paris Cedex 15, France

Dr. Roger Yate Stanier
Professeur à l'Institut Pasteur, Institut Pasteur, Unité de Physiologie Microbienne, 28, rue du Docteur Roux, 75724 Paris Cedex 15, France

Dr. Russell L. Streere
Chief, Plant Virology Laboratory, USDA, A52 Bioscience Building, BARC-West, Beltsville, Maryland 20705

Dr. G. Thompson
National Institute of Mental Health, National Institutes of Health, Building 32A, Room 101, Bethesda, Maryland 20205

Dr. Robert K. Trench
Associate Professor, Department of Biological Sciences, University of California, Santa Barbara, California 93106

Dr. Eugene L. Vigil
Plant Cell Biologist, Department of Botany, University of Maryland, College Park, Maryland 20742

Dr. W. P. Wergin
Research Cytologist, USDA-SEA-AR, BARC-East, Building 177B, Beltsville, Maryland 20705

Dr. Diter von Wettstein
Professor, Department of Physiology, Carlsberg Laboratory, Gl. Carlsberg Vej 10, DK-2500 Copenhagen Valby, Denmark

Dr. Sam Wildman
Emeritus Professor of Biology, Biology Department, University of California, Los Angeles, California 90024

ON ORIGINS

In the sciences hypothesis always precedes law, which is to say, there is always a lot of tall guessing before a new fact is established. The guessers are often quite as important as the fact-finders; in truth, it would not be difficult to argue that they are more important. New facts are seldom plucked from the clear sky; they have to be approached and smelled out by a process of trial and error in which bold and shrewd guessing is an integral part.

H. L. Mencken

Over a year ago I received a telephone call from Roger Stanier, who was visiting at the National Institutes of Health as a Fogarty Fellow. As usual, he had an interesting idea: Why not get a few people together under the auspices of the Fogarty Center for an informal discussion of chloroplast origins? He had made a list of possible participants, had sketched the basic outlines of the program, and had obtained the sponsorship of the Fogarty Center and the use of their conference facilities. It seemed like a fine opportunity. On the one hand, the field had advanced to the point where there was enough information to hazard some guesses about chloroplast origins, on the other, not enough was known to preclude a stimualting and spirited discussion. At this point, Roger was obliged to return to Paris and left the local organization of the meeting in the hands of Peter Condliffe of the Fogarty Center and me.

As we looked over the outline that Roger had left, we realized that we had a splendid opportunity to celebrate not only the origins of chloroplasts but also the origins of the ideas and influence of Roger himself, since the subject matter of the conference was built around his ideas and many of his friends and colleagues would be present. This would be an excellent occasion to honor him. In this way, we engaged in a quiet conspiracy to dedicate this conference to Roger Stanier, a decision that was subsequently ratified unanimously by the participants. Sometime before the conference was to begin, we learned that Roger was to be the first recipient of the Bergey award for his work in microbial taxonomy and that he had elected to receive the award at this meeting (see the Addendum, page 325). The result of all of these events is the present volume.

The various streams that flow into Roger's thinking about microbiology have been developing for some time. His intellectual forebears include such talented comparative microbiologists as Winogradsky and Beijerinck, Kluyver, and, of course, C. B. Van Niel with whom he studied and published. He has made important contributions to a wide variety of areas including his work on simultaneous adaptation of enzymes in the pathway of aromatic biosynthesis; the influence of oxygen on the formation of the photosynthetic apparatus in photosynthetic bacteria (with M. Griffiths, W. P. Sistrom and G. Cohen-Bazire), which led to the realization that carotenoids protect against deleterious photoxidations sensitized by chlorophyll; the localization of bacterial photosynthesis in the chromatophores (with H. K. Schachman and A. B. Pardee); and biochemical evolution, culminating in his work on the cyanobacteria. Perhaps the best indication of the influence of this work is the fact that we now call these organisms cyanobacteria rather than blue-green algae. Since the days of Mereshkowsky and Schimper, interest in the cyanobacteria immediately implies an interest in chloroplasts and Roger Stanier has made important contributions to our thinking about the relations between free-living prokaryotes and eukaryotic organelles.

Once having mentioned some of his important contributions, one cannot easily define Roger's particular way of thinking about scientific problems in a way that gives an adequate impression of the way he integrates knowledge, or to account for how his viewpoint may have originated. One impressive quality is the breadth of his knowledge, understanding, and thinking that leads him to make inspired connections among various fields, which are not at all obvious before he points them out. He is also superbly equipped to make use of the most modern techniques of biochemistry and molecular biology without disdaining the use of the microscope and the petri dish. In other words, despite the timeliness of his work he has not become that sort of narrow molecular chauvinist who feels that certain techniques and areas of investigation are beneath his notice. Rather, all available techniques and information are to be used to the fullest extent possible to answer all questions including taxonomic ones. It takes real talent and taste to be able to work on those larger classical questions that many beginning students tend to regard as quaint or uninteresting, without losing one's credentials as a "modern" scientist working at the flashy cutting edge of cellular biology. Roger Stanier has this talent, and what is more, through his accomplishments, he has made these areas fashionable and exciting for students and colleagues. It is our good fortune that this way of thinking has been preserved and handed on to a new generation of students and teachers in the form of the textbook *The Microbial World* (now in its fourth edition) of which he is a co-author with M. Doudoroff, E. A. Adelberg, and J. Ingraham.

Given the inspiration from Roger Stanier's work, his organization of this meeting, the people he suggested that we gather together, and the timeliness of the subject, it is not difficult to account for the outcome that is recorded in this volume. We have tried to capture the flavor of this meeting by including not only the formal presentations but much of the discussion as well. We hope that the reader will be able to gain some knowledge not only of the questions that are being asked, but also of those small unappreciated facts that, when presented in a Stanier-like synthesis, lead to a new way of thinking about old and persistent problems.

I would be remiss if I did not thank Sophie R. Harrison and Kathy Magnusson for their invaluable help in preparing the manuscript for publication. Others who helped in the organization of the meeting itself are identified on page xv. Finally, we are extremely grateful to Dr. Germaine Cohen-Bazire Stanier for providing the picture of Roger found on page vii and for supplying the freeze-fracture image of a cyanobacterium on which the cover is based.

Jerome A. Schiff

ON THE ORIGINS
OF CHLOROPLASTS

PROLOGUE
OXYGENIC PROKARYOTES
AS PLASTID PRECURSORS

. . . . a Darwinian Man, though well-behaved, at best is only a monkey shaved!

—W. S. Gilbert

Oxygenic Prokaryotes as Plastid Precursors

There is, it is said, an old treatise that describes the natural history of Ireland. In it, there is an entire chapter devoted to the snakes of the island. The entire chapter contains one sentence: "There are no snakes in Ireland." If we devoted ourselves to a discussion of chloroplast origins in the strict sense, we would have as little to say, although we can hardly deny that chloroplasts have had origins. We have not observed the process of chloroplast evolution for ourselves nor do we have any extensive fossil evidence. So, lacking time machines that will take us back to make these observations, we must, like the cosmologists, piece together evidence from contemporary observations and use our ratiocinative talents to suggest a plausible past for this organelle.

Microscopes have been improved over the years to the point where it is possible to observe cells and their internal structures more clearly: the smaller the cells, the larger the microscope. One result of this improved technology has been the realization that cells occur in two forms: prokaryotic cells lack internal organelles delimited by surrounding membranes from the cytoplasm while eukaryotic cells contain membrane-limited organelles such as nuclei, mitochondria, and chloroplasts. By the middle of the nineteenth century, enough was known about the superficial appearance of prokaryotic cells and the chloroplasts of eukaryotic ones in the light microscope to indicate that they were similar in size and structure; this immediately led to the suggestion that they might have common evolutionary origins. As information about the biochemistry, genetics, and molecular biology of cells began to accumulate, this supposition was strengthened. The idea that certain eukaryotic organelles such as the mitochondria and chloroplasts resembled prokaryotic cells received considerable support from evidence that these organelles were capable of division within the eukaryotic cell, had their own DNA genomes, their own genetic systems and their own protein synthesizing machinery. (They also show a great dependence on the rest of the cell for the energy and many of the small molecules and proteins necessary to construct them. The sharing of genetic information from the nuclear and organelle genomes during chloroplast development leads to certain difficulties which must be resolved in theories of chloroplast origins). Biochemical evidence accumulated which showed that the molecules composing the organelles were extremely similar to those found in free-living prokaryotes. Roger Stanier was one of the modern biologists who appreciated the full significance of this similarity; his words are quoted beneath his portrait facing the title page of this book.

Very well, so prokaryotes and chloroplasts are extremely similar in many of their properties. What does this imply about the origins of chloroplasts? As early as the middle of the nineteenth century, suggestions were made that chloroplasts might have originated from the endosymbiotic invasion of a eukaryotic-like host cell by a free-living prokaryote. This hypothesis is still a popular one as you will learn from the pages which follow these. Along the way, another hypothesis has presented itself: that the plastid had an episomal origin, that a detached piece of the main genome of the cell, or an episome, organized an organelle about itself through adaptation and selection over the course of evolution. Between these two widely separated views, a diplomat has interposed him (or her) self by the name of Cluster Clone, who is concerned more with the evolution of molecular mechanisms of cooperation between the organelle and nuclear genomes during evolution than with the initial origins, although it must be admitted that his position is closer to the episomal theory than to the endosymbiotic hypothesis. A discussion of these questions will be found among the contents of this book. Although polarized hypotheses have great heuristic value in science and give us the excitement, courage and interest to do the experiments to find out whether they are true, experience teaches us that neither or both hypotheses may turn out to be true. Biological questions are complex and a hypothesis may not become refined enough for a rigorous test until the results of many experiments have accumulated. As Walt Whitman says "Do I contradict myself? Very well, then, I contradict myself. I am large, I contain multitudes." Therefore, while our two (or three) major hypotheses for chloroplast origins provide the spice and zest for both our meeting and our book, and indeed for future experiments, there is little reason to hold to them beyond their immediate usefulness. Our skeptical scientific spirits should avoid the state of mind that seeks to anathematize colleagues who do not adhere to the same point of view. For after all, as one of my students, Jeffrey Diamond, quoted as an "Oriental Proverb" in his thesis: "When the dust rises thou wilt see whether thou ridest a horse or an ass."

But what direction shall we take through the present haze of dust? Roger Stanier through his extensive work on a particular group of prokaryotes has given us both a direction and a philosophy. He has convinced us that the blue-green algae should be placed among the bacteria based on an impressive amount of biological, molecular biological and biochemical evidence. Blue-green algae are, in fact, oxygenic bacteria closely related to other gram-negative bacteria; for this reason, the name cyanobacteria (or blue-green bacteria) will be found throughout this book. He, and others, have pointed out the close similarities between the chloroplasts of the red algae and the free-living cyanobacteria. Intracellular symbioses of cyanobacteria with eukaryotic cells are known and may represent an intermediate stage in organelle formation, although as you will learn from what follows, this too is an area of controversy. The chloroplasts of the green eukaryotes, such as green algae, *Euglenas* and the multicellular higher plants, are very different; this suggests that they might have had a different precursor than the cyanobacteria. Indeed, the chloroplasts of various groups might have had different progenitors or, in evolutionary language, the chloroplasts probably have polyphylletic origins. Very recently, support for this view has come from the isolation of a green oxygenic prokaryote called *Prochloron* for which a new division, the Prochlorophyta, has been suggested. Organisms of this type may have served as the evolutionary precursors of the present-day chloroplasts of green eukaryotes. It remains to be seen whether this idea is supported by further evidence and whether other free living prokaryotes having the pigmentation and properties of the chloroplasts of other eukaryotic groups (such as the brown and golden-brown algal groups) will turn up. But our present direction through the murky dust lies in the direction of comparing the chloroplasts of modern eukaryotic cells with the cells of free-living oxygenic prokaryotes using every technique at our disposal including pigment analysis, studies of metabolism and biochemistry, the fruitful approaches of molecular biology and careful comparisons of structure and function. For this reason, a considerable portion of this book is concerned with a consideration of comparisons of this sort. In every case, the contributors were encouraged to condense and present research data in such a way that made clear its comparative import. The rapporteurs have carried through the difficult task of providing summaries of this material set in a broader context than was possible in each individual contribution. They have tried to make connections among the views of various contributors and discussants and to add, where possible, additional material from their own work and knowledge.

After a consideration of the comparative aspects and a discussion of the various hypotheses for chloroplast origins, what is left? What is left is perhaps the most important and pervasive generalization that biologists have: evolution through mutation and natural selection.

Evolution will be found to be the guiding precept of this volume. Whatever hypotheses are made to account for plastid origins or for subsequent changes that they and their cells have undergone, must be consistent with our present understanding of evolution and the mechanisms by which mutations occur, are selected, and are established in the population for this is the motive force that organizes the details presented in this volume into a coherent whole. Evolution is our beacon and our guide through the dusty haze of speculations and facts and as we leap into our saddles to confront these problems let us hope that at least some of us will turn out to be riding horses rather than asses.

J.A.S

PART I
PLASTIDS AND THEIR PRECURSORS

Once you have taken the impossible into your calculations,
its possibilities become limitless.

—Bertrand Russell

Diversity of
the Photosynthetic
Prokaryotes

Moshe Shilo

The most widely accepted hypothesis for the origin of the chloroplast is endosymbiosis between a protoeukaryotic cell and a prokaryote possessing a fully
developed machinery for carrying out oxygenic photosynthesis. For this reason
it might be useful, in a discussion of the origin of the chloroplast, to look at
contemporary photosynthetic prokaryotes to recognize the diversity of this group
of organisms, and to describe its physiology and ecology in order to see if we
can find links between any of the presently found types and the progenitor of
the chloroplast.

In spite of the long history of research and study of the photosynthetic prokaryotes, which had its beginnings early in the previous century, most of our
knowledge of this group has been accumulated in the last few years. The fact
that many of the strains have only recently been isolated and grown in axenic
culture, thereby becoming amenable to experimental analysis, may be part of
the explanation for this.

Several excellent reviews by Stanier (1977) and Stanier and Cohen-Bazire
(1977) on the cyanobacteria and by Pfennig (1977, 1978) on the green and purple
photosynthetic bacteria have appeared, and cover the major characteristics of
these groups of organisms in great detail. These publications discuss the taxonomic status of the main groups and some of their most important characteristics,
including: the diversity of structure and function of the photosynthetic mechanisms; their ecological diversity and distribution patterns; insight into their ev-

Address reprint requests to Dr. Moshe Shilo, Professor, Division of Microbial and Molecular
Ecology, Life Sciences Institute, Hebrew University, Jerusalem, Israel.

olution and phylogeny; and a reexamination of the evidence for a polyphyletic origin of the chloroplasts of the different photosynthetic eukaryotes.

Phototrophic prokaryotes have a very long evolutionary history, going back to the Precambrian period more than 3×10^9 years ago. It is not astonishing, therefore, that the phototrophic prokaryotes include an extremely diverse assemblage of organisms, differing widely in their morphological features as well as in their metabolic patterns and predilection for different habitats. The diversity of the phototrophic prokaryotes is supported by data showing an extremely broad spectrum of DNA base composition, with guanine + cytosine mole percents ranging from 29 to 79, only slightly less than for all the prokaryotes together (Herdman et al., 1979a).

In spite of the many publications in recent years on the photosynthetic prokaryotes, the accumulation of new knowledge in this field is still in a dynamic state. Not only have we witnessed the discovery of entire new groups of photosynthetic prokaryotes but these findings have to some extent changed our basic concepts about these organisms. We may mention as examples:

1. The newly discovered nonchlorophyll-mediated photosynthesis of the halobacteria (Stoeckenius, 1978). The work on the mechanisms of bacteriorhodopsin action now makes it necessary to redefine our whole conceptual approach to photosynthesis and has opened a whole new fruitful field of research.

2. The discovery of Prochlorophyta by R. Lewin (1977). This new division of prokaryotes, the members of which contain chlorophylls *a* and *b*, is of special interest in connection with chloroplast origins since *Prochloron* resembles the higher plant and chlorophyta chloroplast more closely than any other prokaryote.

3. The discovery of a new genus of gliding green photosynthetic bacteria of extremely great metabolic versatility, *Chloroflexus aurantiacus*, by Pierson and Castenholz (1974) has considerably broadened our outlook on this suborder of bacteria which formerly included only nonmotile forms.

4. The discovery that many cyanobacteria such as *Oscillatoria limnetica* are capable of shifting between oxygenic photosynthesis, and anoxygenic photosynthesis driven only by system I, now may form a bridge between the two types of photosynthesis, which up to now seemed unlinked (Cohen et al., 1975b).

The photosynthetic prokaryotes can be divided on the basis of their photosynthetic apparatus into five major groups reflecting basic differences in their photochemical mechanisms (Figure 1).

The first group in this scheme, the *halobacteria*, can easily be separated from all the other photosynthetic prokaryotes, since their photosynthesis is not based on a chlorophyll-containing photosynthetic mechanism. The halobacteria have, instead, bacteriorhodopsin in their purple membranes; the light absorbed is used to form ATP or proton gradients to meet cellular metabolic requirements. No antenna pigments are known in this group. The quantum yield of this reaction

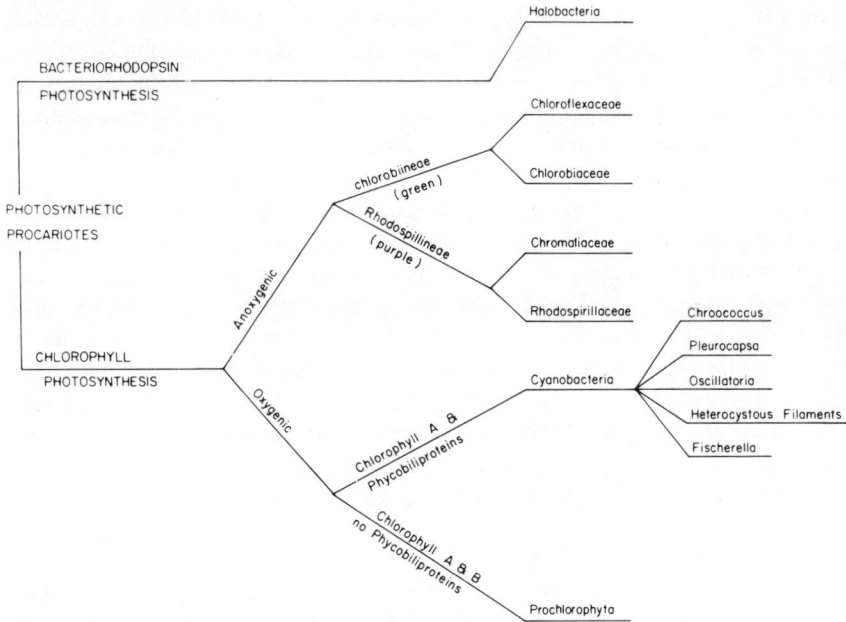

Figure 1. The five major groups of photosynthetic prokaryotes and their interrelationships.

is smaller than that of chlorophyll-sensitized photosynthesis and, up to the time of writing, growth of the organisms in the light in the absence of O_2 has not been possible (Stoeckenius, 1978).

The second and third groups are the phototrophic green and purple bacteria. They carry out anoxygenic photosynthesis only, and possess only a single photosystem (PS I). Reduced sulfur compounds, molecular hydrogen, or simple organic substances can serve as electron donors in this group. The reducing power required for photoassimilation of CO_2 is generally provided by reduced substrates in a pathway linked only indirectly, if at all, with the photosynthetic system. Their photosynthetic pigments are bacteriochlorophylls *a, b, c,* and *d* and a great variety of carotenoids. Light energy absorbed by chlorophyll of the photosystem causes a cyclic electron flow down a redox potential gradient established by electron and/or hydrogen carriers leading the electrons back to the chlorophyll (Pfennig, 1977, 1978).

The last two groups are the cyanobacteria and prochlorophytes. They carry out oxygenic photosynthesis based on two photosystems working in series. The light absorbed maintains an electron flow from water to NADP with evolution of oxygen. They are similar, therefore, in their mode of photosynthesis to the eukaryotic algae and plants.

Despite their possession of both photosystems, certain cyanobacteria can, under suitable conditions, carry out an anoxygenic photosynthesis similar to that found in the green or purple bacteria (Cohen et al., 1975a, 1975b). It is only in recent years that the prokaryotic nature of the cyanobacteria, the largest and most diverse group of the phototrophic prokaryotes, has been fully recognized. The most descriptive approach to the study of this group has now been replaced by the experimental methodology used in the study of other groups of the prokaryotes.

Stanier and his collaborators in Paris, after a concentrated effort to collect and grow in pure culture 200 cyanobacterial strains, have put forward a general comprehensive taxonomic division of the cyanobacteria into five sections based on various patterns of structure and development (Rippka et al., 1979). DNA base composition and genome size were also found by this group to be useful tools for classification of the cyanobacteria (Herdman et al., 1979a, 1979b).

The five sections of cyanobacteria in the Stanier scheme are: (a) unicellular rods or cocci (chroococcal type); (b) the *pleurocapsalean* cyanobacteria, which have a more complex cell cycle; (c) the nonheterocystous filamentous forms; (d) cyanobacteria that form heterocysts of different shape, location, and interspacing; and (e) the most complex cyanobacteria, including branched filamentous forms such as *Chlorogloeocapsis* and *Fischerella*.

The last group is a new division of the photosynthetic prokaryotes, the Prochlorophyta, differing from all the others in their possession of chlorophyll *b* in addition to chlorophyll *a*, and distinct from the cyanobacteria due to the absence of any phycobiliprotein pigments (Lewin, 1977; Thorne et al., 1977; Withers et al., 1978).

Prochloron has up to now been found as an external surface symbiont in the radial grooves surrounding the orifices of colonial ascidians such as *Didemnum cariacem* or internally in the cloacal cavity of *Diplosoma vireus* (Thinh, 1978a, b). The latter type is very common in Eilat, Israel, and has recently been thoroughly investigated there by Duclaux and Lafargue (unpublished). Like all other prokaryotes, they have no membrane-bound organelles and no defined nucleus. Similar to other gram-negative bacteria and the cyanobacteria, *Prochloron* contains muramic acid in its cell wall and a peptidoglycan layer similar to that of cyanobacteria (Moriarty, 1979).

Prochloron has not yet been grown in culture. Its dependence on a symbiotic interaction with the ascidian host could possibly be connected with a requirement for cholesterol, which is an important component of its sterols (Perry et al., 1978).

The photosynthetic green, purple, and blue-green bacteria share a common gram-negative cell wall structure. They all contain lipopolysaccharides in their outer membranes which resemble and yet are distinctly different from the lipopolysaccharides of the Enterobacteriaceae (Drews et al., 1978).

Gas vacuoles are found in several species of the green and the purple bacteria as well as among the cyanobacteria. A typical nonunit membrane is characteristic of the gas vacuoles in all cases (Walsby, 1975).

All types of chloroplasts known, though markedly different in the specific array of the pigments which comprise their photosynthetic machinery, possess only oxygenic photosynthesis based on the combined activity of photosystems I and II. In searching for potential progenitors among the contemporary prokaryotes, we can thus confine ourselves to the cyanobacteria and prochlorophytes.

Diversity in the Structure of the Photosynthetic Apparatus

In the different groups of photosynthetic prokaryotes, the photosynthetic apparatus, including the photochemical reaction centers, the light-harvesting pigments, and the electron transport chains, are functionally integrated in differing spatial patterns.

1. The purple membrane of the halobacteria has been characterized as a uniquely simple structure, shown to function as a light-driven proton pump, creating an electrochemical gradient of hydrogen ions across the membrane. The purple membrane is formed into specialized patches which are integrated within the cytoplasmic membrane of halobacteria (Stoeckenius, 1978).

2. In the green bacteria, reaction centers and electron transport systems are in the cytoplasmic membrane with antenna pigments organized in chlorobium vesicles or chlorosomes, which are arranged peripherally in close contact with the inner surface of the cell membrane (Pfennig, 1977; Remsen, 1978).

3. In the purple bacteria, reaction centers, accessory antenna pigments, and the electron transport system are all contained in extensions of a folded inner membrane forming complex intrusions into the cell interior. A number of different and complex patterns occur within the Rhodospirillaceae (Pfennig, 1977, 1978; Remsen, 1978).

4. The cyanobacteria have compressed, flattened thylakoids with a regular array of aggregates of the phycobilipigments, the phycobilisomes, attached to their external surface (Stanier, 1977; Stanier and Cohen-Bazire, 1977).

The molecular structure of the different phycobiliproteins and their organization and internal order in the phycobilisomes have been established. Allophycocyanin and allophycocyanin-B are presumed to be the site of attachment of the phycobilisomes to the thylakoid membrane, surrounded by phycocyanin subunits which are in turn covered by phycoerythrin subunits, in those organisms containing this phycobiliprotein (Gantt, 1975; Glazer, 1977; Tandeau de Marsac and Cohen-Bazire, 1977; Yamanaka et al., 1978). In addition to the colored phycobiliproteins, a number of non-colored polypeptides were found to be part of the phycobilisome structure (Tandeau de Marsac and Cohen-Bazire, 1977). These polypeptides are likely to play a role in the attachment of the organelle to specific sites on the thylakoid and/or in the positioning of the light-harvesting constituents within the phycobilisome.

The analogy between cyanobacterial cells and chloroplasts can be extended to the proton gradients found inside the thylakoids and in the external membrane of the cyanobacterial cell. Recent measurement carried out by Padan and Schuldiner (1978) showed that upon energization, protons are pumped from the cy-

toplasm into the thylakoidal inner spaces, while at the cytoplasmic membrane
barrier protons are pumped out.

In the cyanobacterium *Gloeobacter violaceus*, which has a unique ultrastruc-
ture, the photosynthetic apparatus is simpler than that in most cyanobacteria
(Rippka et al., 1974). Reaction centers and electron transport system lie in the
cell membrane while a continuous subcortical layer harbors the antenna pigments.
This layer is in close contact with the inner surface of the cell membrane.
Recently, phycobilisomelike structures were isolated even from this organism
(Guglielmi et al., 1979).

5. *Prochloron* is basically very similar to the cyanobacteria in the structure
of its photosynthetic apparatus (Whatley, 1977). However, the thylakoids are
arranged in pairs or in stacks and no phycobilisomes are found. The thylakoids
are not organized into grana, as in higher plant chloroplasts, but form discon-
tinuous bands such as are found in algae or higher plants deficient in certain
nutrients. The reaction center size, 240 chlorophylls per P_{700}, was found to be
half that of the higher green plants (Withers et al., 1978).

Ecological Diversity

Based on their physiological potential, the phototrophic microorganisms have
a typical pattern of distribution covering the entire spectrum of conditions found
in the photic zone of the biosphere. The permanently aerobic layers are dominated
by eukaryotic algae and the higher plants. The green and purple phototrophic
bacteria, on the other hand, inhabit the stable photoanaerobic niches including
anaerobic layers of lakes rich in sulfides, anaerobic sediments, and sulfur springs.
Differing from both of these, cyanobacteria are dominant in ecosystems in which
rapid diurnal fluctuations or seasonal changes of aerobic and anaerobic conditions
alternate. The chemoclines in lakes and shallow marine bays, marshes, man-
groves, and rice fields are examples of typical environments of this type (Padan,
1979).

The daily fluctuations in shallow water environments often involve rapid
changes not only between light and dark periods, but also in O_2, H_2S, pH,
temperature, and salinity.

The ability of many cyanobacteria to shift readily from oxygenic to anoxygenic
photosynthesis (Oren et al., 1977) shows that whereas only a narrow sector of
the spectrum is utilized in the oxygenic type, the entire absorbed spectrum is
utilized in anoxygenic photosynthesis. This limited range of utilization of quantum energies in oxygenic photosynthesis in *Oscillatoria limnetica* and other
cyanobacteria is markedly different from that of most eukaryotic algae and plants
containing chlorophyll *b* in their light-harvesting system. Many cyanobacteria
are, therefore, not only capable of a shift to anoxygenic photoysnthesis, but this
seems to be their preferred pattern of life.

The finding that *O. limnetica* is capable of CO_2 photoassimilation with hydrogen as the electron donor (Belkin and Padan, 1978) now extends the distri-

bution potential of this organism even further into conditions of low redox potential with Eh levels even below the sulfide biotope.

The ability of many cyanobacteria to photosynthesize anoxygenically (Garlick et al., 1977) can now explain the paradox that many nonheterocystous cyanobacteria carry the information for synthesis of the entire enzyme complex for dinitrogen fixation, which is never expressed under aerobic conditions. The process can proceed in nature and possibly plays an important role in the global nitrogen cycle, under conditions in which the organism can grow anaerobically, driven by the energy derived from anoxygenic photosynthesis. Such anaerobic niches suitable for growth and development of cyanobacteria are indeed widespread in anaerobic aquatic layers or sediments, or under rapidly fluctuating conditions.

The extremely great flexibility of the cyanobacteria is further demonstrated by the ability of some of them to exist using limited heterotrophic or mixotrophic means (Stanier and Cohen-Bazire, 1977) in addition to their photoautotrophic way of life. Others have the ability to utilize their internally stored polyglucose reserves for maintenance in the dark by using diverse metabolic patterns, including respiration, fermentation to lactic acid, and anaerobic respiration with elemental sulfur (Oren and Shilo, 1979).

Certain types of cyanobacteria, especially the gas vacuolated planktonic forms capable of bloom formation, are unique in their adaptation to withstand high oxygen concentrations and are resistent to conditions favoring photooxidation (Abeliovich and Shilo, 1972; Eloff et al., 1976). The ability to maintain high levels of superoxidedismutase activity of the manganese type seems to be among the mechanisms responsible for this resistance (Steinitz and Shilo, 1976).

Certain cyanobacteria, such as *Oscillatoria limnetica*, adapt to high oxygen concentrations by increasing their superoxidedismutase levels. When anaerobically grown cells are shifted to aerobic conditions, their resistance to photooxidation increases and concomitantly the iron superoxidedismutase activity of the cells increases many fold (Friedberg et al., 1979).

Phylogenetic Diversity

Application of molecular and genetic techniques now make it possible to construct phylogenetic trees, which test to what extent the empirical taxonomic systems constructed also reflect the degree of relatedness among the various organisms. This approach is important not only in order to obtain a natural rational system for classification, but also because it allows us to look for the molecular fossil record in order to find possible connections between present day phototrophic prokaryotes and eukaryote chloroplasts. A number of methodological approaches have been utilized, including amino acid sequencing of selected proteins such as cytochrome *c* and ferredoxin. The rapid accumulation of sequences from additional organisms and of new protein types allows us to broaden this approach and gain confidence in the results obtained.

A beginning of a phylogenetic grouping of the photosynthetic prokaryotes based on the amino acid sequences and molecular structure of cytochrome c has been attempted by Almassy and Dickerson (1978). Schwartz and Dayhoff (1978) have combined the sequence results for a number of proteins with sequence data from 5S rRNA to obtain an integrated broad overview of biological evolution in the form of a composite phylogenetic tree for several prokaryotes and eukaryotes including several photosynthetic organisms. This may be an important step in approaching a meaningful phylogeny.

Analyses based on 16S rRNA sequencing and cataloging of all of the 500 oligonucleotides obtained after digestion with T_1 ribonuclease have been carried out by Bonen and Doolittle (1976) (see also Doolittle et al., 1979) using several cyanobacteria and chloroplasts from *Porphyra* and *Euglena*; Gibson (1979) has similar results for the green and purple bacteria. Use of this method further confirmed the closeness between the chloroplasts of the rhodophyta and the cyanobacteria and again indicated that the *Euglena* and *Porphyra* chloroplasts seem to have arisen independently. The relation between the cytoplasmic 18S rRNA and the chloroplast 16S rRNA of *Porphyra*, or the cyanobacterial 16S rRNA, was not greater than between unrelated molecules.

Analysis by the above-mentioned method of the 16S rRNA from green and purple photosynthetic bacteria (Tabit et al., 1979) revealed that the photosynthetic bacteria is an extremely ancient group; it also suggested strongly that the transition to nonphotosynthetic energy-generating systems has occurred many times during the course of evolution. DNA–DNA (Herdman et al., 1979) and RNA–DNA (Phillips and Carr, 1975) hybridization have also very effectively been utilized in comparative analysis of the different photosynthetic prokaryotes. Based on the 5S rRNA hybridization, pronounced sequence similarity was found between unicellular cyanobacteria (*Aphanocapsa* 6308) and chloroplast ribosomal RNA from *Euglena gracilis* (Phillips and Carr, 1975).

By optical monitoring of DNA renaturation, Herdman et al. (1979b) provided information of great evolutionary interest. Genome size analysis of more than 118 strains of 25 genera of cyanobacteria show that they fall into four groups which can be interpreted as 2, 3, 4, and 6 multiples of a basic unit of 1.2 × 10^9 daltons. There seems to be a correlation between genome size and organizational or developmental complexity. It is likely that the morphologically more complex cyanobacteria evolved from the morphologically simple ones. The fossil record bears this out (Phillips and Carr, 1977).

Loss of the capacity for cell wall synthesis by the endosymbiont would permanently fix and stabilize its dependence on the host cell. The contemporary example of *Cyanophora paradoxa*, in which the endocyanelle has only a rudimentary peptidoglycan layer, would serve as a demonstration of this phenomenon. The loss of a rigid cell wall structure could possibly also explain the fact that genome multiplicity is often observed in cases of endosymbiosis and in the chloroplast. The lambda 299 particles of *Paramecium* (Soldo and Godoy, 1973), and *Cyanophora paradoxa* (Herdman and Stanier, 1977) can be cited as examples. With the loss of the cell wall and its function in regulated binary fission and genome separation, "relaxed" genome replication may be advantageous over the "stringent" replication control found in the free-living, cell-wall-surrounded cells.

The great differences in pigment composition among chloroplasts of different algal and plant groups have made it difficult to envisage a single endosymbiotic event leading to diverse evolution as the basis for development of all of the various chloroplast types. Margulis (1970) was the first to suggest that the contemporary diversity of algal chloroplast pigments might reflect a polyphyletic origin through endosymbiosis with different types of phototrophic prokaryotes. Raven (1970) suggested that while cyanobacteria could have given rise to the chloroplasts of the Rhodophyta, Cryptophyta, and the anomalous alga *Cyanidium*, a second type of photosynthetic prokaryote containing chlorophyll *b* could have given rise to the green algae and through them to the Bryophytes, Euglenophytes, and all the higher plants. A third "yellow prokaryote" gave rise to the Phaeophyta, Chrysophyta, Xanthophyta, and Pyrrophyta.

Based on pigment composition and the ultrastructural organization of pigment assemblages, Stanier (1974) suggested that the chloroplast of the Rhodophyta *only* was closely similar to typical cyanobacterial cells, while those of other phototrophic eukaryotes contained a completely different set of pigments and must be the result of a polyphyletic evolution of the chloroplast. It seemed that only cyanobacteria may have survived from among the different types of prokaryotic progenitors of the chloroplast to persist among the contemporary types, while the others have become extinct. It was predicted by Stanier that some of these may yet be discovered. The discovery of the Prochlorophyta clearly has confirmed this prediction.

A striking fact observed in present-day symbioses of invertebrates and phototrophic bacteria and algae is that while the phenomenon is widespread in many widely differing animals, the photosynthetic organisms taken up as endosymbionts are usually restricted to very few types. The case of *Symbiodinium microadreatica* is a case in point. It is not impossible, therefore, that the primary endosymbiotic events which have led to the establishment of chloroplasts as organelles may have been restricted to only a few selected photosynthetic prokaryotes, and that possibly, by analogy to present-day symbioses, different protoeukaryotes may have been involved. Thus, even assuming a polyphyletic origin of the different chloroplast types known to us, only a restricted selection of photosynthetic prokaryote types may have been involved.

In addition to the differences found in the antenna pigment composition of various photosynthetic eukaryotic chloroplasts, the location and organization of these pigments may be very different and indicate linkage to different prokaryotic progenitor cells. Thus, while the similarity between the rhodophyta and the cyanobacteria is extremely great in both pigment composition and the aggregation of phycobiliproteins into phycobilisomes attached to thylakoid external surfaces, the chloroplasts of *Cyanidium* and those of the Cryptophyceae, though also

18   M. Shilo

containing phycobiliproteins, are different and may have arisen differently. The cryptophytes have only one phycobiliprotein—phycoerythrin or phycocyanin—but never allophycocyanin (Gantt, 1978). They do not have phycobilisomes and their phycobiliprotein pigments are intrathylakoidal. Sequence studies of the polypeptide chains, immunologic similarity and thermolability of the phycobiliproteins also point to the close connection between the cyanobacteria and the Rhodophyta and a more remote relationship with the Cryptophyceae (MacColl and Berns, 1979).

An approach which could help us to test and establish the polyphyletic origin hypothesis is the comparison of the control mechanisms operative in the biosynthesis and aggregation of the pigments in different photosynthetic prokaryotes during the response of the cells to a change in light intensity. In this connection, the findings of Falkowski and Owens (1978) that in different groups of phototrophic eukaryotes (diatoms and Chlorophyta) two different adaptation strategies are used in response to adaptation to shade conditions may be mentioned. In the diatom *Skeletonema costarum*, light/shade adaption is characterized by a change in the size of the photosynthetic units, but not their number, while in the chlorophyte *Dunaliella tertiolecta*, overall changes in chlorophyll are related to changes in number of the photosynthetic units, but not to changes in their size.

Conclusion

The discovery of new groups of photosynthetic prokaryotes and the change in our approach to the study of their distribution and physiology clearly shows that we can expect interesting further developments in this field and new insights into the problems of their evolution and their relationship to chloroplasts. Plasmids, widely distributed among photosynthetic prokaryotes and carrying information which has not yet been related, for the most part, to functional or structural proteins of their hosts, have introduced a new challenge in addition to their being used as experimental tools. The manipulation and free transfer of genetic materials, including interspecies transfer, which has become possible (Marrs, 1979; Saunders, 1979) in the various photosynthetic bacteria and cyanobacteria, provides us with powerful new tools for understanding these organisms, as well as enabling us to use them for the study of genetic engineering. We are just at the beginning of this development.

References

Abeliovich, A., and Shilo, M. 1972. Photoxidative death in blue-green algae. J. Bacteriol. 111:682–689.
Almassay, R. J., and Dickerson, E. D. 1978. *Pseudomonas* cytochrome C551 at 2.0Å resolution: enlargement of the cytochrome family. Proc. Nat. Acad. Sci. 75:2674.
Belkin, S., and Padan, E. 1978. Sulfide dependent hydrogen evolution in the cyanobacterium *Oscillatoria limnetica*. FEBS Lett 94:291–294.
Bonen, L., and Doolittle, W. F. 1976. Partial sequences of 16S rRNA and the phylogeny of blue-green algae and chloroplasts. Nature 261:669–673.

Cohen, Y., Padan, E., and Shilo, M. 1975a. Facultative anoxygenic photosynthesis in the cyanobacterium *Oscillatoria limnetica*. J. Bacteriol 123:855–861.

Cohen, Y., Jørgensen, B. B., Padan, E., and Shilo, M. 1975b. Sulfide dependent anoxygenic photosynthesis in the cyanobacterium *Oscillatoria limnetica*. Nature 257:489–492.

Doolittle, W. F., Bonen, L., and Fox, G. E. 1979. Phylogenetic relationships among cyanobacteria: Final results of T1 oligonucleotide cataloging of the 16 S ribosomal RNAs of six unicellular and two filamentous cyanobacteria. *In* IIIth Int. Sym. Photosynthetic Prokaryotes—Abstracts. Oxford, E19.

Drews, G., Weckesser, J., and Mayer, H. 1978. Cell envelopes. *In* The Phyotosynthetic Bacteria. R. K. Clayton and W. R. Sistrom (eds.). Pp. 61–77.

Eloff, J. N., Steinitz, Y., and Shilo, M. 1976. Photoxidation of cyanobacteria in natural conditions. Appl. Environ. Microbiol. 31:119–126.

Falkowski P. G., and Owens, T. G. 1978. Effects of light intensity on photosynthesis and dark respiration in six species of marine phytoplankton. Mar. Biol. 45:289–295.

Friedberg, D., Fine, M., and Oren, A. 1979. Effect of oxygen on the cyanobacterium *Oscillatoria limnetica*. Arch. Microbiol. 123:311–313.

Gantt, E. 1975. Phycobilisomes: light harvesting pigment complexes. Bioscience 25:781–788.

Gantt, E. 1978. Phycobiliproteins of Cryptophyceae. *In* The Biochemistry and Physiology of Protozoa. 2nd edition. H. S. Hutner and M. Levandowsky (eds.). Pp 1–12.

Garlick, S., Oren, A., and Padan E. 1977. Occurrence of facultative anoxygenic photosynthesis among filamentous and unicellular cyanobacteria. J. Bacteriol. 129:623–629.

Glazer, A. N. 1977. Structure and molecular organisation of the photosynthetic accessory pigments of cyanobacteria and red algae. Mol. Cell. Biochem. 18:125–140.

Guglielmi, G., Cohen-Bazire, G., and Bryant, D. A. 1979. The structure of *gloeobacter niolaceus* and its Phycobilisomes. *In* IIIth Int. Sym. Photosynthetic Prokaryotes—Abstracts. Oxford, D18.

Herdman, M., Janvier, M., Waterbury, J. B., Rippka, R., Stanier, R. Y., and Mandel, M. 1979a. Deoxyribonucleic acid base composition of cyanobacteria. J. Gen. Microbiol. 111:63–71.

Herdman, M., Janvier, M., Rippka, R., and Stanier, R. Y. 1979b. Genome size of cyanobacteria. J. Gen. Microbiol. 111:73–85.

Herdman, M. and Stanier, R. Y. 1977. The cyanelle; chloroplast or endosymbiotic procaryote? FEMS Microbiol. Lett. 1:7–12.

Lewin, R. A. 1979. Progress in Prochlorophyte Research (1977–1979). IIIth Int. Symp. Photosynthetic Prokaryotes—Abstracts. Oxford, E15.

MacColl R., and Berns, D. S. 1979. Evolution of the Biliproteins. Trends Biochem. Sci. 4:44–47.

Margulis, L. 1970. Origin of Eukaryotic Cells. New Haven and London: Yale University Press.

Marrs, B. 1979. Genetics of Photosynthetic Bacteria. *In* 111th Int. Photosynthetic Prokaryotes—Abstracts. Oxford C1.

Moriarty, D. J. W. 1979. Muramic acid in the cell walls of *Prochloron* (Prochlorophyta). Arch. Microbiol. 120:191–193.

Oren, A., Padan, E., and Avron, M. 1977. Quantum yields for oxygenic and anoxygenic photosynthesis in the cyanobacterium *Oscillatoria limnetica*, Proc. Nat. Acad. Sci. USA. 74:2152–2156.

Oren, A., and Shilo, M. 1979. Anaerobic heterotrophic dark metabolism in the cyanobacterium *Oscillatoria limnetica*: sulphur respiration and lactate fermentation. Arch. Microbiol. Chem. 122:77–84.

Padan, E. 1979. Impact of facultatively anaerobic photoautotrophic metabolism on ecology of cyanobacteria (blue-green algae). *In* Advances in Microbial Ecology. Vol. 3. M. Alexander (ed.). Pp. 1–47.

Padan, E., and Schuldiner, S. 1978. Energy transduction in the photosynthetic membranes of the cyanobacterium (blue-green alga) *Plectonema boryanum*. J. Biol. Chem. 253:3281–3286.

Perry, G. J., Gillian, F. T., and Johns, R. B. 1978. Lipid composition of a prochlorophyte. J. Phycol. 14:369–371.

Pfennig, N. 1977. Phototrophic green and purple bacteria: A comparative, systematic survey. Annu. Rev. Microbiol. 31:275–290.

Pfennig, N. 1978. General physiology and ecology of photosynthetic bacteria. *In* The Photosynthetic Bacteria. R. K. Clayton and W. R. Sistrom (eds.). Pp. 3–18.

Phillips, D. O., and Carr, N. G. 1977. Nucleic acid analysis and the endosymbiotic hypothesis. Taxon 26:3–42.

Pierson, B. K., and Castenholz, R. W. 1974. A phototropic gliding filamentous bacterium of hot springs, *Chloroflexus aurantiacus*, gen. and sp. nov. Arch. Microbiol. 100:5–24.

Raven, P. H. 1970. A multiple origin for plastids and mitochondria. Science 169:641–646.

Remsen, C. C. 1978. Comparative subcellular architecture of photosynthetic bacteria: In The Photosynthetic Bacteria. R. K. Clayton and W. R. Sistrom (eds.). Pp. 31–60.

Rippka, R., Deruelles, J., Waterburry, J. R., Herdman, M., and Stanier, R. Y. 1979. Generic assignments, strains histories and properties of pure cultures of cyanobacteria. J. Gen. Microbiol. 111:1–61.

Rippka, R., Waterbury, J., and Cohen-Bazire, G. 1974. A cyanobacterium which lacks thylakoids. Arch. Microbiol. 100:419–436.

Saunders, V. A. 1979. The molecular biology of photosynthetic prokaryotes: Implications and applications. In IIIth Int. Sym. Photosynthetic Prokaryotes—Abstracts. Oxford E1.

Schwartz, R. M., and Dayhoff, M. O. 1978. Origins of prokaryotes, eukaryotes, mitochondria, and chloroplasts. A perspective is derived from protein and nucleic acid sequence data. Science 199:395–403.

Shilo, M. 1980. Factors that affect distribution patterns of aquatic microorganisms. *In* Microbiology ASM. D. Schlessinger (ed.).

Soldo, A. T., and Godoy, G. A. 1973. Molecular complexity of *Paramecium* symbiont lambda DNA; evidence for the presence of a multicopy genome. J. Mol. Biol. 73:93–108.

Stanier, R. Y. 1974. The Origins of Photosynthesis in Eukaryotes. Evolution in the Microbial World. Symposium 24: M. J. Carlile and J. J. Skehel (eds.). Cambridge, England: Cambridge University Press, 219–240.

Stanier, R. Y. 1977. The position of cyanobacteria in the world of phototrophs. Carlsberg Res. Commun. 42:77–98.

Stanier, R. Y., and Cohen-Bazire, G. 1977. Phototrophic prokaryotes: The cyanobacteria. Annu. Rev. Microbiol. 31:225–274.

Stoeckenius, W. 1978. Bacteriorhodopsin. *In* The Photosynthetic Bacteria. R. K. Clayton and W. R. Sistrom (eds.). Pp. 571–592.

Steinitz, Y., and Shilo, M. 1976. Study of photodynamic damage in cyanobacteria and mechanisms of their resistance to photooxidative death. In Proc. 2nd Int. Sym. Photosynthetic Prokaryotes. G. A. Codd and W. D. P. Stewart (eds.). Dundee; Pp. 18–20.

Tabita, F. R., Gibson, J. L., Whitman, W. B., Martin, M. N., and Robinson, P. D. 1979. The small subunit of ribulose bisphosphate carboxylase/oxygenase: Its role in the molecular regulation of CO_2 fixation and its evolutional significance. In IIIth Int. Sym. Photosynthetic Prokaryotes—abstracts. Oxford, E11.

Tandeau de Marsac, N. and Cohen-Bazire, G. 1977. Molecular composition of cyanobacterial phycobilisomes. Proc. Nat. Acad. Sci. USA. 74:1635–1639.

Thinh, L. V. 1978a. Photosynthetic lamellae of *Prochloron* (Prochlorophyta) associated with the ascidian *Diplosoma virens* (Hartmeyer) in the vicinity of Townsville. Aust. J. Bot. 26:617–620.

Thinh, L. V. 1978b. *Prochloron* (Prochlorophyta) associated with the ascidian *Trididemnum cyclops* Michaelson. Phycologia 18:77–82.

Thorne, S. W., Newcomb, E. H., and Osmond, C. B. 1977. Identification of chlorophyll *b* in extracts of prokaryotic algae by fluorescence spectroscopy. Proc. Nat. Acad. Sci. USA. 74:575–578.

Walsby, A. E. 1975. Gas vesicles. Annu. Rev. Plant Physiol. 26:427–439.

Whatley, J. 1977. The fine structure of *Prochloron*. New Physiol. 79:309–313.

Withers, N. W., Vidaver, W., and Lewin, R. A. 1978. Pigment composition, photosynthesis and fine structure of a non-blue-green prokaryotic algal symbiont (*Prochloron* sp.) in a didemnid ascidian from Hawaiian waters. Phycologia 17:167–171.

Yamanaka, G., Glazer, A. N. and Williams, R. C. 1978. Cyanobacterial phycobilisomes characterisation of the phycobilisomes of *Synechococcus* SP.6301. J. Biol. Chem. 253:8303–8310.

Discussion of Presentation by Dr. Shilo

HALLICK: In construction of family trees for evolution of proteins, it should be kept in mind that two proteins identical in amino acid sequence could have widely different DNA sequences. Ideally one should compare DNA sequences of related proteins to determine evolutionary relationships.

CASTELFRANCO: What is known about the mechanism of repression of oxygenic photosynthesis by H_2S?

SHILO: I don't think a mechanism is known at this time.

GANTT: I would like to comment on the cryptophyte phycobiliprotein aggregation state in regard to its possible phylogenetic relationship to other algae. Both Dr. Glazer and Dr. Shilo have alluded to the lack of phycobilisomes in the cryptophyte algae and to the location of these accessory pigments in the intrathylakoidal spaces (Gantt, 1979). Since only small aggregates of the pigments have been obatined in vitro, it has been assumed that they are incapable of forming large aggregates such as the phycobilisomes existing in the red and blue-green algae. Partly because of this enigmatic characteristic a separate phylogenetic origin has been suggested, in spite of the fact that their amino acid sequence, as pointed out by Dr. Glazer, is like that of red and blue-green algae (Glazer and Apell, 1977). In addition, there is also some immunologic cross-reactivity between phycobiliproteins of these groups (MacColl et al., 1976). It is quite possible that the phycobiliproteins of the three algal groups may indeed have a common origin, but that the binding components responsible for aggregation developed later. These binding components may be specific for the separate algal groups, and even for each species. Certain extra protein bands shown to occur in whole phycobilisomes by Tandeau de Marsac and Cohen-Bazire (1977) appear to be involved in phycobiliprotein binding. In our laboratory we have been able to show, by in vitro recombination studies, that small molecular weight proteins are present in the recombinable fractions, are absent in nonrecombinable fractions, and are never present in purified phycobiliproteins. Binding proteins may also be present in phycobiliprotein aggregation of the cryptophytes, but for some reason they may be very labile in the purification methods thus far employed.

SCHIFF: I take it that there is no possibility that the unique sequences arose by convergent evolution, i.e., that the sequence was so important to the protein's function that it arose independently in two evolutionarily unrelated molecules.

GLAZER: The different phycobiliproteins derived from cyanobacteria and red algae are homologous throughout their polypeptide chain length (Bryant et al., 1978). Aside from cysteinyl residues, which form links to the bilin moieties, the other residues in these proteins contribute to tertiary structure, intersubunit, and interprotein contacts. Such interactions place a high degree of constraint on the variation in amino acid sequences. In general, for proteins which interact with several other macromolecules (e.g., histones which interact with other histones and DNA in nucleosomes or cytochrome c which interacts with both a reductase and oxidase) the amino acid sequences are highly conserved. There are many examples of proteins which have converged to the same function with close similarities in the mechanism of action (e.g., bacterial subtilisins and such pancreatic proteolytic enzymes as chymotrypsin and trypsin). The sequences of such proteins which have converged to the same function show no homologies in amino acid sequence.

References to Discussion

Bryant, D. A., Hixon, C. S., and Glazer, A. N. 1978. Structural studies on phycobiliproteins. III. Comparison of bilin-containing peptides from the β subunits of C-phycocyanin, and phycoerythrocyanin. J. Biol. Chem. 253:220–225.

Gantt, E. 1979. Phycobiliproteins of Cryptophyceae. In Biochemistry and Physiology of Protozoa, Vol. 1. M. Levandowsky and S. H. Hutner (eds.). New York: Academic Press, pp. 121–137.

MacColl, R., Berns, D. S., and Gibbons, E. 1976. Characterization of cryptomonad phycoerythrin and phycocyanin. Arch. Biochem. Biophys. 177:265–275.

Tandeau de Marsac, N., and Cohen-Bazire, G. 1977. Molecular composition of cyanobacterial phycobilisomes. Proc. Nat. Acad. Sci. USA 74:1635–1639.

Additional Comments: Amino Acid Sequences of Biliproteins and the Endosymbiotic Origins of Chloroplasts

DR. ALEXANDER N. GLAZER: In a review examining the hypothesis that chloroplasts had an endosymbiotic origin, Stanier (1974) wrote: "As Margulis (1968) first suggested, the contemporary diversity of algal pigment systems might thus reflect a much more ancient diversity among prokaryotes that performed oxygenic photosynthesis, the type of chloroplast now characteristic of each major algal group having been derived from a specific type of prokaryotic endosymbiont. If the hypothesis of Margulis is correct, it must be assumed that the prokaryotic progenitors of most types of chloroplasts subsequently became extinct as free-living photosynthetic organisms: only cyanobacteria appear to have survived. However, one should perhaps not dismiss out of hand the possibility that other types of oxygen-producing photosynthetic prokaryotes may still be extant."

There was little delay in the fulfillment of the tentative prediction expressed in the last sentence of the above quotation. Lewin (1975) described microscopic unicellular organisms which he had found growing epizoically on calcified colonial ascidians (*Didemnum* sp.). Although Lewin had originally described these organisms as a new cyanobacterial species of *Synechococcus*, further examination revealed that these organisms were lacking in biliproteins and contained both chlorophylls *a* and *b* (Lewin and Withers, 1975). The organisms discovered by Lewin are thus members of a new class of prokaryotes ("prochlorophytes") able to perform oxygenic photosynthesis. They may well be the descendants of the long-sought group of ancestral prokaryotes from which the chloroplasts of green algae and land plants may have been derived.

Already in 1974, the evidence that the light-harvesting systems of cyanobacteria and red algal chloroplasts share a common evolutionary origin was very strong. Biliproteins are the major light-harvesting components of the photosynthetic apparatus of both of these systems. Cyanobacterial and red algal biliproteins were found to be closely related as determined by immunological criteria. Moreover, the supramolecular organization of biliproteins into phycobilisomes was similar in cyanobacteria and red algae (see Stanier, 1974 for a review). More recent amino acid sequence data on the phycobiliproteins confirm the earlier conclusion that these proteins are derived from a common ancestral gene (e.g., see Figure 1). These data support the conclusion that "a cyanobacterial origin for the rhodophytan chloroplast, if not for other structures in the rhodophytan cell, therefore now appears virtually certain" (Stanier, 1974).

Comparison of amino-terminal sequences of biliproteins (allophycocyanins, phycocyanins, and phycoerythrins) of many different cyanobacteria and red algae (see Glazer, 1980, for references to the original sources) demonstrates that certain positions are occupied by either one or at most two amino acid residues in the sequences all the proteins examined (Figure 2). Such constraints on sequence variations are generally the consequence of restrictions imposed

```
 1                                              10
Val Lys Thr Pro Ile Thr Asp Ala Ile Ala Ala Ala Asp Thr Gln Gly
Met                     Glu                             Asn

                     20                                 30
Arg Phe Leu Ser Asn Thr Glu Leu Gln Ala Val Asn Gly Arg Tyr Gln

                                     40
Arg Ala Ala Ala Ser Leu Glu Ala Ala Arg Ala Leu Thr Ala Asn Ala
                                         Ser             Ser

 50                                             60
Gln Arg Leu Ile Asp Gly Ala Ala Gln Ala Val Tyr Gln Lys Phe Pro

                         70                                     80
Tyr Leu Ile Gln Thr Ser Gly Pro Asn Tyr Ala Ala Asp Ala Arg Gly
    Thr Ser     Met Pro         Gln         Ser Ser     Val

                                         90
Lys Ser Lys Cys Ala Arg Asp Ile Gly His Tyr Leu Arg Ile Ile Thr
    Ala                         Tyr                 Met

            100                                     110
Tyr Ser Leu Val Ala Gly Gly Thr Gly Pro Leu Asp Glu Tyr Leu Ile
    Cys                             Met

                             120
Ala Gly Leu Asn Glu Ile Asn Asp Ala Phe Glu Leu Ser Pro Ser Trp
            Glu             Arg Thr     Asp

130                                             140
Tyr Ile Glu Ala Leu Lys Tyr Ile Lys Ala Asn His Gly Leu Ser Gly
    Val         Asn

                         150                                     160
Gln Ala Ala Asn Glu Ala Asn Thr Tyr Ile Asp Tyr Val Ile Asn Ala
                                             Ala

Leu Ser
```

Figure 1. Comparison of cyanobacterial and red algal biliprotein subunits. The continuous sequence is that of the C-phycocyanin α subunit of the filamentous cyanobacterium *Mastigocladus laminosus* (Frank et al., 1978). The residues indicated below this sequence represent replacements in the sequence of the corresponding subunit of the C-phycocyanin of the unicellular rhodophyte *Cyanidium caldarium*. From Troxler and Brown, 1979.

by the three-dimensional structure of the proteins. The pattern of the highly conserved residues may be viewed as a characteristic linear diagram which represents certain features of a particular three-dimensional protein structure.

In this context, it should be noted that biliproteins occur in another group •of eukaryotic algae—the cryptomonads. In these unicellular algae the biliproteins occupy an intrathylakoidal location and have distinctive molecular and aggregation properties (Stanier, 1974). However, the amino-terminal sequence

**Composites of the Amino-terminal Sequences
of the α and β Subunits of Cyanophytan and Rhodophytan Biliproteins**

Composite of the amino-terminal sequence of α-type subunits

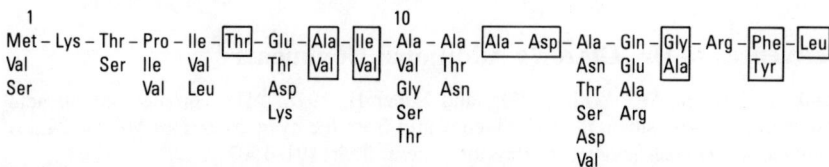

```
1                                        10
Met – Lys – Thr – Pro – Ile –|Thr|– Glu –|Ala|–|Ile|– Ala – Ala –|Ala – Asp|– Ala – Gln –|Gly|– Arg –|Phe|–|Leu|
Val          Ser   Ile   Val        Thr |Val||Val|  Val   Thr           Asn   Glu |Ala|        |Tyr|
Ser                Val   Leu        Asp                   Gly   Asn           Thr   Ala
                               Lys                   Ser                 Ser   Arg
                                                     Thr                 Asp
                                                                         Val
```

Composite of the amino-terminal sequences of β-type subunits

```
Met – Leu – Asp – Thr – Phe –|Thr|– Lys –|Val|–|Val|– Ala – Gln –|Ala – Asp|– Ala – Arg –|Gly|– Glu –|Phe|–|Leu|
Ala    Phe          Ala   Ile  |Ala|  Arg |Ala||Ile|  Val   Ala           Val   Lys |Ala|  Ala |Tyr||Val|
Thr    Tyr          Val              Ala               Asn   Asn           Ser   Gln        Lys
Gly    Glu                                             Gln                 Thr              Asn
                                                       Ser                 Arg
```

**Comparison of the Amino-terminal Sequences
of a Cryptomonad Phycocyanin and a Rhodophytan Phycoerythrin**

Amino-terminal sequence of the β-subunit of *Hemiselmis virescens* phycocyanin

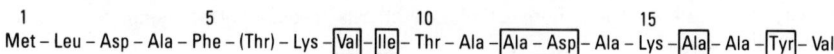

```
1                  5                  10                 15
Met – Leu – Asp – Ala – Phe – (Thr) – Lys –|Val|–|Ile|– Thr – Ala –|Ala – Asp|– Ala – Lys –|Ala|– Ala –|Tyr|– Val
```

Amino-terminal sequence of the β-subunit of *Porphyridium cruentum* β-phycoerythrin

Met – Leu – Asp – Ala – Phe – Thr – Arg – Val – Val – Val – Asn – Ala – Asp – Ala – Lys – Ala – Ala – Tyr – Val

(Residues that are highly conserved in the composite sequences of both α and β subunits are "boxed")

Figure 2. The composite sequences of the α and β subunits of cyanobacterial and red algal biliproteins were constructed by placing the most frequently occurring residue in the continuous sequence and, under the continuous sequence, the alternate residues observed at each position in order of frequency of occurrence.

of the β subunit of a cryptomonad phycocyanin displays the characteristic pattern of conserved residues seen in the cyanobacterial and rhodophytan biliproteins and resembles most closely the amino-terminal sequence of the β subunit of a red algal phycoerythrin (Glazer and Apell, 1977) (Figure 2). In the framework of the endosymbiosis hypothesis two explanations may be offered for this finding. One is that the cyanobacteria and the prokaryotic precursor of the cryptomonad chloroplast diverged, prior to the evolution of eukaryotic cells, from a common prokaryote which carried the ancestral biliprotein gene. A second possibility is that the cryptomonad chloroplast arose from a photosynthetic prokaryote different from the ancestor of the cyanobacteria, but that at a much later stage of evolution it acquired additional characteristics from a red algal endosymbiont (Taylor, 1979).

The data accumulated since 1974 on accessory light-harvesting pigment systems continue to strengthen the case for the endosymbiotic hypothesis for the origins of chloroplasts.

References to Dr. Glazer's Additional Comments

Frank, G., Sidler, W., Widmer, H., and Zuber H. 1978. "The complete amino acid sequence of both subunits of C-phycocyanin from the cyanobacterium *Mastigocladus laminosus*. Hoppe-Seyler's Z. Physiol. Chem. 359:1491–1507.

Glazer, A. N. 1980. Structure and evolution of photosynthetic accessory pigment systems with special reference to phycobiliproteins. *In* The Evolution of Protein Structure and Function. UCLA Forum in Medical Sciences. Vol. 21. S. Sigman and M. A. B. Brazier (eds.). New York: Academic Press.

Glazer, A. N., and Apell, G. A. 1977. A common evolutionary origin for the biliporteins of cyanobacteria, rhodophyta and cryptophyta. FEMS Lett. 1:113.

Lewin, R. A. 1975. A marine *Synechocystis* (Cyanophyta, Chroococcales) epizoic on ascidians. Phycologia 14:153–160.

Lewin, R. A., and Withers, N. W. 1975. Extraordinary pigment composition of a prokaryotic alga. Nature 256:735–737.

Margulis, L. 1968. Evolutionary criteria in thallophytes: a radical alternative. Science 161:1020–1023.

Stanier, R. Y. 1974. The origins of photosynthesis in eukaryotes. Symp. Soc. Gen. Microbiol. 24:219.

Taylor, F. J. R. 1979. Symbionticism Revisited: A Discussion of the Evolutionary Impact of Intracellular Symbioses. Proc. R Soc. Lond. B. Biol. Sci. 204:267–286.

Troxler, R. F., and A. S. Brown. 1979. Amino Acid Sequence of the Phycocyanin α Subunit from the Alga, *Cyanidium caldarium*. Fed. Proc. 38:510(abs.).

Evolution of Photosynthetic Catalysts in Cyanobacteria

David W. Krogmann

Cyanobacteria are the most primitive of the oxygen-producing photosynthetic organisms and are separated from higher plants by an exceedingly long evolutionary distance. However, the cyanobacteria possess a photosynthetic apparatus which is very similar in most of its details to the one found in higher plant chloroplasts (Krogmann, 1977). The path of light-driven electron transport (Figure 1), the proton gradient for energy coupling, and even the protein assembly of the coupling factor for ATP synthesis are identical or nearly so in cyanobacteria and spinach chloroplasts. There are a few important points of difference, however. In cyanobacteria, the photosynthetic membranes are diffused throughout the cell; in eukaryotic organisms, the photosynthetic membranes are gathered together within a chloroplast envelope membrane and are often organized into grana stacks. The cyanobacteria share with the red algae, but not with other eukaryotic plants, an unusual type of organization of the major light-absorbing pigments of photosystem 2. Phycobiliproteins replace the chlorophyll-*b*-enriched pigment protein of the green algae and higher plants as light-harvesting antennae of photosystem 2, and the phycobiliproteins are not integrated into the membranes but rather are organized as special structures—phycobilisomes—on the surface of the thylakoids

An interesting point of evolutionary divergence has been revealed by studying the proteins which serve as electron donors to photosystem 1. In cyanobacteria

Supported by Grant No. PMC 01956-A-02 from the Metabolic Biology Program of the National Science Foundation.

Address reprint requests to Dr. David Krogmann, Professor, Department of Biochemistry, Purdue University, West Lafayette, Indiana 47907.

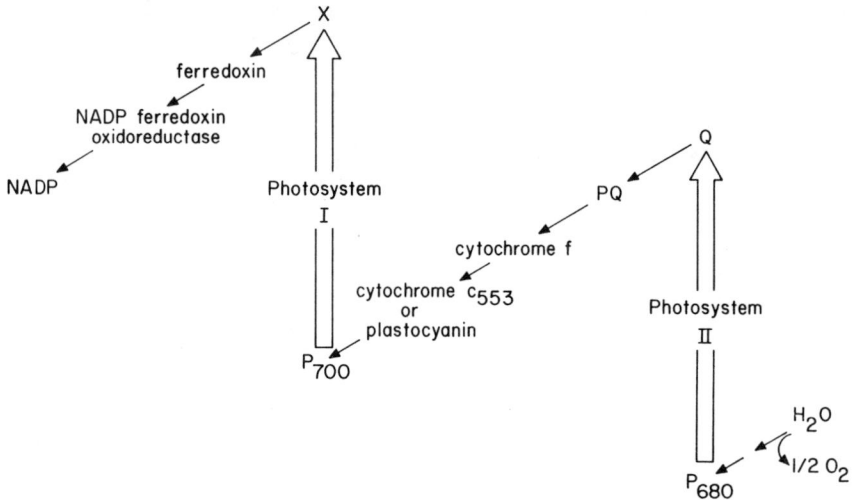

Figure 1. The photosynthetic electron transport sequence.

and in many eukaryotic algae and higher plants, P_{700} is reduced via the copper protein, plastocyanin. When cyanobacteria and many of the eukaryotic algae are grown under conditions of limited availability of copper, plastocyanin is replaced by cytochrome c_{553} (Wood, 1978). Since these two different types of molecules are functionally interchangeable, it seems likely that they will show points or domains of similarity corresponding to the sites of interaction with reducing and oxidizing reactants in the membrane. Both plastocyanin and cytochrome c_{553} are small (approximately 10,000 daltons), stable, and easy to isolate. Plastocyanin shows interesting similarities to the azurins—copper proteins in the respiratory chain of the pseudomonads. Cytochrome c_{553} is similar to other respiratory cytochromes c and seems quite like the cytochrome c_{552} of *Chromatium.* The first indication of an evolutionary alteration in this region of the photosynthetic electron transport chain came from measurements of the isoelectric points of plastocyanins. Spinach plastocyanin has an isoelectric point of 3 while the plastocyanin from the cyanobacterium *Anabaena variabilis* has an isoelectric point of 8.4. Extensive measurements of the nuclear magnetic resonance spectra of plastocyanins from a variety of higher plants and from *A. variabilis* indicate that the tertiary structures of these molecules must be nearly identical (Markley et al., 1975; Freeman et al., 1977). When Freeman and his colleagues solved the structure of a plastocyanin by X-ray crystallographic techniques (Coleman et al., 1978), this structure could be used to identify the region of charge alteration. Figure 2 shows the Freeman structure of plastocyanin. On the left the primary structure of plastocyanin from spinach is superimposed and the charged residues are marked. On the right the primary structure of *A. variabilis* is used to identify the charged areas. The two molecules differ in that a band of positive charges

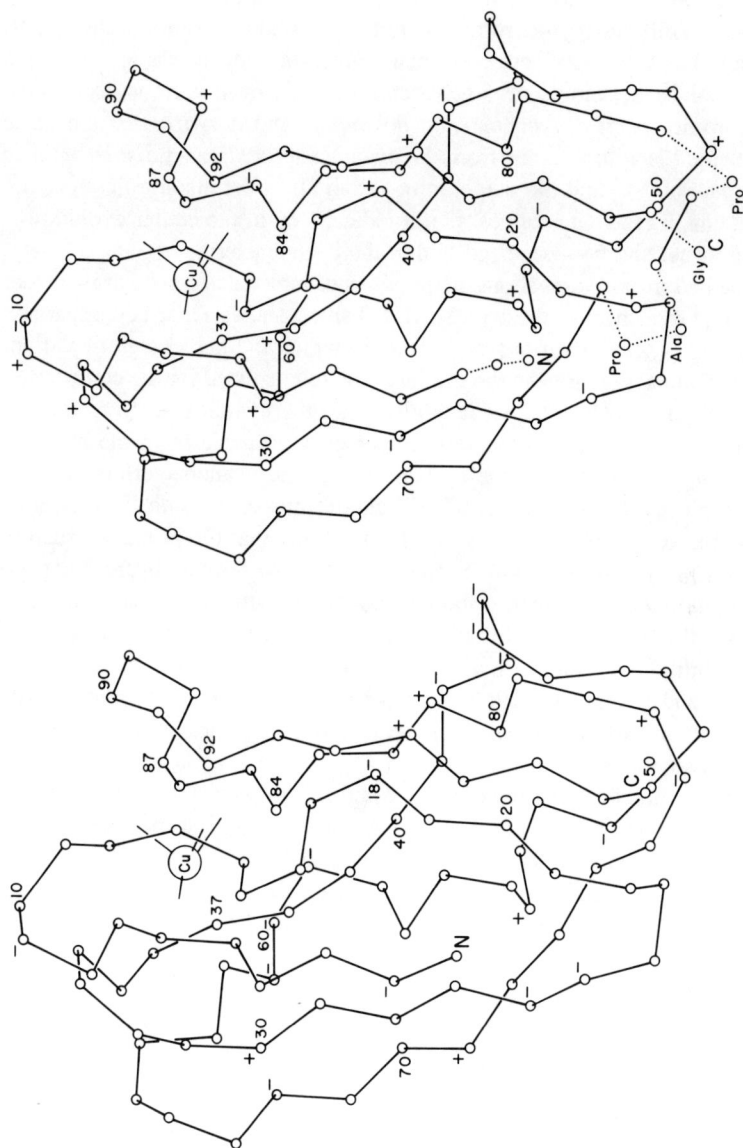

Spinach plastocyanin

A. variabilis plastocyanin

Figure 2. Tertiary structures of plastocyanins showing the distribution of charged residues.

around the center of the plastocyanin from *A. variabilis* is replaced by a band of negative charges in the molecule from spinach.

The alteration of net charge seen in plastocyanins is also observed in cytochrome c_{553}. While this cytochrome has not been isolated from a higher plant, the molecule has been obtained from many eukaryotic algae. Table 1 shows a compilation of isoelectric points of cytochromes c_{553} from various organisms. The cytochromes from *Aphanizomenon flos-aquae* and *A. variabilis* are basic proteins, while the cytochromes from the cyanobacteria *Microcystis aeruginosa* and *Spirulina maxima* and the cytochromes from all of the eukaryotic algae are acidic proteins. These data suggest that a radical step in molecular evolution— the change in net charge—occurred within the cyanobacteria.

More detailed information about the relations of molecules can be drawn from the amino acid sequences of these molecules. Table 2 shows a matrix comparison of cytochrome c_{553} from various organisms in which the percentage of variant residues in identical positions in the primary structure is listed for the comparison of one cytochrome with another. It is clear that the cytochrome from the cyanobacterium *S. maxima* is more similar to the cytochrome from the red alga *Porphyra teneria* than it is to the cytochrome of the cyanobacterium *A. flos-aquae*. There is evidence that a similar situation may occur with another gene product, ferredoxin. Takruri et al. (1978) have found that the primary structure of ferredoxin of *Spirulina* is more similar to that of the ferredoxin from the red alga *Porphyra umbilicalis* than to the ferredoxins of some other cyanobacteria. Nonetheless, the evolution of ferredoxin has not greatly altered its isoelectric point or its ability to interact with adjacent electron carriers in the photosynthetic chain. Susor and Krogmann (1966) demonstrated that ferredoxin from *A. variabilis* was functionally interchangeable with spinach ferredoxin in NADP photoreduction assays.

A functional interchangeability of plastocyanins from *A. variabilis* and spinach was not found by Tsuji and Fujita (1972). Spinach chloroplast membrane preparations did not respond with restored photosytem 1 activity on addition of *A. variabilis* plastocyanin as they did on addition of spinach plastocyanin. The study of P_{700} interaction with plastocyanin is difficult since, at the time of writing,

Table 1. Isoelectric Points of Cytochromes c_{553}

Source	pI
Aphanizomenon flos-aquae (Cyanophyta)	9.3
Anabaena variabilis (Cyanophyta)	8.9
Euglena gracilis (Euglenophyta)	5.5
Microcystis aeruginosa (Cyanophyta)	5.3
Spirulina maxima (Cyanophyta)	5.1
Porphyridium cruentum (Rhodophyta)	4.3
Porphyra tenera (Rhodophyta)	3.9
Bumilleriopsis (Chlorophyta)	3.8

Table 2. Matrix of Differences in Amino Acid Sequence of Cytochromes c_{553}

	S. maxima[a]	M. aeruginosa[a]	A. flos-aquae[a]	P. tenera[a]
S. maxima	0	46	53	47
M. aeruginosa	46	0	44	46
A. flos-aquae	53	44	0	56
P. tenera	47	46	56	0

[a]Percentage of variant residues.

P_{700} is indissolubly linked to a rather complex membrane fragment. Interaction of P_{700} with plastocyanin may be greatly influenced by adhering constituents of the photosynthetic membrane in the P_{700} preparation. A highly resolved P_{700} preparation from spinach chloroplasts was used by Davis et al. (1979) to obtain some interesting insights concerning the interaction of protein electron donors with P_{700}. The assay used was to measure the rate of dark reduction of photooxidized P_{700} by various plastocyanins or cytochromes present in varying concentrations. Magnesium ion was known to facilitate the interaction of spinach plastocyanin with this P_{700} enriched membrane fragment preparation. These experiments established the affinity of spinach P_{700} for various electron donors and the influence of magnesium ions on these interactions in the reconstructed system. The results are summarized in Table 3. The first four entries in the table show that spinach plastocyanin and cytochromes c_{553} from eukaryotic algae (all with isoelectric points below 5) show a modest affinity for P_{700} which is greatly improved by the presence of magnesium ions. The magnesium ion effect can

Table 3. Affinities of Various Electron Donors for a Spinach P_{700} Preparation

Elected donor	K_m without Mg^{2+}	K_m with 20 mM Mg^{2+}
Spinach plastocyanin	10.0 μM	1.6 μM
Bumilleriopsis cytochrome c_{553}	5.0 μM	0.25 μM
Porphyra tenera cytochrome c_{553}	4.0 μM	0.67 μM
Porphyridium cruentum cytochrome c_{553}	1.25 μM	0.23 μM
Spirulina maxima cytochrome c_{553}	0.50 μM	0.71 μM
Microcystis aeruginosa cytochrome c_{553}	0.37 μM	0.42 μM
Euglena gracilis cytochrome c_{553}	0.33 μM	0.20 μM
Anabaena variabilis plastocyanin	0.09 μM	0.20 μM
Anabaena variabilis cytochrome c_{553}	0.09 μM	0.25 μM
Aphanizomenon flos-aquae cytochrome c_{553}	0.05 μM	0.08 μM

Source: Davis et al. (1979).

be duplicated by high concentrations of monovalent cations or by very low concentrations of polycations so this seems to be a facilitation due to the suppression of repulsive anion charges on the P_{700} enriched particle. At the proper magnesium ion concentration, soluble spinach plastocyanin reacts with the P_{700} particle at a stoichiometry similar to the one found in vivo where both constituents are embedded in the chloroplast membrane. Cytochromes with isoelectric points near 5 (from two cyanobacteria and one eukaryote) are hardly affected by magnesium ions in their interaction with P_{700}. The proteins with isoelectric points above 8—plastocyanin or cytochrome from *A. variabilis* and the cytochrome from *A. flos-aquae*—have a high affinity for the P_{700} particle and magnesium ions inhibit these interactions.

Cytochrome f is the other reaction partner with plastocyanin or cytochrome c_{553} in the photosynthetic electron transfer sequence shown in Figure 1. Ho and Krogmann (1980) have purified cytochrome f from spinach, from the eukaryotic red alga *Porphyridium cruentum,* and from prokaryotic cyanobacteria *S. maxima* and *A. flos-aquae*. This cytochrome is larger (mol wt 35,000 to 37,000) and shows only small variations in its isoelectric point. Cytochromes f do show some reaction specificity in that spinach cytochrome f is more rapidly oxidized by spinach plastocyanin than by *S. maxima* cytochrome c_{553} while *S. maxima* cytochrome f is more rapidly oxidized by the cytochrome c_{553} of *S. maxima* than by spinach plastocyanin. Full primary structures of these cytochromes f are needed but the N terminal regions of these molecules from prokaryotes and eukaryotes are quite similar.

Several interesting lines of speculation arise from the data described here. First, there has been a radical evolution in the ionic charge character of the proteins mediating electron flow between cytochrome f and P_{700}. This radical change in isoelectric point has occurred among the cyanobacteria, and the isoelectric point of either plastocyanin or cytochrome c_{553} from a given genus of cyanobacteria may be useful in classifying that genus in terms of its proximity to the eukaryotes. Second, the radical change in charge of these molecules has not altered their tertiary structure but may have led to an alteration of the site in the photosynthetic membrane through which they react with P_{700}. The photosynthetic membranes may have adjusted to the evolution of these electron donors by evolving an additional electron carrier or some special binding elements which have been removed during the purification of P_{700} enriched particles used in the studies cited here. The analysis of proteins in P_{700} enriched particles from spinach by Bengis and Nelson (1977) and the kinetic studies on P_{700} reduction and plastocyanin oxidation in eukaryotic cells by Bouges-Bocquet and Delosme (1978) suggest that there may be a special element or an additional intermediate in this region of the electron transfer sequence. It will be interesting to make a detailed comparison of P_{700} and its neighboring molecules in eukaryotes and in primitive prokaryotes to see how the evolution of an electron carrier is accommodated.

References

Bengis, C., and Nelson, N. 1977. Subunit structure of chloroplast photosystem 1 reaction center. J. Biol. Chem. 252:4564–4569.

Bouges-Bocquet, B. and Delosme, R. 1978. Evidence for a new electron donor to P_{700} in Chlorella pyrenoidosa. FEBS Lett. 94:100–104.

Coleman, P. M., Freeman, H. C., Guss, J. M., Murata, M. Norris, V. A. Ramshaw, J. A., and Venkatappa, M. P. 1978. X-ray crystal structure analysis of plastocyanin at 2.7 ÅA resolution. Nature 272:319–324.

Davis, D. J., Krogmann, D. W., and San Pietro, A. 1980. Electron donation to photosystem 1. Plant Physiol. 65:697–702.

Freeman, H. C., Norris, V. A., Ramshaw, J. A., and Wright, P. E. 1977. High resolution proton magnetic resonance studies of plastocyanin. IN Proc. 4th Int. Photosynthesis. D. O. Hall, J. Coombs, and T. W. Goodwin (eds.). London: Biochemical Society.

Ho, K. K., and Krogmann, D. W. 1980. Cytochrome f from spinach and cyanobacteria: purification and characterization. J. Biol. Chem. 255:3855–3861.

Krogmann, D. W. 1977. Blue-green algae. In Encyclopedia of Plant Physiology, New Series, Vol. 5. A. Trebst and M. Avron (eds.). Berlin, Heidelberg, New York: Springer-Verlag.

Markley, J. L., Ulrich, E. L., Berg, S. P. and Krogmann, D. W. 1975. Nuclear magnetic resonance studies of the copper binding site of blue copper proteins: oxidized, reduced and apoplastocyanin. Biochemistry 14:4428–4432.

Susor, W. A., and Krogmann, D. W. 1966. TPN photoreduction with cell-free preparations of Anabaena variabilis. Biochim. Biophys. Acta 120:65–72.

Takruri, I., Haslett, B. G., Boulter, D., Andrew, P. W., and Rogers, L. J. 1978. The amino acid sequence of ferredoxin from the red alga Porphyra umbilicalis. Biochem. J. 173:459–466.

Tsuji T., and Fujita, Y. 1972. Electron donor specificity observed in photosystem 1 reactions of membrane fragments of the blue-green alga Anabaena variabilis and higher plant Spinacia oleracea. Plant Cell Physiol. 13:93–99.

Wood, P. M. 1978. Interchangeable copper and iron proteins in algal photosynthesis. Eur. J. Biochem. 87:9–19.

Discussion of Presentation by Dr. Krogmann

SCHIFF: Sandmann and Böger (1980) have also shown the induction of plasto-cyanin by growth on copper and the formation of c-type cytochrome in the absence of copper in the green alga Scenedesmus; Wood (1978) has found this phenomenon in Chlamydomonas. Since you find this in cyanobacteria, perhaps these represent adaptations of certain organisms to copper availability regardless of phylogenetic position. There are organisms which as far as we know are committed to plastocyanin (e.g., higher plants) or to cytochrome c (e.g., Euglena). (See Freyssinet et al. (1980) for a list of papers concerned with cytochromas c and plastocyanin in photosynthetic electron transport.)

KROGMANN: Perhaps, but many of the organisms which seem committed to plastocyanin, like spinach, have not been examined for the cytochrome under conditions of copper deficiency. Copper deficiency is necessary to see this cytochrome in algae and cyanobacteria, and may be required for cytochrome appearance in other organisms, even higher plants, if the gene for the cytochrome has persisted.

COHEN: Copper-containing enzymes are pretty rare in prokaryotes and the existence of copper-containing plastocyanin in the cyanobacteria might be considered as a possible evolutionary marker. If copper enzymes evolved late only under conditions in which copper was formed in a soluble state it can be imagined that the earliest cyanobacteria might have lacked a Cu-containing plastocyanin. Is it possible that there are cyanobacteria lacking such an enzyme which may reflect such an early origin?

KROGMANN: That is a very attractive possibility. One simply must grow more cyanobacterial species under conditions of copper sufficiency and look for a plastocyanin. Plastocyanin is present in *Anabaena variabilis* which is among the more primitive cyanobacteria if primitiveness is truly reflected in the isoelectric point of cytochrome$_{533}$. Both the plastocyanin and cytochrome$_{533}$ are basic proteins in *A. variabilis,* in contrast to the acidic proteins in eukaryotes.

References to Discussion

Freyssinet, G., Harris, G. C., Nasatir, M., and Schiff, J. A. 1980. Isolation and immunological estimation of chloroplast cytochrome c-552 from *Euglena gracilis. In* Methods in Chloroplast Molecular Biology. M. Edelman, R. B. Hallick, and N.-H. Chua (eds). Amsterdam: Elsevier (in press).

Sandmann, G., and Böger, P. 1980 Physiological factors determining formation of plastocyanin and plastidic cytochrome c-553 in *Scenedesmus.* Planta 147:330–334.

Wood, P. M. 1978. Interchangeable copper and iron proteins in algal photosynthesis. Eur. J. Biochem. 87:9–19.

Diversity of Chloroplast Structure

Shimon Klein

The subject of diversity in chloroplast structure includes the genetically fixed differences between the structures of the mature chloroplasts in the various groups of plants, the extent to which the "genotypic" wild-type pattern of chloroplast structure can be affected by internal or external factors, and the morphogenetic changes which occur in those plastids which undergo ontogenetic development.

The variation in chloroplast structures among the various groups of lower and higher plants, and the value and implication for phylogenetic considerations of this diversity, has been discussed elsewhere (Klein and Cronquist, 1967; Gibbs, 1970; Dodge, 1973; Kirk and Tilney-Bassett, 1967). Also, the effects of mutations on plastid structure and the variations induced by environmental factors have been reviewed.

In regard to plastids, it has been suggested that ontogenesis seems to repeat phylogenesis (Klein and Cronquist, 1967). If indeed phylogeny is a series of ontogeneses (Simpson and Beck, 1965), and what is passed on from one generation to the next is a blueprint for a developmental mechanism, then early steps in ontogenetic pathways may have a bearing on the question of origin and evolution of plastids. Because of this, it may be appropriate to approach the subject of diversity in chloroplast structure on an ontogenetic basis. I would like, therefore, to start with a short summary of the morphogenetic changes which occur during plastid ontogenesis in higher plants. This will permit a discussion of whether there are indeed separate ontogenetic pathways (e.g., in darkness and in light) leading to different structures or whether, as has been suggested by

Address reprint requests to Dr. Shimon Klein, Professor of Botany, Department of Botany, The Hebrew University, Jerusalem, Israel.

Whatley (1977), the formation of different structures under different conditions could be reconciled with the concept of a common, although "plastic," developmental pathway. A clarification of this point may be useful in comparing plastid ontogeny in higher plants with that in algae.

In this chapter I will discuss morphological changes in chloroplast structure mainly with respect to the thylakoid system and related, or apparently related, structures.

Morphogenetic Changes During Ontogeny of Plastids in Higher Plants

Figure 1 shows a proplastid in a primary bean leaf 2-3 days after the onset of germination, while the leaf is still enclosed in the cotyledons. Leaf proplastids at this early stage already contain well-developed starch grains. Comparable proplastids from leaf primordia, for example, may still be devoid of starch. The proplastid is surrounded by the two envelopes and contains some few prothylakoids which have generally been presumed to originate from the inner envelope, either through invaginations, or by fusion of blebs originating from them. The numerous earlier examples, on which this idea is based, have been summarized by Kirk and Tilney-Bassett (1967) and by Rosinski and Rosen (1972). However, it has been stressed by Douce and Joyard (1978) that no convincing evidence to support this idea has been produced up to now. The apparent inner envelope-prothylakoid connections could easily be artifacts caused by section geometry (serial sections or tilting devices have usually not been used in these investigations). Also, fixation artifacts cannot be excluded. It is relevant that in an investigation based on serial sectioning of plastids which contain a peripheral reticulum, no connections were found to exist between this structure and the thylakoid system (Sprey and Laetsch, 1978). Also, the differences between chemical composition of the envelope membranes and developing thylakoids (Douce and Joyard, 1978) and the size and distribution of particles in them (Sprey and Laetsch, 1978) make it questionable whether the envelopes serve as direct structural precursors for the thylakoid system. This, however, does not imply that the envelopes do not participate in thylakoid assembly. A portion of the thylakoid polypeptides and some of their lipids are synthesized outside the plastids. Galactolipids are synthesized in the envelopes. Rather than being direct structural precursors of thylakoids, it is plausible that invaginations and blebs formed by the inner envelopes facilitate the transport of these compounds to, and their incorporation into, the developing thylakoids.

In most of the electron micrographs of young proplastids, small unattached prothylakoids can be recognized. The chloroplast envelopes are morphologically permanent structures. It appears, then, that with respect to thylakoid membranes in higher plants, the largely discarded concept of their continuity may still be valid.

During early proplastid development there is an increase in the prothylakoid

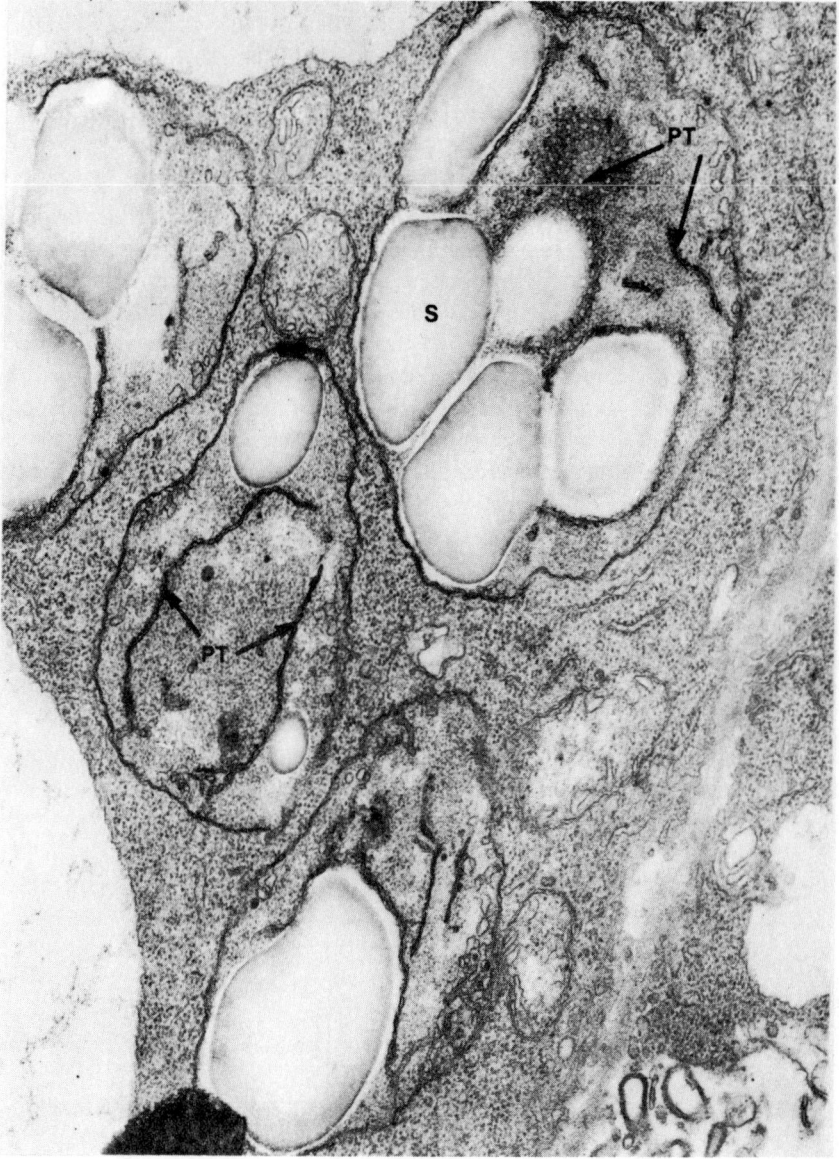

Figure 1. Proplastids in a primary bean leaf, 3 days after imbibition of the seed. PT = prothylakoids, S = starch grains. Note perforations in those PTs that were sectioned tangentially. (×37,000) *Source:* Klein and Schiff, 1972.

membranes, which appear as perforated structures. At this stage starch grains may be formed, frequently in close connection with the growing membranes, which may also surround them. Such proplastids in primary bean leaves contain a small amount of protochlorophyll(ide) [Pchl(ide)] with a predominant absorption peak in vivo of 630-635 nm (P_{635}) (Klein and Schiff, 1972).

At this early stage of plastid development lighting conditions might cause a diversity in chloroplast structure. In light-grown seedlings the prothylakoids frequently develop directly into chlorophyll[ChL]-containing, photosynthetically active thylakoids (Kirk and Tilney-Bassett, 1967) while in dark-grown seedlings tubular complexes appear (Figure 2). How these tubular complexes form is not yet clear. Weier and Brown (1970) and Bradbeer et al. (1974) have suggested that they are formed through fusion and condensation of the previously formed perforated thylakoids—in which case composition of the latter should be similar or identical to that of the tubular complex. Klein and Schiff (1972) have suggested that the formation of the tubular complexes requires continuous synthetic processes and starts with bridge formations between the existing prothylakoids. This would allow for differences in the chemical composition of the complexes and the prothylakoids. Then the complex enlarges, due to an increase in the number of tubules, and the organization of the complex, which at the beginning is rather loose and irregular, becomes light and regular (Weier and Brown, 1970; Klein and Schiff, 1972) (Figure 3). The complex continues to increase in size for a

Figure 2. Proetioplasts in a primary leaf from a 4-day-old bean seedling, germinated in the dark. Note the tubular complexes which are beginning to be formed (intact arrow) and the stacking of the thylakoids (broken arrow). (× 29,000)

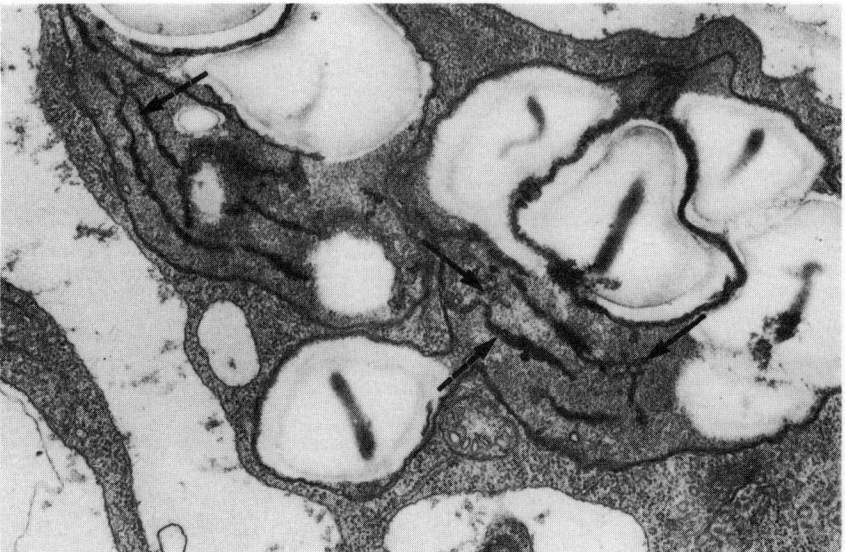

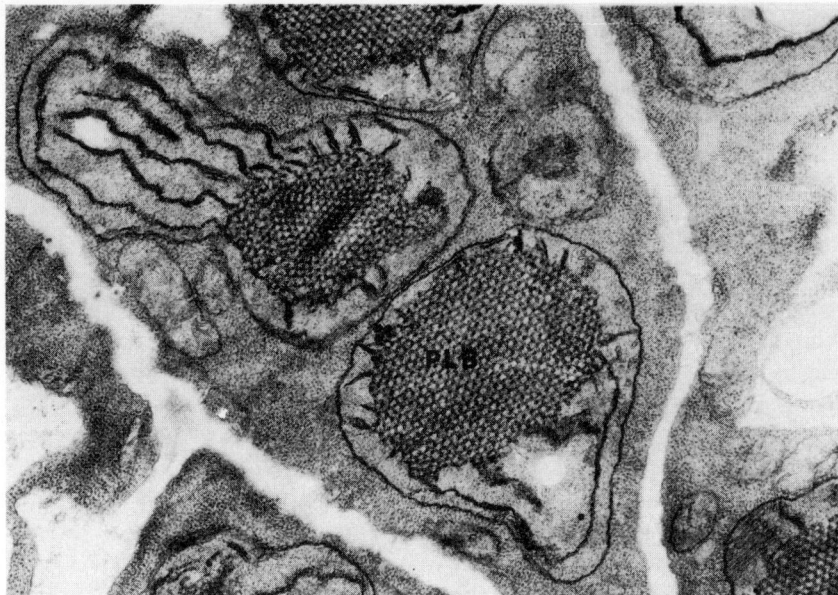

Figure 3. Etioplasts in a primary bean leaf from a 7-day-old seedling grown in the dark. PLB = prolamellar body. ($\times 26{,}000$)

number of days in darkness, and this is paralleled by a concomitant increase in Pchl(ide) content of the plastids, P_{650} now becoming the predominant form (Klein and Schiff, 1972; Lancer et al., 1976).

For a more detailed discussion of the occurrence and architecture of the tubular structures, the "prolamellar bodies" (PLBs), than is possible here, the reader is referred to the reviews of Kirk and Tilney-Basset (1967), Rosinski and Rosen (1972), and Gunning and Jagoe (1967).

During the earliest development of the proplastid in the dark, most of the Pchl(ide) appears to be nonphotoconvertible. In the bean, after 2 days of dark germination only about 40% of the Pchl(ide) is converted by saturating light flashes to Chl(ide), P_{635} being transformed to Chl_{671} without a Shibata shift. With the accumulation of P_{650}, the percentage of active Pchl(ide) increases up to 80% and the photoconversion is followed by a Shibata shift (Klein and Schiff, 1972; Lancer et al., 1976). Obviously, the in vivo form of Pchl(ide) P_{650}, which is commonly encountered in work with older plant materials, is not the predominant form of Pchl(ide) in young proplastids.

Previous fluorescence-microscopic investigations (Boardman and Anderson, 1964), combining fluorescence and negative contrast electron microscopy (Kahn, 1968), and autoradiography of etioplast sections containing ^{3}H-Pchl(ide) (Lafleche et al., 1972), suggested that the pigment may be located primarily in the PLBs. The observations noted above concerning the concomitant increase in

Chl(ide) content in relation to PLB growth pointed in the same direction (Klein and Schiff, 1972). Because of this, relationships between the molecular properties of the Pchl(ide) holochrome and the peculiar structure of the PLBs were postulated (Boardman et al., 1978).

When such etiolated leaves are exposed to relatively strong white light, the PLBs rapidly lose their crystalline character, and the tubules, while still interconnected, become irregularly arranged. In some cases the light requirements for this structural change were found to be similar to those for the Pchl(ide) to Chl(ide) conversion (Eriksson et al., 1961; Klein et al., 1964) and the two may coincide. This too, suggested that light-induced changes in the Pchl(ide) holochrome, the Pchl(ide) to Chl(ide) conversion, the Shibata shift, or phytylation may be causally related to the structural changes in the PLB membranes (Boardman et al., 1978; Treffry, 1978).

After this reorganization, the PLBs start to diminish in size, while the presumably attached prothylakoids begin to grow, frequently forming closed, apparently spherelike structures, which appear as girdles in the sections. This is especially prominent when normal development is interfered with, e.g., by lowering the temperature or by treatment with certain inhibitors such as aminooxyacetate, which is an inhibitor of transamination (Figure 4).

Figure 4. Etiochloroplasts in a cell from the second leaf of a 9-day-old dark-grown maize seedling, which was exposed for 8 hr to white light while being treated with aminooxyacetate. (× 17,500)

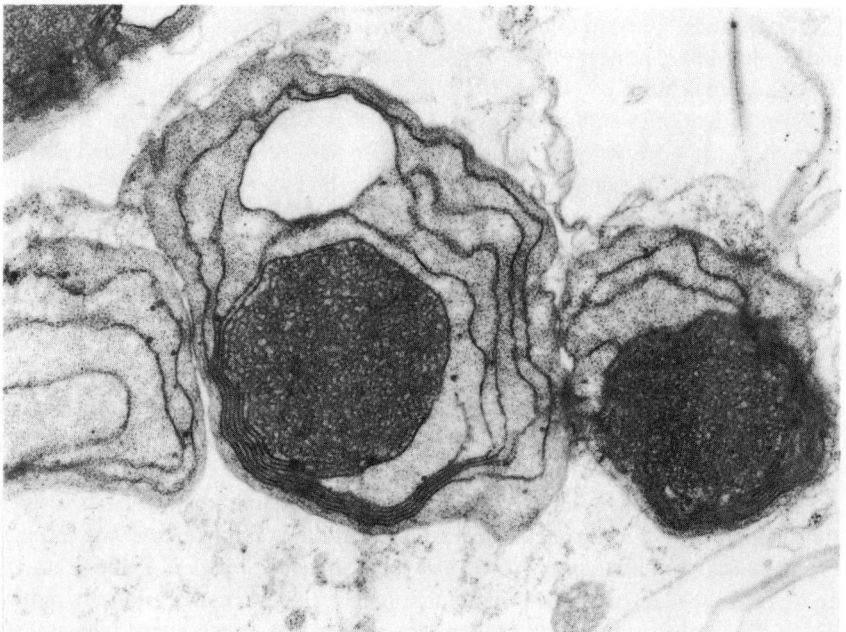

With the continuing disappearance of the PLBs the thylakoids become non-perforated and are arranged in a parallel manner. It has been calculated that the entire increase in membrane surface of the enlarging primary thylakoids prior to grana formation could be accounted for by that of the PLB tubules (Gunning and Jagoe, 1967).

These events coincide roughly with the lag period in Chl synthesis. It is during this period that the light-induced changes in the polypeptides of the photosynthetic membranes become apparent and the photosystems become assembled and active. Only then, approximately at the beginning of rapid Chl accumulation, is massive grana formation initiated. This general chain of events occurs under continuous light, but can be interfered with by changes in light, temperature, and other external conditions, leading to the formation of agranal plastids, of those with different amounts of stacking, or to the appearance of giant grana. Mutants may also possess plastids, the final development of which is stopped at practically any of the light-induced developmental stages (v. Wettstein, 1976).

PLBs were looked for and found predominantly in dark-grown etiolated tissues where they, together with the prothylakoids, represent the final and culminating structures. Since, as mentioned above, plastid development in light-grown seedlings may proceed directly from proplastids to chloroplasts, the concept of a dual pathway for plastid development in higher plants evolved.

The above observations also led to the assumption that "the prolamellar body has an extensive role in normal plastid development" (Rosinski and Rosen, 1972) by actively participating in the light-induced thylakoid formation of etioplasts. This again implied that thylakoid formation in illuminated etioplasts, involving the PLB, differs from that in light-grown plants where a direct prothylakoid to thylakoid transition occurs (Leech, 1977).

Most of the present information on light-induced chloroplast development in higher plants is derived from studies of the development of etioplasts into chloroplasts. The relevance of the information thus obtained to the light-induced development of proplastids into chloroplasts—the process usually occurring in nature—can be questioned. Are there indeed two pathways with significant differences between them? As pointed out, this concept is based mainly on interpretations which relate the formation of the prolamellar body to conditions of darkness, and which assume that Pchl(ide) accumulation occurs in the prolamellar bodies. Despite findings which lend themselves to such an interpretation, the concept itself is still questionable.

Rosinsky and Rosen (1972) stated that the concept of the PLB as a structure developing only in the dark is not in accord with all the available evidence. By then, PLBs had been known to occur as transient structures in differentiating meristematic cells of light-grown seedlings, and were not uncommonly found together with stroma thylakoids and grana. It was shown by Whatley (1977) that in the bean, early stages in chloroplast development (including PLB formation) are similar in darkness and in light, the existing differences being more of a quantitative than a qualitative nature. Similar results were reported by Platt-Aloia and Thomson (1977), and by Rascio et al. (1976), who made the interesting finding that etioplast formation was a common step in chloroplast development in meristematic cells of younger maize leaves, while in comparable cells of older leaves this step was bypassed. This would indicate that the formation of PLBs might depend on the state of cell differentiation rather than being directly related to light or dark conditions.

PLBs which were formed in the dark persist under weak white or red light, or are formed de novo or recrystallized under these conditions (Treffry, 1978). Treatment of bean leaves with δ-aminolevulinic acid, a precursor of Chl, allows prolonged persistence of the PLBs under exposure to light and causes their recrystallization (Klein et al., 1974). Under all these conditions Pchl(ide) is converted to Chl(ide), but the level of Pchl(ide) remains relatively high. Thus, even if darkness is not a necessary requirement for PLB formation, it could still be maintained that there exists a close relationship between PLBs and Pchl(ide) accumulation. This too has become doubtful. A complete Pchl(ide) to Chl(ide) conversion occurs in etiolated bean leaves under strong red light, but paracrystalline PLBs remain (Treffrey, 1973). On the other hand, the cotyledons of dark-grown pine seedlings contain Chl, but no Chl(ide) or Pchl(ide). Nevertheless, in addition to stacked and nonstacked thylakoids, relatively well-developed PLBs are found in the plastids (Michel-Wolwertz and Bronchart, 1974).

A more direct approach to determining the properties of PLBs, and to ascertaining whether they are really the sites of protopigment deposition, is by the analysis of PLBs which have been separated from prothylakoids. Such a separation has been achieved recently (Lütz, 1978; Wellburn and Hampp, 1979). According to Lütz (1978), PLBs have a significantly higher lipid/protein ratio than prothylakoids. Apparently the lipophilic part of the PLBs can be accounted for almost entirely by two steroids (Lütz, 1978; Kesselmeier and Budzikiewic, 1979), and the protein part seems to be made up mainly of low-molecular-weight polypeptides. Since more than 20 to 40 thylakoid polypeptides are already membrane-bound in the etioplasts (Nielsen, 1975) this implies that these polypeptides are part of the prothylakoids. Also, PLB tubules may not display osmotic activity (Bogorad et al., 1971) and it is questionable whether the PLB tubules can be regarded as true biological membranes. An analysis of separated PLB and prothylakoid fractions by Lütz and Klein (1979) indicates that most if not all of the Pchl(ide) and/or converted Chl(ide) is located in the prothylakoids, and that the PLBs may be rather poor in these pigments. This analysis agrees with the findings of Wrischer (1978), who could not obtain histochemical evidence for Chl-dependent light reactions in PLBs after the conversion of Pchl to Chl, while the prothylakoids in the same etioplasts reacted strongly. Wellburn and Hampp (1979) did not find any positive associations of existing or developing photochemical activities with PLB structures proper. These activities seem to be located entirely in the developing prothylakoids.

All this strongly indicates that PLBs can be formed in light as well as in darkness and that they accumulate whenever the developmental state of the cell or external conditions do not allow complete assembly of the photosynthetic membranes. Similar structures are formed in the light and in darkness, and they can thus be fitted into a single morphogenetic pathway (Whatley, 1977). The PLBs differ in their chemical makeup from prothylakoids and they seem to be poor in pigments and polypeptides. They appear not to be directly involved in the development and assembly of thylakoids, except for a possible transfer of β-carotene and NADP-dependent oxyreductase to the developing prothylakoids (Wellburn and Hampp, 1979).

The formation of the prolamellar bodies is thus not indicative of a different developmental pathway, but rather of a disturbance of the tightly integrated activities required for normal thylakoid assembly. It is therefore compatible with the concept of a single morphogenetic pathway for chloroplast development (Wheatly, 1977).

If thylakoid development in illuminated etioplasts does not involve the pro-lamellar bodies, this process may indeed be similar to that occurring during the development of the proplastid into a chloroplast, and information gained from developmental studies with illuminated etioplasts may be relevant to the more normal mode of chloroplast development. Obviously, further evidence would be required to verify the apparent noninvolvement of the PLB in thylakoid formation. The mechanism responsible for the increase in primary thylakoids, concomitant with the disappearance of the prolamellar body, in illuminated isolated etioplasts when cultured on relative simple media should certainly be studied in greater detail (Wellburn and Wellburn, 1973; Wrischer, 1973; Kohn and Klein, 1976). If the involvement of the prolamellar body in thylakoid de-velopment can be shown to be minimal, the concepts concerning its functions have run a full cycle. When PLBs were discovered, they were assumed to be abnormal, pathological structures (Rosinsky and Rosen, 1972). Then a more active and important role was ascribed to them. Present evidence again tends to favor the original, primarily intuitive, interpretation. This concept of a single pathway for chloroplast development in higher plants, along with the relegation of the prolamellar body to a rather unimportant and/or as yet unknown role, is attractive because it agrees well with the available information concerning the development of chloroplast structures in the algae.

Morphogenetic Changes During Ontogenesis of Algal Plastids

In the majority of higher plants, division of plastids and their transmission occurs predominantly during the proplastid stage, although cases of division of mature plastids are known (Kirk and Tilney-Basset, 1967). In contrast, in most of the algae the mature chloroplast divides by fission (Bisalputra, 1974), forming two smaller chloroplasts which grow to their final size without any pronounced morphogenetic changes. Information about occurrence in algae of proplastids and their morphogenetic development into chloroplasts is rather limited. In gen-eral, it has been described for some red and green algae. It may occur in multicellular algae, concomitantly with cell maturation (Bouck, 1962; Brown and Weier, 1968); in siphonous algae with the proplastids located below the growing tip (Borowitzka and Larkum, 1974; Borowitzka, 1976); during carposporogenesis (Kugrens and West, 1974); and after light exposure of certain dark-grown unicellular algae, either wild-type or mutants, in which Chl synthesis has become light-dependent (Ben Shaul et al., 1964a; Klein et al., 1972; Ohad et al., 1967b; Bryan et al., 1967). Plastid ontogenesis and fission of mature plastids may also occur in the same algae (Bouck, 1962; Bisalputra, 1972).

In the unicellular algae with light-dependent Chl synthesis, proplastids are formed due to dedifferentiation of the chloroplasts when the cells are grown organotrophically for a number of generations in the dark (Ben Shaul et al., 1964b; Ohad et al., 1967a). The degree of dedifferentiation obtainable seems to differ in various organisms, as indicated by the differences in proplastid size, in pigment content and form, in amount and arrangement of the prothylakoids, and in enzyme activity. There is no apparent correlation among these parameters; the relatively large proplastids in the *Chlamydomonas reinhardii* Y-1 mutant contain few prothylakoids but they are associated with small amounts of Chl and Pchl (Ohad et al., 1967a). The smaller *Euglena gracilis* proplastids have a much more elaborate prothylakoid system, but lose all their Chl and contain only small amounts of Pchl(ide) (Klein et al., 1972). Incidentally, the relatively few unicellular algae—among them *Euglena, Chlamydomonas,* and *Scenedesmus*—in which it has been possible to study chloroplast development have contributed spectacularly toward an understanding of the development of the photosynthetic apparatus. The amount of information these few organisms have yielded is impressive compared with what we know about chloroplast development in the various other classes of algae.

Despite the large structural differences among the mature chloroplasts in the various groups of algae, proplastids in multicellular, siphonous, or dedifferentiated dark-grown unicellular algae are rather similar. They are spherical or ameboid, surrounded by envelopes, and contain inner membranes which are frequently arranged as single peripheral girdles (Bouck, 1962; Brown and Weier, 1968; Klein et al., 1972; Friedberg et al., 1971). Here too it has been claimed that the prothylakoids may arise from the envelopes, but the evidence is no more convincing than it is for higher plants. It may be significant that among the published micrographs of sections through early proplastids of *Lomentaria bayleyana* (Bouck, 1962), the few proplastids which are devoid of thylakoid membranes have a diameter three to four times less than the other proplastids in the same cell, all of which contain primarily girdle-like thylakoids. This suggests that the smaller membrane-free sections may be tangential ones, in which the thylakoids were bypassed. Despite earlier reports (Ben Shaul et al., 1964a), *Euglena* proplastids (Figure 5) contain 2 to 3 or even more well-defined girdle thylakoids, and there is no evidence that they originate from the envelopes (Klein et al., 1972). In *Chlamydomonas* Y-1 no envelope invaginations occurred during

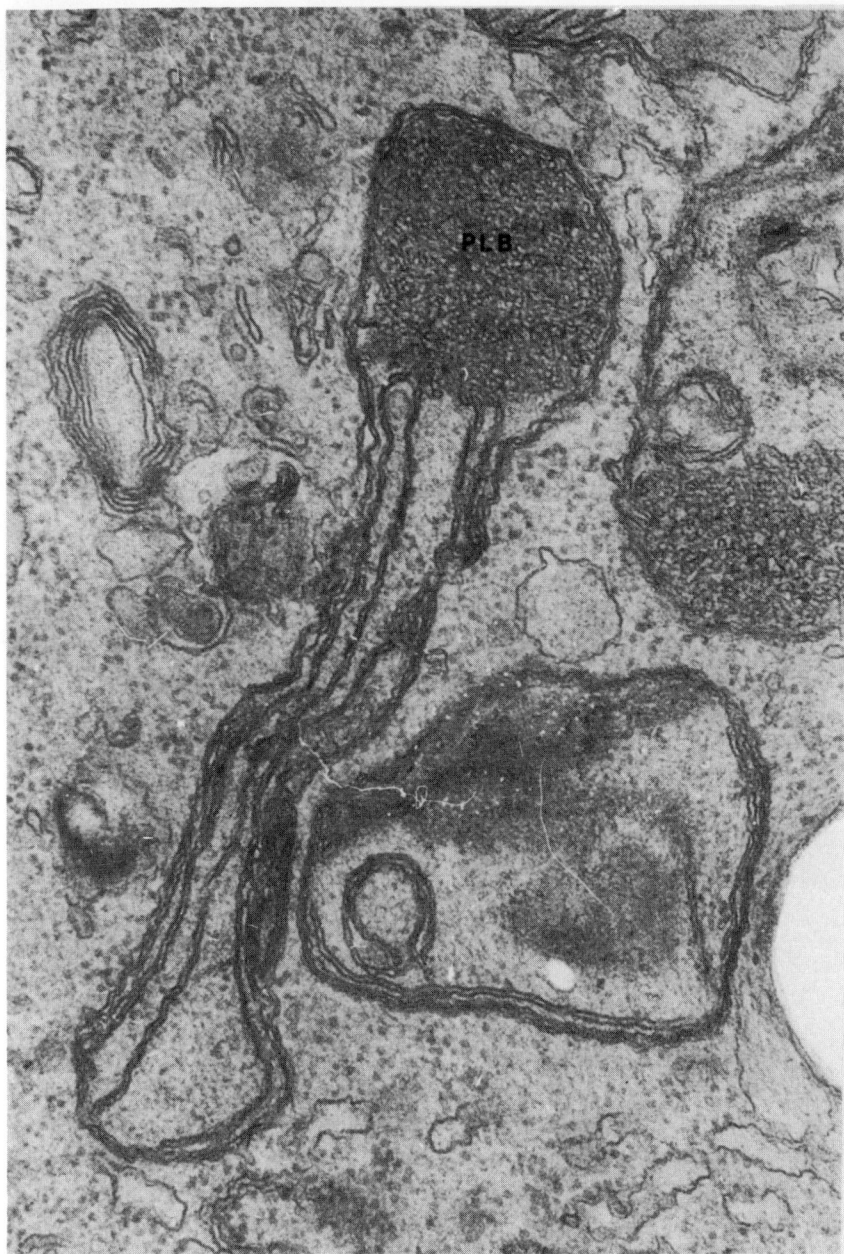

Figure 5. Proplastid from a dark-grown cell of *Euglena gracilis* var. *bacillaris*. (×84,000) *Source:* Klein et al., 1972.

Table 1. PLB-like Complexes in Cyanobacteria and Algal Plastids

Cyanobacteria (blue-green algae)	Permanent	Transient	Dark	Light	Pattern	Reference
Anabaena		+		+	Regular	Lang and Rae (1967)
Green algae						
Chlorella vulgaris C-10		+	+			Bryan et al. (1967)
Chlorella pyrenoidosa		+	+			Budd et al. (1969)
Chlamydomonas reinhardii Y-1		+	+		Irregular	Friedberg et al. (1971)
Caulerpa lamouroux (17 species)	+			+	Regular	Calvert et al. (1976)
Euglenas						
Euglena gracilis		+	+		Irregular	Schiff (1970) Klein et al. (1972) Salvador et al. (1971)
Red algae						
Batrachospermum moniliformae		+	+		Irregular	Sheath et al. (1979)

rapid thylakoid formation (Ohad et al., 1967b); by using a goniometer stage, apparent connections between the inner envelope and growing thylakoids could be shown to be nonexistent (Ohad et al., 1977).

Tubular complexes, similar to PLBs in higher plants, may develop during proplastid formation, or may be formed under certain conditions in mature plastids. Table 1 lists some of those algae in which plastids with tubular complexes were found that are similar to the PLBs in higher plants. The first observation was that by Lang and Rae (1967) of a paracrystalline complex of interconnected tubules in the prokaryote cyanobacterium *Anabaena*. The conditions which led to the formation of these complexes are not described, but it has been suggested that their formation may have been due to self-shading of the filaments in the dense cultures (Sheath et al., 1979). In *Euglena gracilis* (Schiff, 1970; Salvador et al., 1971) and the *Chlamydomonas* mutant Y-1 (Friedberg et al., 1971) irregular tubular complexes develop during organotrophic growth in the dark. They resemble the PLBs of higher plants that have been illuminated for a short time. Similar environmental conditions lead to the formation of comparable tubular complexes in the *Chlorella vulgaris* C-10 mutant (Bryan et al., 1967) and *Chlorella pyrenoidosa* (Budd, 1967). For the first time in a Rhodophyte, a similar structure was found in plastids of dark-grown cultures of *Batrachospermum monoliformae* (Sheath et al., 1979).

In these unicellular algae, the formation of the tubular complexes can be considered to be a response to darkness and to conditions unfavorable for complete membrane biosynthesis. However, in all species of *Caulerpa* and *Halimedes* a tubular paracrystalline complex occurs as a permanent plastid structure, independent of light conditions. This tubular complex occurs at one side of the plastid (Dawes and Rhamstine, 1967; Calvert and Dawes, 1976); between it and the inner envelope there is a cuplike arrangement of concentric membrane pairs. The membranes of the concentric lamellar system (CLS) are interconnected with each other and with the inner plastid envelope (Borowitzka and Larkum, 1974) and have been compared to the peripheral reticulum in some higher plant chloroplasts (Borowitzka, 1976). The spacing of the tubules in the central complex is consistent with that in the PLBs of higher plants, while the dimensions of the tubules themselves are similar but not directly comparable (Calvert and Dawes, 1976). From the other side of the tubular complex, thylakoid bands extend into the interior of the plastid. It is not clear whether the bands are connected to the complexes; Borowitzka (1976) did not find evidence for such transitions.

In *Caulerpa*, evolution appears to have progressed from species with large chloroplasts with peripheral girdle thylakoid bands and pyrenoids surrounded by starch sheaths, through intermediate stages into species with small chloroplasts without pyrenoids or girdle thylakoids, and with rudimentary starch grains only (Calvert et al., 1976). All these different types of chloroplasts develop from similar proplastids. In these the CLS and the central tubular complex are already present at very early stages of development, together with one or two peripheral girdle thylakoids. Starch grains are formed very early and thylakoid growth is initiated in close proximity to the tubular complex (Borowitzka and Larkum, 1974; Borowitzka, 1976).

The thylakoids then become closely appressed and bands are established. During development of these proplastids into the mature chloroplasts of the more primitive types, pyrenoids are formed, while in the more advanced species the girdle bands become transformed into straight bands.

Borowitzka (1976) suggested that the tubular center is involved in the organization of the thylakoids. It is obvious, however, that in *Caulerpa* the "thylakoid organizing body" is not a direct structural precursor to the thylakoids. Its membranes differ from the latter in their staining properties, their solubility in 1% SDS, and they do not show Chl fluorescence.

In dark-grown illuminated *Euglena gracilis* cells the developmental sequence leading from proplastids with PLBs and girdle thylakoids to chloroplasts with straight thylakoid bands and pyrenoids is very similar to that described above (Klein et al., 1972). After a rather prolonged lag period, during which Chl accumulates—apparently in the preexisting single thylakoids—these start to enlarge and new ones are formed as outgrowths from older thylakoids. These early membranes become appressed and peripheral girdle bands and central straight bands are formed. Later, the girdles open up, pyrenoids appear and the plastids attain their mature form. During this process the PLBs gradually disappear, but their eventual and gradual disappearance does not seem to be directly or closely related to any of the structural or chemical events involved in light-induced thylakoid assembly and growth. In *Euglena* proplastids, the existing small amount of Pchl(ide) is predominantly in the 635 nm form, and, as in young proplastids of higher plants, is converted to Chl_{673} by light, without a large Shibata shift. Although in *Euglena* proplastids 1 to 3 fluorescent spots can be observed, which have been associated with the existence of PLBs (Klein et al., 1972), the Pchl(ide) to Chl(ide) conversion is not paralleled by any synchronous structural changes in the PLB, and subsequent Chl accumulation is not correlated with recognizable changes in the PLB. Also, there is no indication of a pigment transfer from the PLB to the thylakoids. All this might indicate that, in this alga as in higher plants, the PLB may not be the main site for Pchl(ide) deposition. In this respect *Euglena* resembles *Chlamydomonas* Y-1. In dark-adapted illuminated cells of this mutant, thylakoid assembly and growth had to be discussed (Ohad, 1977) independently of changes occurring in the PLB and, as in the other algae and the higher plants, no direct relation appears to exist between these two events. In *Chlamydomonas* Y-1 too Pchl(ide) exists apparently in the 630–632 nm form and the action spectrum for plastid development is identical with the absorption spectrum for this in vivo form (Ohad and Drews, 1974).

Thus, among widely different algae, paracrystalline or nonparacrystalline tubular complexes occur either as intrinsic or early-formed parts of the proplastids, or are induced under conditions which favor dedifferentiation. As in higher plants they may be transient or permanent chloroplast features, perhaps indirectly involved in membrane assembly, but they seem not to be direct structural thylakoid precursors; development of thylakoids appears to be unaffected by their presence or absence.

Rhodophytes (red algae) contain plastids with unfused single thylakoids. In representatives of this group, for example *Lomentaria bayleana* (Bouck, 1962) or *Batrachospermum* (Brown and Weier, 1968) plastid ontogeny is similar to early development in chlorophytes and euglenophytes. In the developing proplastids peripheral girdle thylakoids exist and/or are formed which then give rise to the more centrally located thylakoids. In the chrysophyte *Ochromonas danica* the same pattern of development has been described (Gibbs, 1962); it is followed by band formation due to the fusion of existing thylakoids and formation of new ones from those formed previously.

Plastid Origins

Ideas concerning the origin of chloroplasts and possible plastid precursors have been based mainly on similarities between certain generic, physiological, and structural properties of cyanobacteria and those of mature chloroplasts in algae and higher plants. The possibility of a polyphylletic origin has been discussed in relation to different pathways for carotenoid synthesis (Stanier, 1974).

Similarities between early stages in the development of chloroplasts in *Euglena* and higher plants has been pointed out by Klein et al. (1972) and Lancer et al. (1976). The early structural plastid development in *Euglena* resembles that in those other algal groups which possess plastids with widely diverse structures at maturity. The similar pathway for early plastid development, apparently common for plastids both in higher and lower plants, could indicate ancient mechanisms, while the diversity in mature plastid structure may well be due to the expression during cell differentiation of the different nuclear genomes in the various groups of plants. We suggest therefore that proplastids may be closer to possible plastid precursors than the mature chloroplast, and that a more detailed investigation of their properties and mechanisms may be a valuable tool in elucidating questions of ancestry and poly- or monophyletic origins.

References

Beale, S. I. 1977. Biosynthesis of photosynthetic pigments—pathways and regulation. *In* Photosynthesis '77. D. O. Hall, J. Coombs, and T. W. Goodwin (eds.). Proc. 4th Internat. Cong. Photosynthesis, pp. 507–516.

Ben Shaul, Y., Schiff, J. A., and Epstein, H. T., 1964a. Studies of chloroplast development in *Euglena*. VII. Fine structure of the developing plastid. Plant Physiol. *39*:231–240.

Ben Shaul, Y., Schiff, J. A., and Epstein, H. T., 1964b. Studies on chloroplast development in *Euglena*. X. The return of the chloroplast to the proplastid condition during dark adaptation. Can. J. Bot. *43*:129–136.

Bisalputra, T. 1974. Plastids. *In* Algal Physiology and Biochemistry. W. D. Stewart (ed.). Oxford: Blackwell Scientific Publications, pp. 124–160.

Boardman, N. K., and Anderson, J. M. 1964. Studies on the greening of dark grown bean plants. I. Formation of chloroplasts from proplastids. Aust. J. Biol. Sci. 17:86–92.

Boardman, N. K., Anderson, J. M. and Goodchild, D. J. 1978. Chlorophyll-protein complexes and structure of mature and developing chloroplasts. In Current Topics in Bioenergetics, Vol. 8. D. R. Sanadi and L. P. Vernon (eds.). New York: Academic Press, pp. 36–101.

Bogorad, L., Falk, R. M., Forger, J. M. III, and Lockshin, A. 1971. Properties of etioplast membranes and membrane development in maize. In Autonomy and Biogenesis of Mitochondria and Chloroplasts. N. K. Boardman, A. W. Linnane, and R. M. Smillie (eds.). Amsterdam: North-Holland Publishing Co., pp. 85–91.

Borowitzka, M. A. 1976. Some unusual features of the ultrastructure of the chloroplasts of the green algal order Caulerpales and their development. Protoplasma 89:129–147.

Borowitzka, M. A., and Larkum, A. W. D. 1974. Chloroplast development in the Caulerpean alga Halimeda. Protoplasma 81:131–144.

Bouck, G. B. 1962. Chromatophore development, pits and other fine structure in the red alga, Lomentaria baileyana (Harv.) Farlow. J. Cell Biol. 12:553–564.

Bradbeer, J. W., Ireland, H. M. M., Smith, J. W., Rest, J., and Edge, H. J. W. 1974. Plastid development in primary leaves of Phaseolus vulgaris. VI. Development of growth in continuous darkness. New Phytol. 73:263–270.

Bronchart, R., 1970. Une nouvelle interpretation de la structure du corps prolamellaire basée sur des observations en cryodécapage. C. R. Acad. Sci. Ser. D. 270:1789–1791.

Brown, D. L., and Weier, T. E. 1968. Chloroplast development and ultrastructure in the freshwater red alga Batrachospermum. J. Phycol. 4:199–206.

Bryan, G. W., Zadylak, A. H., and Ehret, C. F. 1967. Photoinduction of plastids and of chlorophyll in a Chlorella mutant. J. Cell Sci. 2:513.

Budd, T. W., Tjostem, J. L., and Duysen, A., 1969. Ultrastructure of Chlorella pyrenoidosa as effected by environmental changes. Am. J. Bot. 56:540–545.

Calvert, H. E. and Dawes, C. J., 1976. Ontogenetic membrane transitions in plastids of the coenocytic alga Caulerpa (Chlorophyceae). Phycologia 15:37–40.

Calvert, H. E., Dawes, C. J., and Borowitzka, M. A. 1976. Phylogenetic relationship of Caulerpa (Chlorophyta) based on comparative chloroplast ultrastructure. J. Phycol. 12:149–162.

Dawes, C. J., and Rhamstine, E. C. 1967. An ultrastructural study of the giant green algal coenocyte Caulerpa prolifera. J. Phycol. 3:117–126.

Dodge, J. D. 1973. The Fine Structure of Algal Cells. London, New York: Academic Press.

Douce, R., and Joyard, J. 1978. Importance of the envelope in chloroplast biogenesis. In Chloroplast Development. G. Akoyunoglou and J. H. Argyroudi-Akoyunoglou (eds.). Amsterdam: Elsevier-North Holland Biomedical Press, pp. 283–295.

Eriksson, S., Kahn, A., Walles, B. and v. Wettstein, D., 1961. Zur Makromolekularen Physiologie der Chloroplasten. III. Ber. Deutsch. Bot. Ges. 79:222–232.

Friedberg, J., Goldberg, J., and Ohad, I. 1971. A prolamellar like structure in Chlamydomonas reinhardii. J. Cell Biol. 50:268–275.

Gibbs, S. P., 1962. Chloroplast development in Ochromonas danica. J. Cell Biol. 15:343–361.

Gibbs, S. P. 1970. The comparative ultrastructure of the algal chloroplast. Ann. NY Acad. Sci. 175:413–781.

Granick, S., and Sassa, S. 1971. δ-Aminolevulinic acid synthetase and the control of heme and chlorophyll synthesis. *In* Metabolic Regulation, Vol. 5. H. J. Vogel (ed.). New York: Academic Press, pp. 77–141.

Gunning, B. E. S., and Jagoe, M. P. 1967. The prolamellar body. *In* The Biochemistry of Chloroplasts, Vol. 2. T. W. Goodwill (ed.). London, New York: Academic Press, pp. 655–676.

Hampp, R., and Wellburn, A. R., 1978. Development of photochemical activities in preparations of unresolved internal membranes, enriched prolamellar bodies, and prothylakoid vesicles during etioplast chloroplast transitions. Ber. Deutsch. Bot. Ges. *91*:551–561.

Kahn, A. 1968. Developmental physiology of bean leaf plastids. II. Negative contrast electron microscopy of tubular membranes in prolamellar bodies. Plant Physiol. 43:1769–1780.

Kesselmeier, J., and Budzikiewic, H. 1979. Identification of saponins as structural building units in isolated prolamellar bodies from etioplasts of *Avena sativa* L. Z. Pflanzenphysiol. *91*:333–344.

Kirk, J. T. O., and Tilney-Basset, R. A. E. 1967. The Plastids. London: W. H. Freeman and Co.

Klein, R. M., and Cronquist, A. 1967. A consideration of the evolutionary and taxonomic significance of some biochemical, micromorphological, and physiological characters in the thallophytes. Q. Rev. Biol. 42:105–331.

Klein, S., Bryan, G., and Bogorad, L. 1964. Early stages in the development of plastid fine structure in red and far-red light. J. Cell Biol. *22*:433–442.

Klein, S., Manori, I., Ne'eman, E., and Katz, E. 1974. The effect of cycloheximide and δ-aminolevulinic acid on structural development of chloroplasts. *In* Proc. 3rd Int. Cong. Photosynthesis, pp. 2089–2095.

Klein, S., and Schiff, J. 1972. The correlated appearance of prolamellar bodies, protochlorophyll(ide) species and the Shibata shift during development of bean etioplasts in the dark. Plant Physiol. 49:619–626.

Klein, S., Schiff, J. A., and Holowinsky, A. W. 1972. Events surrounding the early development of *Euglena* chloroplasts. II. Normal development of fine structure and the consequences of preillumination. Dev. Biol. 28:253–273.

Kohn, S., and Klein, S. 1976. Light induced structural changes during incubation of isolated maize etioplasts. Planta *132*:169–175.

Kugrens, P., and West, J. A. 1974. The ultrastructure of carposporogenesis in the marine hemiparasitic red alga *Erythrocystis saccata*. J. Phycol. *10*:137–139.

Lafleche, D., Bove, J. M., and Duranton, J. 1972. Localization and translocation of the protochlorophyllide holochrome during the greening of etioplasts in *Zea mays* L. J. Ultrast. Res. 40:250–214.

Lancer, H. A., Cohen, C. E., and Schiff, J. A. 1976. Changing ratios of phototransformable protochlorophyll and protochlorophyllide of bean seedlings in the dark. Plant Physiol. 57:369–374.

Lang, N. J., and Rae, P. M. M. 1967. Structures in a blue green alga resembling prolamellar bodies. Protoplasma *64*:67–74.

Leech, R. M. 1977. Etioplast structure and its relevance to chloroplast development. Biochem. Soc. Trans. 5:81–84.

Lütz, C., 1978. Separation and comparison of prolamellar bodies and prothylakoids of etioplasts from *Avena sativa* L. *In* Chloroplast Development. G. Akoyunoglou and J. H. Argyroudi-Akoyunoglou (eds.). Amsterdam: Elsevier-North-Holland Biomedical Press, pp. 481–488.

Lütz, C., and Klein, S. 1979. Biochemical and cytological observations on chloroplast development. IV. Chlorophylls and saponins in prolamellar bodies and prothylakoids separated from etioplasts of etiolated *Avena sativa* L. leaves. Planta (in press).

Michel-Wolwertz, M. R., and Bronchart, R. 1974. Formation of prolamellar bodies without correlative accumulation of protochlorophyllide or chlorophyllide in pine cotyledons. Plant Sci. Lett. *2:*45–54.

Nielsen, N. C. 1975. Electrophoretic characterization of membrane proteins during chloroplast development in barley. Eur. J. Biochem. *50:*611–623.

Ohad, I., 1977. Biogenesis of chloroplast membranes in algae. *In* Bioenergetics of Membranes. L. Packer et al. (eds.). Amsterdam: Elsevier/North-Holland Biomedical Press, pp. 61.

Ohad, I., Bar-Nun, S., Cohen, D., Gershoni, J. M., Gurevitz, M., and Lavintman, N. 1977. Stepwise synthesis and assembly of photosynthetic membranes in algae. Proc. 4th Int. Cong. Photosyn. 517–526.

Ohad, I., and G. Drews. 1974. Action spectrum for the synthesis of chlorophyll and chloroplast membrane proteins of cytoplasmic origin. In Proc. 3rd Int. Cong. Photosynthesis: 1907–1912.

Ohad, I., Siekevitz, P., and Palade, G. E. 1967a. Biogenesis of chloroplast membranes. I. Plastid dedifferentiation in a dark grown algae mutant *(Chlamydomonas reinhardii)*. J. Cell Biol. *35:*521–552.

Ohad, I., Siekevitz, P., and Palade, G. E. 1967b. Biogenesis of chloroplast membranes. II. Plastid differentiation during greening of a dark grown algal mutant *(Chlamydomonas reinhardii)*. J. Cell Biol. *35:*521–551.

Platt-Aloia, K. A. and Thomson, W. W. 1977. Chloroplast development in young sesame plants. New Phytol. *78:*599–605.

Rascio, N., Orsenigo, M., and Arbott, D. 1976. Prolamellar body transformation with increasing cell age in the maize leaf. Protoplasma *90:*253–263.

Rosinski, J., and Rosen, W. G. 1972. Chloroplast development: Fine structure and chlorophyll synthesis. Q. Rev. Biol. 47:160–191.

Salvador, G., Lefort-Tran, M., Nigon, V., and Holowinsky, A. 1971. Structure et évolution du corps prelamellaire dans les proplastes d'*Euglena*. Exp. Cell Res. 64:458–462.

Schiff, J. A., 1970. Developmental interactions among cellular compartments in *Euglena*. Symp. Soc. Exp. Biol. 24:227–230.

Sheath, R. G., Hellebust, J. A., and Sawa, T., 1979. Effects of low light and darkness on structural transformation in plastids of the Rhodophyta. Phycologia 18:1–12.

Simpson, G. G., and Beck, W. S. 1965. *Life*. New York: Harcourt, Brace, and World.

Sprey, B., and Laetsch, W. M. 1978. Structural studies of peripheral reticulum in C_4 plant chloroplasts of *Portulaca oleracea* L. Z. Pflanzenphysiol. *87:*37–53.

Stanier, R. Y., 1974. The origin of photosynthesis in eukaryotes. *In* Evolution in the Microbial World. Twenty-fourth Symp. Soc. Gen. Microbiol. London: Cambridge Univ. Press, pp. 219–240.

Treffry, T. 1973. Chloroplast development in etiolated peas: Reformation of prolamellar bodies in red light without accumulation of protochlorophyllide. J. Exp. Bot. *24:*185–196.

Treffry, T., 1978. Biogenesis of the photochemical apparatus. Int. Rev. Cytol. *52:*159–196.

v. Wettstein, D. 1976. Genetic regulation of membrane synthesis in chloroplasts as studied with lethal gene mutants. *In* Membranes and Disease. L. Bolis, J. F. Hoffman, and A. Leaf (eds.). New York: Raven Press, pp. 123–130.

Weier, T. E., and Brown, D. L. 1970. Formation of the prolamellar body in 8 day, dark grown seedlings. Am. J. Bot. 57:267–275.

Wellburn, A. R., and Hampp, R. 1979. Appearance of photochemical function in prothylakoids during plastid development. Biochim. Biophys. Acta 547:380–397.

Wellburn, A. R., and Wellburn, F. A. M., 1973. Developmental changes of etioplasts in isolated suspensions and in situ. Am. Bot. 37:11–19.

Whatley, J. M. 1977. Variations in the basic pathway of chloroplast development. New Phytol. 78:407–420.

Wrischer M., 1973. Ultrastructural changes in isolated plastids. I. Etioplasts. Protoplasma 78:291–303.

Wrischer, M. 1978. Ultrastructural localization of diaminobenzidine photooxidation in etiochloroplasts. Protoplasma 97:85–92.

Cyanelles

Robert K. Trench

The term *cyanelle,* introduced by Pascher (1929), refers to symbiotic cyanobacteria and has no taxonomic significance. With few exceptions, very little is
known about the systematic affinities of symbiotic cyanobacteria. In instances
where a symbiosis can be demonstrated unambiguously by the isolation and
culture of the cyanelles, taxonomic affinities between cyanelles and free-living
cyanobacteria can be more readily established than in instances where the apparent integration between the cyanelles and the host renders the association
obligate.

Cyanobacteria are phototrophic prokaryotes, and many occur in intercellular
or intracellular associations with a broad range of invertebrate hosts, the associations being randomly distributed from the Protozoa through the Urochrodata
(Table 1). Despite this broad distribution of the phenomenon, offering a wide
range of potential experimental systems, information on functional aspects of
the interactions between symbiotic cyanobacteria and their respective hosts is
very sparse indeed.

A prokaryote containing chlorophylls *a* and *b* but no phycobilins has been
recently described (Lewin 1976, 1977). This organism, named *Prochloron,* has
been assigned by the isolator to a group separate from the cyanobacteria called
the prochlorophytes. I am not fully convinced that this separation is justified but
will use the suggested nomenclature in the interests of uniformity.

Preparation of this chapter was supported by National Science Foundation Grant No. PCM
78–15209.

Address reprint requests to Dr. Robert Trench, Associate Professor, Department of Biological
Sciences, University of California, Santa Barbara, California 93106.

Table 1. Phyletic Distribution of Cyanobacterial and Prochlorophyte Symbioses

Host taxon	Host identity	Cyanobacterial or prochlorophyte symbiont	Habitat
Protozoa	*Cyanophora paradoxa*	*Cyanocyta korschikoffiana*	Fresh water
	Paulinella	?	Fresh water
	Glaucocystis	?	Fresh water
	Glaucosphera	?	Fresh water
	Gloeochaete	?	Fresh water
	Cryptella	?	Fresh water
Porifera	*Ircinia*	*Aphanocapsa*	Marine
	Demospongia	?	Marine
Urochordata	*Didemnum*	*Prochloron*	Marine
	Trididemnum	*Prochloron*	Marine
	Diplosoma	*Prochloron*	Marine

With the recent resurgence of the serial endosymbiosis hypothesis for the origin of eukaryotic cell organelles (Margulis, 1970, 1976; Schnepf and Brown, 1971; Stanier, 1974; Taylor, 1974, 1979; Phillips and Carr, 1977), the concept that plant chloroplasts might have arisen, possibly with multiple origins (Raven, 1970), from endosymbioses between phototrophic prokaryotes such as cyanobacteria and heterotrophic hosts has gained increased credence. The search for extant examples of possible "bridge" associations has progressed apace, and with respect to symbioses involving cyanelles, has concentrated on a very few select consortia.

Within the known spectrum of intracellular symbioses involving phototrophic organisms, it is possible to perceive consortia based on organisms which are functionally autonomous *(facultative)* and can survive after separation, as well as those which appear to be only partly autonomous *(obligate)*, and do not survive isolation. As pointed out before, integrative processes in symbiotic associations may progress to the point where distinguishing between an obligate cytobiont and a de facto organelle becomes very difficult indeed (Trench et al., 1978; Trench, 1979, 1980; Taylor, 1979).

In the context of the evolution of chloroplasts, the main theme of this chapter will be to analyze the criteria used to define and thereby distinguish phototrophic prokaryotic cytobionts and chloroplasts. I shall confine the discussion to those symbioses involving cyanelles which have been analyzed from an ultrastructural as well as a functional point of view.

Cyanelles and Chloroplasts

As stated before, cyanelles are phototrophic cyanobacteria. The morphological and biochemical characteristics of cyanobacteria have recently been extensively reviewed (Stanier and Cohen-Bazire, 1977). Cyanobacteria, being organized on a prokaryotic level, lack nuclei and other membrane-bound organelles. Genetic information in the form of DNA is usually confined to a region termed the *nucleoplasm* through which the chromosome, when seen in the electron microscope, is usually dispersed. The cyanobacterial chromosome has been shown to be covalently closed circular DNA (Lau and Doolittle, 1979). Some other characteristics of the DNA of a few select cyanobacteria are given in Table 2 (see also recent reviews of Herdman et al., 1979a, b).

The cyanobacteria display oxygenic photosynthesis. The photosynthetic apparatus is usually located in a series of flattened membranous sacs termed *thylakoids*, which are the sites of the light-harvesting pigments, the photochemical reaction centers, and the photochemical electron transport system. Characteristic photosynthetic pigments usually include chlorophyll *a*, β-carotene, a few xanthophylls, and the phycobiliproteins allophycocyanin, phycocyanin, and/or phycoerythrin (Table 3). The phycobiliproteins are organized into structural and functional units termed *phycobilisomes* which are interthylakoidally attached and are the principal light-harvesting system.

Other inclusions within the cyanobacterial cell are the 70S ribosomes which are composed of ribosomal RNA with subunit structure typical of prokaryotes belonging to the 23S, 16S, and 5S classes (Loening, 1968), "polyhedral bodies" known to be predominantly ribulose diphosphate carboxylase (Shively et al., 1973), polyarginyl-polyaspartic acid (cyanophycin) granules, so-called polyphosphate granules, and osmiophilic droplets.

The cyanobacterial cell is bounded by a unit membrane. External to this plasmalemma is a prokaryotic cell wall similar in fine structure to that seen in gram-negative bacteria, and typically peptidoglycan in composition containing diaminopimelic and muramic acids.

Chloroplasts are membrane-bound organelles found within the cytoplasm of those eukaryotic cells which demonstrate oxygenic photosynthesis. A mature chloroplast contains characteristic arrays of membranous sacs, the thylakoids, associated with which are the light-harvesting and photochemical reaction centers and the photochemical electron transport system. Among the chloroplasts of red algae, the same spectrum of light harvesting pigments is encountered as found in the cyanobacteria (Table 3). Morphological and biochemical aspects of the photosynthetic apparatus are also remarkably similar.

The chloroplast is usually limited by a double membrane system (three membranes are present in dinoflagellates) termed the *envelope*. Within the chloroplast, the covalently closed circular DNA comprising the chromosome is found in the stroma, and the physical and many chemical properties of chloroplast DNA closely resemble those of cyanobacterial DNA (Table 2). The same applies to chloroplast ribosome RNA's, being of the 70S class with 23S, 16S, and 5S. Other particles within the chloroplast may include osmiophilic droplets and starch.

The initiation of polypeptide synthesis in cyanobacteria and chloroplasts requires formylmethionyl-tRNA in contrast to the eukaryotic cytoplasmic system. Protein synthesis by chloroplasts and cyanobacteria is inhibited by the same

Table 2. Comparison of Select Cyanobacterial and Chloroplast DNA

	Buoyant density (g/cm³)	Genome size (daltons × 10⁻⁹)	Chromosome morphology	Guanine + cytosine (mol %)	No. of genome copies
Chloroplasts (algae)					
Chlamydomonas reinhardii	1.695[e]	0.194	Circular	35.7	82
Chlorella pyrenoidosa	1.687[e]	0.210	Circular	27.6	19
Euglena gracilis	1.685[a,e]	0.094–0.180	Circular	25.5	67–80
Cyanobacteria (free-living)					
Anacystis marina	1.728[a]		Circular	69.0	
A. nidulans	1.715[a]	2.27[c]	Circular	56.0	
A. cylindrica		2.47[c]	Circular		
Gleocapsa alpicola	1.694[a]			35.0	
Cyanobacteria (symbiotic)					
Cyanocyta korschikoffiana	1.716[a,b]		Circular	57.0	60
	1.695[d]	0.117		35.7	

[a] Edelman et al. (1967).
[b] In his analysis of *C. korschikoffiana*, L. Vernon (personal communication) also found a buoyant density of 1.716 and a guanine + cytosine value of 57%.
[c] Herdman and Carr (1974).
[d] Herdman and Stanier (1977).
[e] Kirk and Tilney-Bassett (1978).

Table 3. Comparison of Photosynthetic Pigments among Cyanobacteria and
Rhodophyta[a,b]

	Cyanobacteria	Rhodophyta
Chlorophylls		
a-Chlorophyll	+	+
Carotenoids		
α-carotene	−	+
β-carotene	+	+
zeaxanthin	+	+
lutein	−	+
myxoxanthophyll	+	−
echinenone	+	−
carotenoid glycosides	+	−
Phycobiliproteins		
allophycocyanin	+	+
phycocyanin	+	+
phycoerythrin	+	+

[a]*Source:* From Stanier (1974) and Kirk and Tilney-Bassett (1978).
[b]A single photosynthetic prokaryote, *Prochloron,* has been shown to possess chlorophyll *b* (Thorne et al., 1977; Withers et al., 1978).

spectrum of antibiotics. Finally, there is remarkable homology between cyanobacterial phycobilins and those of red algal chloroplasts (Bogorad, 1965; Berns,
1967; Glazer et al., 1971, 1976).

From the brief comparison given above, it is apparent that cyanobacteria and
chloroplasts possess many features in common, which are supportive of the
concept of common ancestry. However, these similarities could be effectively
argued as supporting both the serial endosymbiosis theory and the alternative
cluster-clone hypothesis (Uzzell and Spolsky, 1974; Bogorad, 1975), since the
evidence based on similarities provides very little direct insight into possible
evolutionary mechanisms in the origin of eukaryote organelles.

Raven (1970) proposed a multiple symbiotic origin for chloroplasts (see also
Lee, 1972) and remarked on the absence of extant oxygenic prokaryotes containing chlorophyll *b,* which potentially could have been the stock from which
the green algal chloroplasts arose. Subsequently, Lewin (1976, 1977) discovered
a phototrophic prokaryote intercellularly associated with certain didemnid ascidians, which did not contain phycobilins but did contain chlorophylls *a* and
b. Lewin's proposal to create a new division, the *Prochlorophyta,* has been
critically appraised (Antia, 1977). The existence of a prokaryote with light-
harvesting pigments similar to that found in the chlorophytes suggests that the
present diversity of plant pigments is a reflection of an early divergence among
photosynthetic prokaryotes (Stanier and Cohen-Bazire, 1977), but of itself does
not resolve the symbiotic origin of green algal chloroplasts.

There are several phototrophic protists which have defied classification based
on standard phycological criteria. Among these are *Cyanophora paradoxa, Glau-*

cocystis nostochinearum, Glaucosphera vacuolata, Gloeochaete wittrockiana, Paulinella chromatophora, and Cyanidium caldarium. All these organisms share a history of being regarded as a symbiosis between an apochlorotic host and a cyanobacterium (cyanelle) on the one hand and a de facto plant cell (red alga) with chloroplasts on the other (Schnepf and Brown, 1971; Chapman, 1974; Herdman and Stanier, 1977; Stanier and Cohen-Bazire, 1977; Kremer et al., 1978). Drawing on the serial endosymbiosis theory, others have suggested that these protists represent transition stages or bridges between symbiotic associations and plant cells with authentic chloroplasts (Holton et al., 1968; Ikan and Seckbach, 1972; Fredrick, 1976; Aitken and Stanier, 1979). The significance of the latter suggestion is far-reaching in that it implies that integrative processes resulting in the transformation of a once totally autonomous genetic unit into a semiautonomous organelle are currently in progress (Trench, 1980). The validity of such hypotheses based on our current understanding of the organisms involved will be the focal point of the remaining discussion.

Presumptive and de facto Chloroplasts

Given a protist, organized on the eukaryotic level, containing lamellated inclusions and capable of photosynthesis, how can one determine whether one is dealing with a photosynthetic protist (with de facto chloroplasts) or a symbiotic protist with cyanelles? Past analyses of several such systems from both cytological and biochemical standpoints have not resolved the problem. An example illustrating this is C. caldarium, which has been classified as a cyanophyte, a chlorophyte, a rhodophyte, and a cryptophyte (Allen, 1959; Silva, 1962; Chapman, 1974) as well as a symbiosis based on a cyanobacterium and an apochlorotic eukaryote (Fredrick, 1976; Kremer et al., 1978). Based on pigment composition and cytology, it is very difficult to distinguish between the organization seen in C. caldarium (Mercer et al., 1962; Seckbach, 1972; Edwards and Mainwaring, 1973) and a well-recognized red alga such as Porphyridium cruentum (Gantt and Conti, 1965) (see Figures 1 and 2). Chapman (1974) argued for placing C. caldarium among the red algae but Kremer et al. (1978), based on the analysis of ^{14}C-labeled photosynthetic products, believe that C. caldarium represents a symbiosis since the heterosides typical of rhodophytes are not synthesized by this organism. Instead free glucose and fructose, more characteristic of cyanobacteria, are produced. Similarly, Ikan and Seckbach (1972) and Holton et al. (1968), based on their analyses of the fatty acids, concluded that C. caldarium resembled cyanobacteria rather than red algae (see also Fredrick, 1968, 1971).

In the case of C. paradoxa the situation is somewhat clearer. Initially the cyanelles were thought to be free of a peptidoglycan wall, but a reduced wall has now been demonstrated at both the ultrastructural and biochemical levels (Pickett-Heaps, 1972; Trench et al., 1978; Aitken and Stanier, 1979). In addition, the cyanelles are found within vacuoles in the host cytoplasm. This situation is very similar to that seen in Paulinella chromatophora and Gloeochaete wittrock-

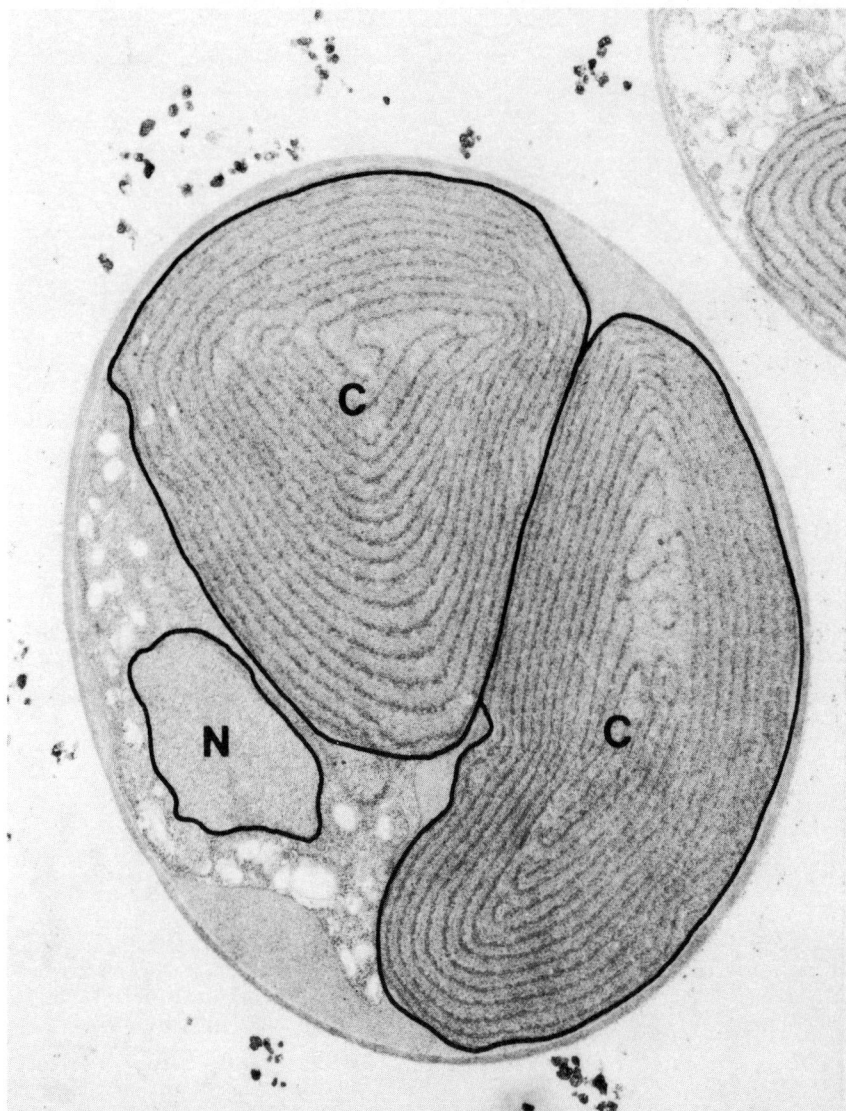

Figure 1. Transmission electron micrograph of the enigmatic "alga" *Cyanidium caldarium*. N = nucleus; C = chloroplast or cyanelle. The phycobilisomes attached to the thylakoid membranes were removed during $KMnO_4$ fixation. Magnification approximately $\times 30,000$. Photograph kindly provided by Dr. M. Edwards.

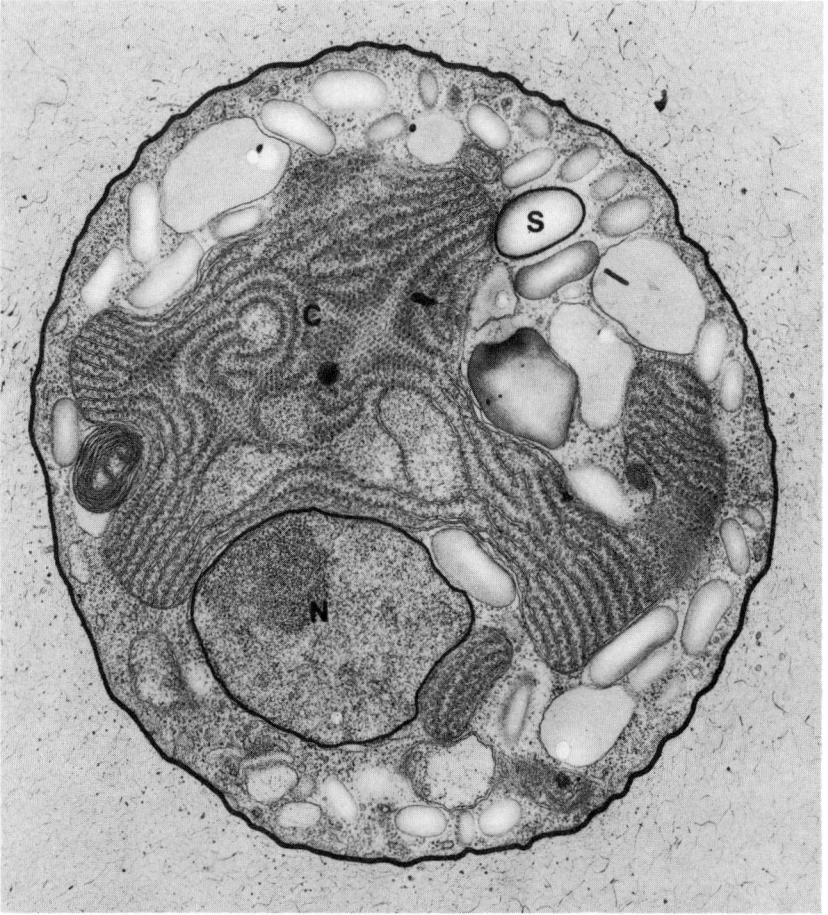

Figure 2. Transmission electron micrograph of the red alga *Porphyridium cruentum*. N = nucleus; C = chloroplast (note phycobilisomes attached to the thylakoid membranes); S = starch. Magnification approximately × 30,000. Photograph kindly provided by Dr. E. Gantt.

iana (Kies, 1974, 1976). In the case of the cyanelles in *Glaucocystis,* the wall around the cyanelles is regarded as absent, with concomitant absence of diaminopimelic acid (Holm-Hansen, et al., 1965; Echlin, 1967; Schnepf and Brown, 1971), but the cyanelles are clearly enclosed within host vacuoles (Schnepf, Koch, and Deichgräber, 1966).

To my knowledge, chloroplasts are never found within vacuoles in the cytoplasm of plant cells unless such vacuoles represent secondary lysosomes. The concept that the outermost chloroplast envelope membrane is homologous with the phagosome which enclosed the symbiotic precursor of the chloroplast (Schnepf and Brown, 1971) is at best speculative, and could only become a realistic view if it was demonstrated that cyanelles were indeed "on the way to becoming chloroplasts." Nonetheless, it is intriguing that the outermost membrane of the chloroplast envelope, as well as the outermost membrane of the mitochondrion, possesses properties which are distinct from those of the inner envelope membrane.

If the assumption is made that cyanelles represent transition stages between symbiotic cyanobacteria and de facto chloroplasts, what criteria could be used to estimate the extent of the transition or when the transition is complete? It is apparent that the observed presence or absence of a peptidoglycan wall in cyanelles is subject to the resolution of this structure at the ultrastructural or biochemical level. But the reduction or complete loss of cell walls is a phenomenon repeatedly encountered in various intracellular symbioses involving plant cells (Muscatine et al., 1975; Trench, 1979, 1980). Many of these associations are facultative, that is, the algae can be readily cultured away from the host. Success in culturing cyanelles away from their hosts is subject to the vicissitudes of culture techniques and our current inability to grow many cyanelles outside the host cytoplasm indicates that the association is probably obligate. However such failure, of itself, sheds very little light indeed on the basis for the obligatory nature of the association.

One possible approach to determining the degree of biochemical integration existing between cyanelles and their hosts' cytoplasm, particularly in light of possible transitions to becoming chloroplasts, would be to assay the extent to which biosynthetic processes in the cyanelles are regulated by, or actually occur in the host cytoplasm. Genetic experiments aimed at determining whether characteristics of the cyanelles are inherited in a Mendelian or non-Mendelian manner could demonstrate where the genetic information directing the expression of such characters resided, as has been done with chloroplasts (Sager, 1977). In this manner better analogies could be drawn with chloroplast-plant cell systems, where many components of the chloroplast have been shown to be synthesized outside the chloroplast (Table 4), some completely under nuclear control (Kirk and Tilney-Bassett, 1978). Unfortunately, many of the intracellular symbioses involving cyanelles are based on hosts which exhibit only asexual reproduction, precluding genetic analysis. Fortunately, there are other approaches which could potentially yield similar information. One organism with which an in-depth experimental approach has been made is *C. paradoxa* (Edelman et al., 1967; Reisfeld and Edelman, 1976; Codd and Stuart, 1977; Herdman and Stanier, 1977; Siebens and Trench, 1978; Trench et al., 1978; Trench and Ronzio, 1978; Trench and Siebens, 1978; Aitken and Stanier, 1979), and the remainder of my remarks will be confined to this organism (Figures 3 and 4).

Cyanophora paradoxa: A Symbiosis in Transition?

Herdman and Stanier (1977), starting with the premise that discrimination between chloroplasts and cyanelles could not be made by ultrastructural analysis alone, determined the genome size of the cyanelles in *C. paradoxa*. Their anal-

Table 4. Sites of Synthesis of Some Chloroplast Proteins[a]

Chloroplast protein	Organism	Antibiotic inhibitor[b]	Presumed site of synthesis
RuBP carboxylase			
large subunit	*Pisum sativum*	CAP	Chloroplast
		Lincomycin	Chloroplast
small subunit	*Pisum sativum*	CYX	Cytoplasm
Ferredoxin	*Chlamydomonas*	CYX	Cytoplasm
	Euglena gracilis	CYX	Cytoplasm
Ferredoxin-NADP			
reductase	*Chlamydomonas*	CYX	Cytoplasm
Cytochrome f553	*Euglena gracilis*	CYX	Cytoplasm
Cytochrome *f*	*Phaseolus vulgaris*	CAP	Chloroplast
ALA synthetase			
ALA dehydrase	*Nicotina*	CYX	Cytoplasm
Porphobilinogenase	*tabacum*		
Chlorophyllase			
Fatty acid synthetase	*Euglena gracilis*	Spectinomycin	Chloroplast
chlorophyll-protein			
complex I	*Pisum sativum*	CAP	Chloroplast
complex II	*Pisum sativum*	CYX	Cytoplasm

[a]Data from Kirk and Tilney-Bassett, 1978.
[b]CAP = chloramphenicol; CYX = cycloheximide.

yses, based on DNA renaturation kinetics, gave a value of 0.117×10^9 daltons. In addition, they reported a DNA base ratio (G + C) of 35.7% and a buoyant density of 1.695 g/cm^3. This value of kinetic complexity is about 5-10% of that found in free-living cyanobacteria, and generally in the range of the chloroplast genome size, though even smaller than some (Table 2). The base ratio and buoyant density values are identical with those found in chloroplasts of *Chlamydomonas*, and are in disagreement with the values obtained by Edelman et al. (1967) and L. Vernon (personal communication). Based on their observations, Herdman and Stanier (1977) concluded that the cyanelles in *C. paradoxa* are actually chloroplasts, the peptidoglycan cell wall being regarded as a vestige of their recent evolutionary past (Aitken and Stanier, 1979).

The chloroplast genome ($0.094-0.230 \times 10^9$ daltons) with a chromosomal contour length of 44 μm, is believed to code for the synthesis of about 126 proteins of 40 kd. If the genome size of the cyanelles in *C. paradoxa* was the same as that of chloroplasts, it would be reasonable to assume that the cyanelles would be able to direct the synthesis of only about 126 proteins. This would presuppose that the remainder of the cyanellar proteins are derived from synthetic processes in the host cytoplasm in a manner analogous to chloroplasts (see Kirk and Tilney-Bassett, 1978). This possibility has not been tested to any great extent.

One approach to investigating the extent to which the cyanelles in *C. paradoxa*

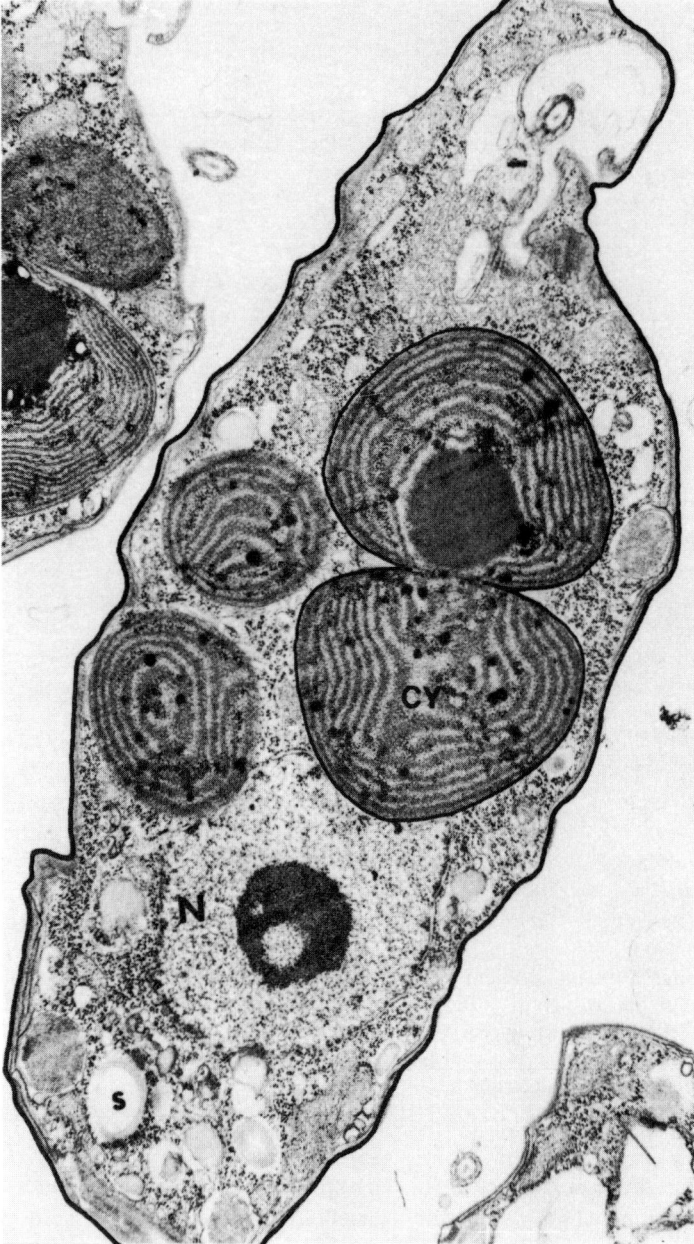

Figure 3. Transmission electron micrograph of *Cyanophora paradoxa* showing the disposition of the endosymbiotic cyanobacteria. One cyanobacterium is depicted undergoing binary fission. N = nucleus; S = starch; C = cyanelle. Magnification approximately ×15,000.

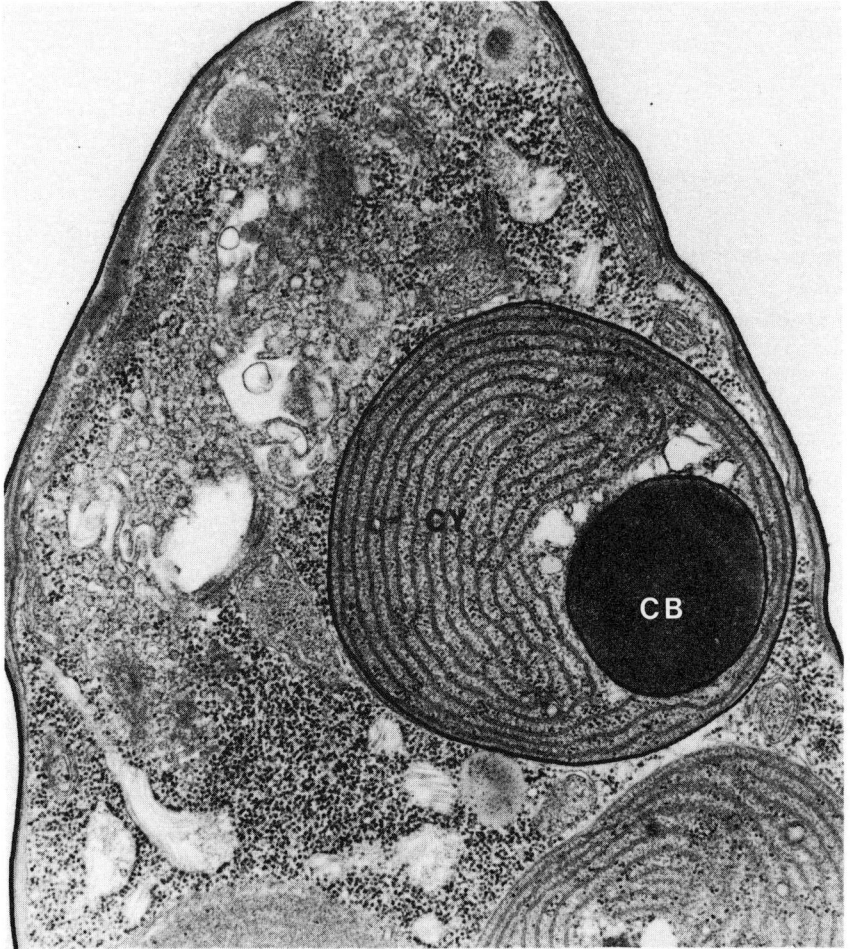

Figure 4. Transmission electron micrograph of *C. paradoxa* showing further details of the morphology of the association. The cyanelle is enclosed within a host vacuolar membrane. Cyanellar ribosomes appear as small black dots; by comparison, host ribosomes are larger. The peptidoglycan wall is barely discernible. CY = cyanelle; CB = carboxysome. Magnification approximately ×32,000.

are dependent on the host cytoplasm for the expression of their genetic information is to determine whether or not the cyanelles are able to synthesize all of their necessary components independent of the host's nuclear-cytoplasmic machinery. One protein that might lend itself to such experimental analysis is the ubiquitous enzyme ribulose bisphosphate carboxylase-oxygenase responsible for photosynthetic CO_2 reduction. In chloroplasts of higher plants, algae, and some cyanobacteria, this protein has a subunit structure involving polypeptides of distinct size classes, the large subunits being about 50 kd and the small subunits about 12-15 kd (Rutner and Lane, 1967; Kawashima, 1969). In plant cells the large subunits are known to be synthesized on chloroplast ribosomes (Criddle et al., 1970; Kawashima, 1970) and the small subunits on cytoplasmic ribosomes. RuBP-Carboxylase from the cyanelles of *C. paradoxa* shows a quaternary structure very similar to the same enzyme found in chloroplasts (Reisfeld and Edelman, 1976; Codd and Stuart, 1977). To my knowledge no attempts have been made to determine the sites of RuBP-carboxylase subunit synthesis in *C. paradoxa*. The significance of such observations is obvious.

A protein-synthesizing system probably analogous to that forming RuBP-carboxylase in chloroplasts involving c-phycocyanin may exist in *C. paradoxa*. In their analyses of the phycobilins in *C. paradoxa*, Trench and Ronzio (1978) found that allophycocyanin was a homodimer while c-phycocyanin was a heterodimer, the large subunit polypeptide being about 15 kd and the small subunit about 13 kd (Figure 5). Electronic absorption and fluorescence emission spectral analyses characterized this pigment as c-phycocyanin as opposed to r-phycocyanin (see Chapman, 1973).

The antimetabolite rifampicin (40 µg/ml) inhibited the incorporation of photosynthetically fixed ^{14}C into both phycobilins in vivo in *C. paradoxa*. However, several experiments showed that cycloheximide (3 µg/ml) preferentially inhibited the incorporation of ^{14}C into the small subunit of c-phycocyanin (Figure 6). Chloramphenicol (300 µg/ml) inhibited incorporation of ^{14}C into the large subunit. Experiments (Trench, unpublished) using ^{35}S produced the same results. Independent evidence (Trench and Siebens, 1978) showed that rifampicin completely inhibited cyanellar r-RNA synthesis; such inhibition might well be responsible for the loss of c-phycocyanin synthesis in the presence of rifampicin. However, cycloheximide showed no effect on cyanellar r-RNA synthesis, which implies that the effect of cycloheximide on the synthesis of the small subunit of c-phycocyanin might be exerted at the level of the host's cytoplasmic ribosomes, and therefore outside the cyanelles (Trench, 1980). These observations do not show that the small subunit polypeptide of c-phycocyanin is synthesized on the host cytoplasmic ribosomal system. It is possible that some other protein involved in the synthesis of the molecule by the cyanelles is derived from the host cytoplasm.

Several studies have produced evidence indicating that the synthesis of chloroplast r-RNA and chlorophyll are inhibited when cycloheximide inhibits cytoplasmic protein synthesis (Gassman and Bogorad, 1967a, b; Ingle, 1968; Detchon and Possingham, 1973). In an attempt to explore the possible existence of an analogous situation in *C. paradoxa*, Trench and Siebens (1978) analyzed the synthesis of cyanellar r-RNA and chlorophyll in vivo under the influence of a variety of antibiotics.

Incorporation of ^{14}C into chlorophyll *a* by cells in log phase of growth was found to be inhibited by cycloheximide more than by either chloramphenicol or rifampicin. Rifampicin inhibited the incorporation of ^{33}P into both the 23S and

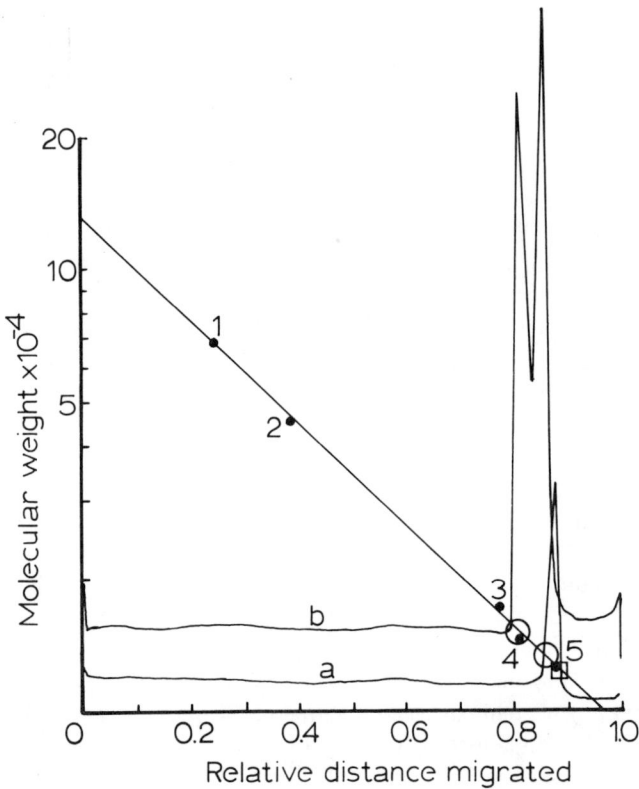

Figure 5. Densitometer scans of SDS-polyacrylamide gel electrophoretograms of (a) allophycocyanin (AP) and (b) c-phycocyanin (CP) from *C. paradoxa*. The scans are superimposed on a semilogarithmic plot of the molecular weight of the subunits of AP and CP against the relative migration distance of the polypeptides in polyacrylamide gels. Solid points represent molecular weight markers: 1. Bovine serum albumin; 2. ovalbumin; 3. myoglobin; 4. lysozyme; 5. cytochrome *c*. Open circles represent the heavy and light subunits of CP. The open square represents the subunits of AP. Data from Trench and Ronzio (1978). Reproduced by permission of the Royal Society of London.

16S cyanellar r-RNAs. Chloramphenicol, in contrast, produced a more marked inhibition of the synthesis of the 16S r-RNA than of the 23S r-RNA. Cyclo-heximide showed no demonstrable effect on the synthesis of cyanellar r-RNA (Figure 7). These data are consistent with the interpretation that some component of chlorophyll synthesis in the cyanelles is dependent on protein synthesis in the host cytoplasm. What this component might be remains unresolved. By contrast, cyanellar r-RNA appears to be totally independent of host cytoplasmic protein synthesis, a situation which is in marked contrast to the chloroplast, and suggests less dependence of the cyanelles on the host cytoplasm than that demonstrated by chloroplasts with respect to the synthesis of r-RNA.

Additional information was provided by studies of the degradation patterns of cyanellar r-RNA. Siebens and Trench (1978) found that the degradation pattern of r-RNA from the cyanelles of *C. paradoxa* resembled much more closely the patterns observed in chloroplast r-RNA than r-RNA from free-living cyanobacteria. When chloroplast r-RNA was degraded, the 23S r-RNA cleaved toward the middle and produced fragments of 0.67 and 0.53 $\times$ 10^6 daltons, while the 16S r-RNA was stable (Leaver and Ingle, 1971). By contrast, 23S r-RNA from several cyanobacteria cleaved toward the end and produced fragments of 0.9 and 0.2 $\times$ 10^6 daltons (Szalay et al., 1972; Grierson and Smith, 1973). Figure 8 shows that when the 23S r-RNA from the cyanelles of *C. paradoxa* was degraded, it cleaved toward the middle and produced fragments of 0.61 and 0.43 $\times$ 10^6 daltons. The 16S r-RNA remained stable.

Figure 6. Densitometer scans of SDS-polyacrylamide gel electrophoretograms of c-phycocyamin from *C. paradoxa* showing the distribution of ^{14}C (broken lines) in the subunit polypeptides in (a) untreated controls and (b) cycloheximide-treated cells.

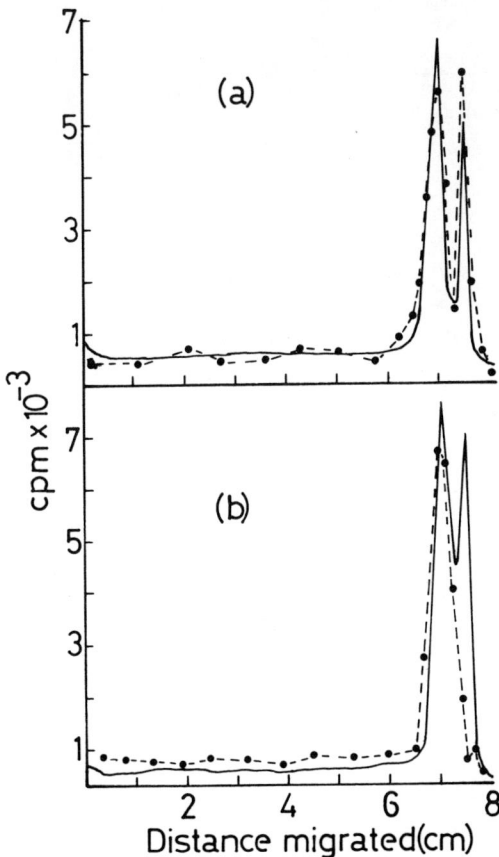

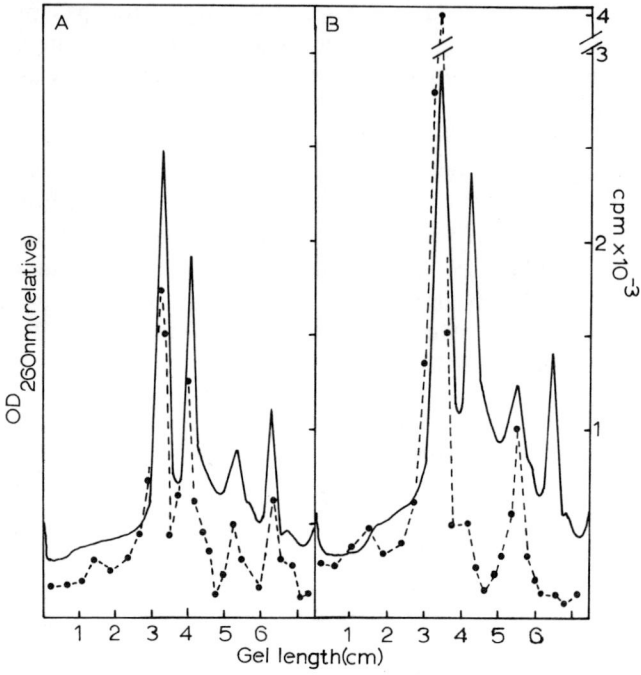

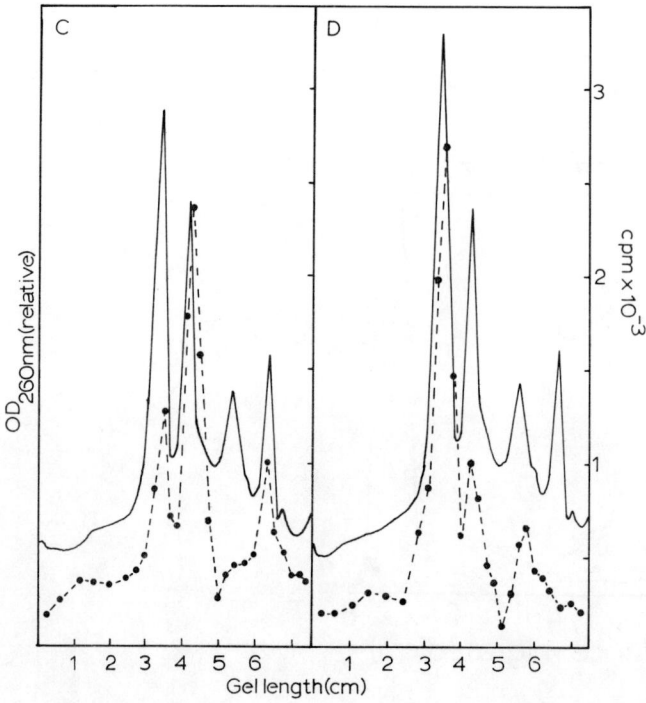

Figure 7. Densitometer scans (-) and pattern of incorporation of ^{33}P (---) into r-RNA in *C. paradoxa* and its endosymbiotic cyanelles. Densitometer peaks, from left to right, correspond to 25S, 23S, 18S, and 16S r-RNA. (a) Control culture to which no antibiotics were added; (b) cells treated with 40 μg/ml rifampicin; (c) cells treated with 3 μg/ml cycloheximide; and (d) cells treated with 300 μg/ml D(-)chloramphenicol. Data from Trench and Siebens (1978), reproduced by permission of the Royal Society of London.

These data obtained from studies with *C. paradoxa* tend to support the idea that in some respects the cyanelles demonstrate characteristics intriguingly similar to those of chloroplasts. In other respects, however, the cyanelles appear to be independent of the host cytoplasm, more so than would be expected of an organelle with a highly reduced genome.

From the discussion it is apparent that it is not possible to form any real conclusion on the origin of chloroplasts. The data presently available can in most instances be effectively used as arguments in support of either the endosymbiotic origin or an origin by cluster and cloning of genes. The potential beauty of the endosymbiosis theory is that it is readily testable because of the many extant examples of apparently tightly integrated symbioses between prokaryotes and eukaryotes. The fact, best illustrated by the controversy over *C. caldarium* and

Figure 8. Densitometer scans of polyacrylamide gels after electrophoretic separation of r-RNA from the cyanelles of *C. paradoxa*. The numbers associated with the peaks represent the molecular weight × 10^{-6} daltons. (a) r-RNA extracted and separated in the presence of 1 mM MgCl$_2$. The RNAs of 0.61 and 0.43 × 10^6 daltons are derived from breakdown of the 23S species; (b) intact 23S and 16S r-RNA isolated and separated in the presence of 25 mM MgCl$_2$; (c) r-RNA from *E.coli* used as standard. Data reproduced from Siebens and Trench (1978) by permission of the Royal Society of London.

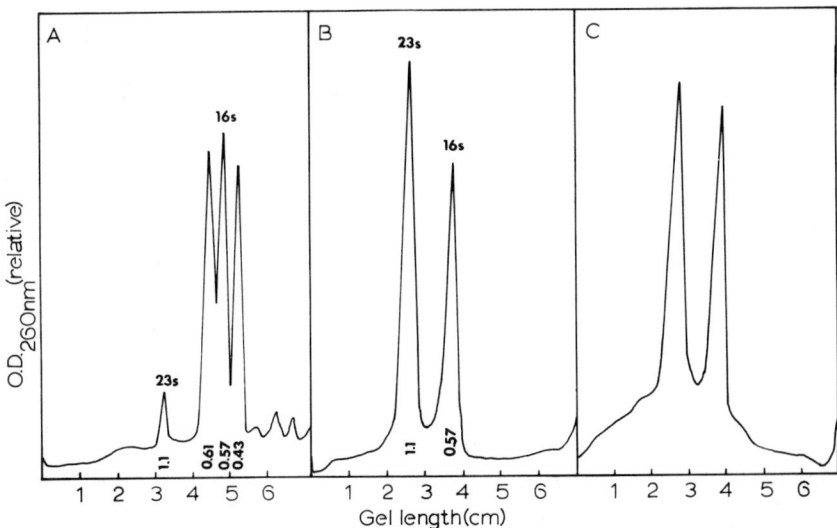

C. paradoxa, that it is so difficult to define and so distinguish between a photosynthetic prokaryotic cytobiont and a de facto chloroplast is in itself a strong
argument in favor of a symbiotic origin for chloroplasts. However, our knowledge
of the details of many of these symbioses involving cyanelles is so meager that
we should exercise extreme caution in "creating" organelles from cytobionts.

I am grateful to Mrs. Helena Kirk for her assistance with the literature search.

References

Aitken, A., and Stanier, R. Y. 1979. Characterization of peptidoglycan from the cyanelles of *Cyanophora paradoxa.* J. Gen. Microbiol. 112:219–223.

Allen, M. B. 1959. Studies with *Cyanidium caldarium,* an anomalously pigmented chlorophyte. Arch. Microbiol. 32:270–277.

Antia, N. J. 1977. A critical appraisal of Lewin's Prochlorophyta. Br. Phycol. J. 12:271–276.

Berns, D. S. 1967. The immunochemistry of biliproteins. Plant Physiol. 42:1569–1586.

Bogorad, L. 1965. Studies of phycobiliproteins. Rec. Chem. Prog. 26:1–12.

Bogorad, L. 1975. Evolution of organelles and eukaryotic genomes. Science 188:891–898.

Chapman, D. J. 1973. Biliproteins and bile pigments. *In* The Biology of Blue Green Algae. N. G. Carr and B. A. Wilton (eds.). Berkeley and Los Angeles: Blackwell Scientific Publications.

Chapman, D. J. 1974. Taxonomic status of *Cyanidium caldarium,* the Porphyridiales and Goniotrichales. Nova Hedwigia 25:673–682.

Codd, G. A., and Stuart, W. D. P. 1977. Quaternary structure of the D-ribulose 1,5-diphosphate carboxylase from the cyanelles of *Cyanophora paradoxa.* FEMS. Lett. 1:35–38.

Criddle, R. S., Dan, B., Kleinkoff, G. E., and Huffaker, R. C. 1970. Differential synthesis of ribulose diphosphate carboxylase subunits. Biochem. Biophys. Res. Commun. 41:621–627.

Detchon, P., and Possingham, J. V. 1973. Chloroplast ribosomal ribonucleic acid synthesis on cultured spinach leaf tissue. Biochem. J. 136:829–836.

Echlin, P. 1967. The biology of *Glaucocystis nostochinearum.* I. The morphology and fine structure. Br. Phycol. Bull. 3:225–239.

Edelman, M., Swinton, D., Schiff, J. A., Epstein, H. T., and Zeldin, B. 1967. Deoxyribonucleic acid of the blue-green algae (Cyanophyta). Bacteriol. Rev. 31:315–331.

Edwards, M. R., and Mainwaring, J. D., Jr. 1973. Ultrastructural localization of phycocyanin in the acidophilic, thermophilic alga, *Cyanidium caldarium. In* 31st Ann. Proc. Electron Microscopy Soc. Amer. C. J. Arceneaux (ed.).

Fredrick, J. F. 1968. Glucosyltransferase isoenzymes in algae. III. The polyglucosides and enzymes of *Cyanidium caldarium.* Phtyochemistry 7:1573–1576.

Fredrick, J. F. 1971. Storage polyglucan-synthesizing isoenzyme patterns in the Cyanophyceae. Phytochemistry 10:395–398.

Fredrick, J. F. 1976. *Cyanidium caldarium* as a bridge alga between Cyanophyceae and

Rhodophyceae: Evidence from immunodiffusion studies. Plant Cell Physiol. 17:317–322.

Gantt, E., and Conti, S. F. 1965. The ultrastructure of *Porphyridium cruentum*. J. Cell Biol. 26:365–381.

Gassman, M., and Bogorad, L. 1967a. Control of chlorophyll production in rapidly greening bean leaves. Plant Physiol. 42:774–780.

Gassman, M., and Bogorad, L. 1967b. Studies on the regeneration of protochlorophillide after brief illumination of etiolated bean leaves. Plant Physiol. 42:781–784.

Glazer, A. N., Cohen-Bazire, G., and Stanier, R. Y. 1971. Comparative immunology of algal biliproteins. Proc. Nat. Acad. Sci. USA. 68:3005–3008.

Glazer, A. N., Apell, G. S., Hixon, C. S., Bryant, D. A., Rimon, S., and Brown, D. M. 1976. Biliproteins of cyanobacteria and Rhodophyta: Homologous family of photosynthetic accessory pigments. Proc. Nat. Acad. Sci. USA. 73:428–431.

Grierson, D., and Smith, H. 1973. The synthesis and stability of ribosomal RNA in blue-green algae. Eur. J. Biochem. 36:280–285.

Herdman, M. and Carr, N. G. 1974. Estimation of the genome size of blue-green algae from DNA renaturation rates. Arch. Microbiol. 99:251–254.

Herdman, M., and Stanier, R. Y. 1977. The cyanelle: chloroplast or endosymbiotic prokaryote? FEMS Lett. 1:7–12.

Herdman, M., Janvier, M., Waterbury, J. B., Rippka, R., and Stanier, R. Y. 1979a. Deoxyribonucleic acid base composition of cyanobacteria. J. Gen. Microbiol. 111:63–71.

Herdman, M., Janvier, M., Rippka, R., and Stanier, R. Y. 1979b. Genome size of cyanobacteria. J. Gen. Microbiol. 111:73–85.

Holm-Hansen, O., Prasad, R., and Lewin, R. A. 1965. Occurrence of a ε-diaminopimelic acid in algae and flexibacteria. Phycologia 5:1–14.

Holton, R. W., Blecker, H. H., and Stevens, T. S. 1968. Fatty acids in blue-green algae: Possible relation to phylogenetic position. Science 160:545–547.

Ikan, R., and Seckbach, J. 1972. Lipids of the thermophilic alga *Cyanidium caldarium*. Phytochemistry 11:1077–1082.

Ingle, J. 1968. Synthesis and stability of chloroplast ribosomal RNAs. Plant Physiol. 43:1448–1454.

Kawashima, N. 1969. Comparative studies on fraction I protein from spinach and tobacco leaves. Plant Cell Physiol. 10:31–40.

Kawashima, N. 1970. Non-synchronous incorporation of ^{14}C into amino acids of two subunits of fraction I protein. Biochem. Biophys. Res. Commun. 38:119–124.

Kies, L. 1974. Elektronenmikroskopische Untersuchungen an *Paulinella chromatophora* Lauterborn, einer Thekamöbe mit blaugrünen Endosymbionten (Cyanellen). Protoplasma 80:69–89.

Kies, L. 1976. Untersuchungen zur Feinstruktur und taxonomischen Einordnung von *Gloeochaete wittrockiana*, einer apoplastidalen capsalen Alge mit blaugrünen Endosymbionten (Cyanellen). Protoplasma 87:419–446.

Kirk, J. T. O., and Tilney-Bassett, R. A. E. 1978. The Plastids. Amsterdam, New York, Oxford: Elsevier-North Holland Biomedical Press.

Kremer, B. P., Feige, G. B., and Schneider, H. A. W. 1978. A new proposal for the systematic position of *Cyanidium caldarium*. Naturwiss. 65:157.

Lau, R. H., and Doolittle, W. F. 1979. Covalently closed circular DNA in closely related unicellular cyanobacteria. J. Bacteriol. 137:648–652.

Leaver, C. J. and Ingle, J. 1971. The molecular integrity of chloroplast ribosomal ribonucleic acid. Biochem. J. 123:235–243.

Lee, R. E. 1972. Origin of plastids and the phylogeny of algae. Nature 237:44–46.

Lewin, R. A. 1976. Prochlorophyta as a proposed new division of algae. Nature 261:697–698.

Lewin, R. A. 1977. *Prochloron:* type genus of the Prochlorophyta. Phycologia 16:217.

Loening, U. E. 1968. Molecular weights of ribosomal RNA in relation to evolution. J. Mol. Biol. 38:355–365.

Margulis, L. 1970. Origin of Eukaryotic Cells. New Haven and London: Yale University Press.

Margulis, L. 1976. Genetic and evolutionary consequences of symbiosis. Exp. Parasitol. 39:277–349.

Mercer, F. V., Bogorad, L., and Mullens, R. 1962. Studies with *Cyanidium caldarium.* I. The fine structure and systematic position of the organism. J. Cell Biol. 3:393–403.

Muscatine, L., Pool, R. R., Jr., and Trench, R. K. 1975. Symbiosis of algae and invertebrates: Aspects of the symbiont surface and host symbiont interface. Trans. Am. Micros. Soc. 94:450–469.

Pascher, A. 1929. Studien über Symbiosen. I. Über einige Symbiosen von Blaualgen in Einzellern. Jahrb. Wissensch. Bot. 71:386–462.

Phillips, D. O. and Carr, N. G. 1977. Nucleic acid analysis and the endosymbiotic hypothesis. Taxon. 26:3–42.

Pickett-Heaps, J. 1972. Cell division in *Cyanophora paradoxa.* New Phytol. 71:561–567.

Raven, P. H. 1970. A multiple origin for plastids and mitochondria. Science 169:641–646.

Reisfeld, A., and Edelman, M. 1976. Ribulose diphosphate carboxylase from *Cyanophora paradoxa.* Isr. J. Bot. 25:97.

Rutner, A. C., and Lane, M. D. 1967. Non-identical subunits of ribulose diphosphate carboxylase. Biochem. Biophys. Res. Commun. 28:531–537.

Sager, R. 1977. Genetic analysis of chloroplast DNA. Adv. Genet. 19:287–337.

Seckbach, J. 1972. On the fine structure of the acidophilic hot-spring alga *Cyanidium caldarium.* Microbios 5:113–142.

Schnepf, E., and Brown, R. M., Jr. 1971. On the relationships between endosymbiosis and the origin of plastids and mitochondria. *In* The Origin and Continuity of Cell Organelles. J. Reinert and H. Ursprung (eds.). New York, Heidelberg, Berlin: Springer Verlag.

Schnepf, E., Koch, W., and Deichgräber, G. 1966. Zur Cytologie und Taxonomischen Einordnung von *Glaucocystis.* Arch. Mikrobiol. 55:149–174.

Shively, J. M., Ball, F. L., and Kline, B. W. 1973. Electron microscopy of the carboxysomes (polyhedral bodies) of *Thiobacillus neopolitans.* J. Bacteriol. 116:1405–1411.

Siebens, H. C., and Trench, R. K. 1978. Aspects of the relation between *Cyanophora paradoxa* (Korschikoff) and its endosymbiotic cyanelles *Cyanocyta korschikoffiana* (Hall and Claus). III. Characterization of ribosomal ribonucleic acids. Proc. R. Soc. B. 202:463–472.

Silva, P. C. 1962. Classification of Algae. *In* Physiology and Biochemistry of Algae. R. A. Lewin (ed.). New York: Academic Press.

Stanier, R. Y. 1974. The origins of photosynthesis in eukaryotes. Symp. Soc. Gen. Microbiol. 24:219–240.

Stanier, R. Y., and Cohen-Bazire, G. 1977. Phototrophic prokaryotes: the cyanobacteria. Annu. Rev. Microbiol. 31:225–274.

Szalay, A., Munsche, D., Wollgiehn, R., and Parthier, B. 1972. Ribosomal ribonucleic acid precursor ribonucleic acid in *Anacystis nidulans*. Biochem. J. 129:135–140.

Taylor, F. J. R. 1974. Implications and extensions of the serial endosymbiosis theory of the origin of eukaryotes. Taxon 23:229–258.

Taylor, F. J. R. 1979. Symbiotism revisited: a discussion of the evolutionary impact of intracellular symbiosis. Proc. R. Soc. B. 204:267–286.

Thorne, S. W., Newcomb, E. H., and Osmond, C. B. 1977. Identification of chlorophyll *b* in extracts of prokaryotic algae by fluorescence spectroscopy. Proc. Nat. Acad. Sci. USA. 74:575–578.

Trench, R. K. 1979. The cell biology of plant animal symbiosis. Annu. Rev. Plant Physiol. 30:485–531.

Trench, R. K. 1980. Integrative mechanisms in mutualistic symbioses. *In* Cellular Interactions. C. B. Cook, E. Rudolph, and P. W. Pappas (eds.). Columbus: Ohio State University Press.

Trench, R. K., and Ronzio, G. S. 1978. Aspects of the relation between *Cyanophora paradoxa* (Korschikoff) and its endosymbiotic cyanelles *Cyanocyta korschikoffiana* (Hall and Claus). II. The photosynthetic pigments. Proc. R. Soc. B. 202:445–462.

Trench, R. K., and Siebens, H. C. 1978. Aspects of the relation between *Cyanophora paradoxa* (Korschikoff) and its endosymbiotic cyanelles *Cyanocyta korschikoffiana* (Hall and Claus). IV. The effects of rifampicin, chloramphenicol and cyaloheximide on the synthesis of ribosomal ribonucleic acids and chlorophyll. Proc. R. Soc. B. 202:473–482.

Trench, R. K., Pool, R. R., Jr., Logan, M., and Engelland, A. 1978. Aspects of the relation between *Cyanophora paradoxa* (Korschikoff) and its endosymbiotic cyanelles *Cyanocyta korschikoffiana* (Hall and Claus). I. Growth, ultrastructure, photosynthesis and the obligate nature of the association. Proc. R. Soc. B. 202:423–443.

Uzzell, T., and Spolsky, C. 1974. Mitochondria and plastids as endosymbionts: a revival of special creation? Am. Sci. 62:334–343.

Withers, N. W., Alberte, R. S., Lewin, R. A., Thornber, J. P., Britton, G., and Goodwin, T. W. 1978. Photosynthetic unit size, carotenoids and chlorophyll-protein composition of *Prochloron* sp., a prokaryotic green algae. Proc. Nat. Acad. Sci. USA. 75:2301–2305.

Discussion of Presentation by Dr. Trench

SHILO: Could genome multiplicity, which is also found in lambda particles of *Paramecium* by Soldo and Godoy (1973), be a loss of stringent DNA replication in a cell which has lost its cell wall and rigidity? How rapidly are parts of the genome lost, when (as in endosymbiosis) they are not functionally required any more?

TRENCH: I must emphasize that the cyanelles in *C. paradoxa* have not lost the cell wall, so I do not really have an answer. I do not think that anyone knows how rapidly parts of the genome are lost, or whether there really is a major reduction in genome size in the cyanelles, despite the report of Herdman and Stanier (1977).

SCHMIDT: There is nitrate reduction in cyanelles and nitrate-dependent oxygen evolution according to Bothe and Floener (1978).

TRENCH: The cyanelles do function in the production of organic matter, which ultimately becomes available to the host.

BARNETT: Does *Cyanophora* grow in the dark? Are there strain differences among *C. paradoxa* cultures?

TRENCH: No—the organism is an obligate phototroph. Whether there are strain differences has not, to my knowledge, been determined. Most reports cite Luigi Provasoli (of Haskins Laboratories at Yale University) or Starr's collection (Starr, 1978) as their source. (Dr. Starr and his collection have moved to The University of Michigan at Ann Arbor).

SCHIFF: I noticed on one of your slides that you show *Euglena* cytochrome c-552 as being synthesized on cytoplasmic ribosomes based on cycloheximide inhibition. In fact recent data indicate that the synthesis of this enzyme is sensitive to both 80S and 70S inhibitors (Freyssinet et al., 1979). Either different parts of the molecule (heme, polypeptide, etc.) are made in different compartments or the synthesis (or lack of synthesis) of one regulates the synthesis of the other.

TRENCH: The data I presented were from Kirk and Tilney-Bassett (1978). They actually present data showing that certain components of plastids are sensitive to both 80S and 70S inhibitors.

References to Discussion

Bothe, D., and Floener, L. 1978. Physiological characterization of *Cyanophora paradoxa*, flagellate containing cyanelles in endosymbiosis. Z. Naturforsch. 33C:981–987.

Freyssinet, G., Harris, G. C., Nasitir, M., and Schiff, J. A. 1979. Events surrounding the early development of *Euglena* chloroplasts 14. Biosynthesis of cytochrome c-552 in wild type and mutant cells. Plant Physiol 63:908–915.

Herdman, M., and Stanier, R. Y. 1977. The cyanelle: chloroplast or symbiotic prokaryote? FEMS Lett. 1:7–12.

Kirk, J. T. O., and Tilney-Bassett, R. A. E. 1978. *The Plastids*. Amsterdam, New York, Oxford: Elsevier-North Holland Biomedical Press.

Soldo, A. T., and Godoy, G. A. 1973. Molecular complexity of *Paramecium* symbiont lambda DNA; evidence for the presence of a multicopy genome. J. Mol. Biol. 73:93–108.

Starr, R. C. 1978. The culture collection of algae at the University of Texas at Austin. J. Phycol. 14:47–100.

Establishment of Photosynthetic Eukaryotes as Endosymbionts in Animal Cells

Leonard Muscatine

Eukaryotic algae symbiotic with animal cells are relevant to an evaluation of the serial endosymbiosis theory of origin of organelles because these contemporary associations illustrate the ease and frequency with which such algae establish endosymbioses in nature and the impressive selective advantage which the association confers on both partners (Raven, 1970; Flavell, 1972). Some idea of the breadth of associations is given in a review by Trench (1979), who lists associations between representatives of six algal classes and as many invertebrate phyla. In addition, unlike prokaryotic algae in association with heterotrophs, eukaryotic symbionts can be separated from their hosts and recombined with relative ease. As such they represent material favorable for experimental studies on initiation and maintenance of an endosymbiosis. Contemporary studies of the origins of organelles (see as described, for example, in various papers in this volume) have tended to focus on structural and functional homologies drawn between present day free living and symbiotic prokaryotes and organelles such as plastids and mitochondria. Relatively little attention has been given to dynamic cellular processes involved in the establishment of symbioses and how these might bear on an evaluation of the serial endosymbiosis theory. With this in mind, I have attempted in this paper to review and describe generally a range of cellular events observed when a foreign, self-reproducing phototroph becomes established as an hereditary endosymbiont within a heterotrophic cell. I have

Preparation of this chapter was supported by National Science Foundation Grant No. PCM-78-27380.

Address reprint requests to Dr. Leonard Muscatine, Professor of Biology, Department of Biology, University of California, Los Angeles, California 90024.

drawn on information from a few eukaryotic endosymbioses, particularly the *Chlorella-Hydra viridis* association, which provide the most comprehensive set of available data. Comparative aspects of the establishment of algae-invertebrate symbioses are reviewed by Karakashian (1975), Muscatine et al. (1975), Trench (1979, 1980), Cook (1980), and Jolley and Smith (1980). The term *symbiosis* is used here in the general sense, but reference is made almost exclusively to mutualistic biotrophic associations.

Establishment of Endosymbionts: Hypothetical Evolutionary Sequence

The range of phenomena to be considered may be illustrated by the hypothetical evolutionary sequence of cellular events in the establishment of hereditary mutualistic biotrophic endosymbiosis (e.g., see Dubos and Kessler, 1963; Karakashian and Karakashian, 1965; Margulis, 1970; Flavell, 1972). Potential symbionts and host cells, pursuing widely divergent lifestyles, must have come into frequent contact with the result that the symbiont was phagocytized by the host cell and was probably contained in a vacuole. Thus contained, the foreign symbiont must have resisted exocytosis and digestive attack and eventually must have come into growth synchrony with the host cell, with provision for transmission to its offspring. With the symbiont and host brought together in a close and protracted association, copies of genetic information present in both partners manifested redundancy. As a result of selection pressure, the symbiont, for example, may have lost redundant genetic information, but its survival would not have been affected because its needs would then be satisfied by the host cells. Further loss of redundant genes would increase the dependence of symbiont on host, increase the specificity of the association, and define the level of partner integration. Extant associations may thus be regarded as representing a spectrum of degrees of evolution from recent, facultative, minimally integrated associations to ancient, obligate, and well-integrated ones.

Establishment of Endosymbionts: Contemporary Sequence

The Association of Hydra viridis and Chlorella

Hydra viridis is a small freshwater metazoan (Cnidaria: Hydridae) and, like all hydras, consists of two tissue layers and about six cell types. In addition, this species contains about 1.5×10^5 algal cells. The algae are found exclusively in the endodermal layer in digestive cells. The numbers of algae per cell vary with the position of the digestive cell along the body tube and with the maintenance conditions (Pardy and Muscatine, 1973; Pardy, 1974). In hydra fed daily and maintained at about 17°C on a 12:12 light:dark photoperiod, the cells from the stomach and budding zone each contain about 18 algae. The algae are situated at the base of each digestive cell. Details of the fine structure of *Chlorella* symbiotic with *Hydra* and *Paramecium* are given by Oschman (1967); Karakashian et al. (1968); Karakashian (1970); and Vivier et al. (1967) Maintained in a defined medium (Muscatine and Lenhoff, 1965) and fed daily on *Artemia* larvae, *Hydra* grows rapidly by asexual budding with a generation time of approximately 1.5 days, giving rise to clones with similar genetic, nutritional, and developmental histories. Algae and host may readily be separated and the algae may be artificially reintroduced into aposymbiotic hosts. The latter are obtained by photobleaching (Pardy, 1976b) or from algae-free eggs.

Until the systematics of green hydras and their algae are firmly established, we designate strains by the geographic origin of the host. Most of our work has been carried out on *H. viridis* (Fla.) [Florida strain]. The algae in virtually all strains of *Hydra* thus far studied seem to be indistinguishable from members of the genus *Chlorella*. From time to time symbiotic algae from *Paramecium bursaria* (strain NC64A) and free-living *Chlorella vulgaris* (strain 397) as well as from other free-living and symbiotic strains have been used in reinfection experiments (Jolley and Smith, 1980).

Methods of Study

Methods used to study symbiotic *Chlorella* in *Hydra* are given by Park et al. (1967), Pardy and Muscatine (1973), Muscatine et al. (1975), Pool (1979), Cook (1980), Jolley and Smith (1980), and McNeil (1980, 1981). Algae are harvested from stocks of green hydra by gentle homogenization and centrifugation. Algae thus freshly isolated, as well as other cultured algal cell types and an assortment of nonliving particles (e.g., heat-killed algae; latex spheres) can then be introduced by microinjection into the gut of aposymbiotic hydra. Subsequently the hydras are macerated and the numbers of algae or other injected particles taken up by each digestive cell are observed directly. The total surface area of interiorized plasma membrane may be computed. In addition, the percentage of total digestive cells which acquire algae or particles can be ascertained. The morphology of the course of uptake can be determined by scanning and transmission electron microscopy. Other methods are described in the text below.

The relevance of artificial initiation of the symbiosis to that which occurs in nature is noteworthy at this point. Whereas Hamann (1882) (cited in Kanaev, 1952 [1969]), Brien and Reniers-Decoen (1950), Valkanov (1950), and Stagni (1974) describe transmission of algae via eggs of green hydra, more recent work of Thorington et al. (1979) suggests that in addition, juvenile hydra hatch from eggs in the aposymbiotic condition and then acquire algae by ingestion. The algae are situated in abundance on the surface of the egg capsule, having been released from the hydra during oogenesis. In addition, Goetsch (1924) demonstrated that algae associated with small crustaceans can effect reinfection after the crustaceans have been ingested by hydra. Therefore, reinfection by algae entering the gut of aposymbiotic hydra is not without precedent in nature.

The Reinfection Process in *Hydra viridis*

For convenience, we refer to a series of arbitrary phases in the reinfection process. These are phagocytosis (including contact and adhesion events), digestion resistance, migration, repopulation (i.e., restoration of the normal number of algae per cell), and transmission to offspring. Except for the last, these events are depicted in Figures 1a–j. Recognition of specific algae as normal symbionts also takes place during the foregoing processes but it is not yet known how or where recognition is manifested.

Contact and Phagocytosis

When aposymbiotic hydra are injected with freshly isolated symbiotic algae, fixed within 10 min and examined by electron microscopy, algae are seen in the coelenteron (Fig. 1a) and contacting the digestive cell surface (Fig. 1b) which manifests abundant microvilli (Muscatine et al., 1975; McNeil, 1980). The potential for electrostatic attraction between algal and digestive cell surfaces is suggested by the studies of McNeil (1980). He demonstrated that anionized

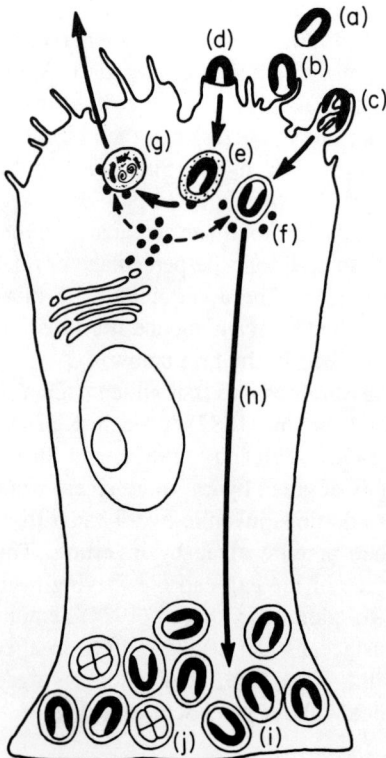

Figure 1. Schematic composite drawing showing the fate of freshly isolated symbionts (or free-living *Chlorella vulgaris*) after being presented to an aposymbiotic *Hydra viridis* (Fla.) digestive cell.

ferritin binds to endoderm and cationized ferritin to ectoderm, but not vice versa, inferring that the digestive cell surface bears a net positive charge. Further, in whole-cell electrophoresis studies, freshly isolated symbionts exhibited a much higher electrophoretic mobility toward the cathode than free-living *Chlorella,* demonstrating that the symbionts have a relatively strong net negative surface charge. These features appear to be ideal for electrostatic attraction of symbiotic algae to *Hydra viridis* digestive cells.

Using scanning and transmission electron microscopy, McNeil (1981) also demonstrated that the morphological response of digestive cells to a range of particle types covered a much broader spectrum than had previously been suspected for a nutritive phagocyte. Thus, unchallenged digestive cells, whose distal surface architecture is manifested by sparse microvilli and assorted small folds, respond to challenge with freshly isolated symbionts by producing abundant microvilli which engulf the symbiotic algae (Fig. 1c). In contrast, particles such as heat-killed symbionts, latex spheres, and free-living *Chlorella* do not evoke microvilli formation. Instead, they evoke the formation of cylindrical extensions of the cell membrane, termed by McNeil "funnels," each of which engulfs individual challenge particles (Fig. 1d). These observations have important implications for the definition of phagocytic recognition.

All particles described above are interiorized by *Hydra* digestive cells but to different extents (McNeil, 1981). For example, examination of digestive cells five hr after challenge, when phagocytosis is usually completed, reveals that particles such as freshly isolated symbionts enter more than 90% of the digestive cells of the gastric region (Pool, 1979). They enter in modest numbers, ranging from about 6–10 algae per cell, and effect the interiorization of about 1600 μm^2 of cell surface membrane (Muscatine et al., 1975; McNeil, 1981). In contrast, other particles (such as heat-killed symbionts, latex spheres, free-living *Chlorella,* and other strains of symbiotic algae such as NC64A from *Paramecium bursaria*) infect a much smaller percentage of digestive cells (5–10%) and enter in fewer numbers (2 to 3 per cell) (Muscatine et al., 1975; McNeil, 1981). This apparent dichotomy will be treated further in a later section.

Factors Affecting Phagocytosis of Algae

Membrane limitation. Whereas the normal carrying capacity of digestive cells is about 18 algae per cell, the aposymbiotic hydra take up only 6–10 symbiotic algae per cell even though the supply of algae is adequate for uptake in greater numbers. Intracellular space does not seem to be a limiting factor since digestive cells from green hydra which already contain their full normal complement of algae will take up supernumerary algae but only in the same small numbers. It has been hypothesized (Pardy and Muscatine, 1973) that the uptake of normal symbiotic algae is limited by the availability of plasma membrane for phagocytosis and the time required for synthesis of new membrane. Consistent with this hypothesis is the observation (Figure 2) that when a series of hydra are

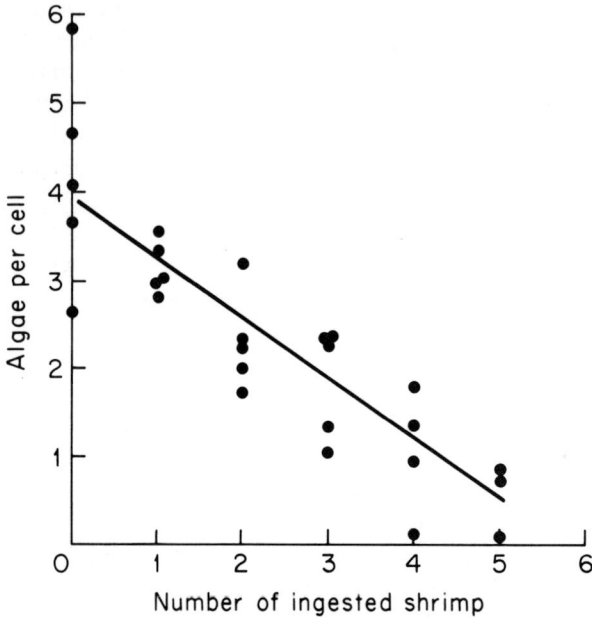

Figure 2. The effect of food ingestion on the endocytosis of algae by *Hydra viridis* (Fla.) digestive cells. The least squares regression line is significant at the 5% level ($r = -0.87$). *Source:* Reproduced with permission of The Faculty Press; from C. B. Cook et al., Cytobios. 23:17–31 (1979).

fed increasing numbers of *Artemia nauplii* and then injected with algae 7 hr later after uneaten food is regurgitated, the uptake of algae is drastically reduced and in inverse proportion to the number of *Artemia nauplii* previously ingested (Cook et al., 1979). In addition, fewer digestive cells are infected in the more well-fed hydra. These data are interpreted to mean that digestive cells utilize similar limiting membrane constituents in the phagocytosis of food and algae (Cook et al., 1979). In other experiments, where aposymbiotic hydra are injected with algae at intervals after an initial injection, the capacity of the digestive cells to take up the normal numbers of algae is restored gradually over a 24-hr period (D'Elia, Cook and Muscatine, unpublished). Presumably during this time the putative limiting constituents are resynthesized.

Surface charge. The extent of uptake is influenced by treatment of algae with various substances believed to coat the algal surface and modify the net external surface charge. Thus, the extent of uptake of freshly isolated symbionts is reduced by treatment with poly-D-lysine, protamine sulfate, and ferric chloride. All of these bear a net positive charge. Uptake is not reduced and in some cases may even be enhanced by treatment with poly-D-glutamic acid or bovine serum albumin, polypeptides with a net negative charge (McNeil, 1980).

Antibodies. Pool (1979) has demonstrated that, in contrast to freshly isolated *Hydra* symbionts (designated F/F), when symbionts from *P. bursaria* (strain NC64A) grown in culture (NC/L) are injected into *H. viridis* (Fla.), relatively few cells are infected (<10%) and the number of algae per cell averages 2–3. However, if NC64A algae are allowed to proliferate in *Hydra* for several months and are then removed and introduced into a second set of aposymbiotic hydra, the infectivity of these algae (now designated NC/F) is increased and resembles that of the normal hydra symbiosis. This increased infectivity is believed to be the result of changes in the antigenicity of the algae. Complement fixation tests show that they acquire surface antigenic sites similar to F/F algae and have fewer NC64A antigenic sites. In addition to these results, Pool also showed that when F/F and NC/F algae are treated with F/F antisera and then injected into *H. viridis,* their infectivity is significantly decreased, but treatment with antiserum from cultured NC64A (NC/L) had no effect on infectivity. Pool (1979) interpreted these data to mean that F/F antiserum masks specific phagocytic recognition sites and that the inhibiting effect on uptake is not due merely to a nonspecific effect of exposure to antiserum.

Enzymes and lectins. Ingestion of freshly isolated *H. viridis* (Fla.) algae was significantly reduced after host endoderm cells were exposed in situ to trypsin and concanavalin-A. The inhibitory effect of trypsin was not observed when, prior to use, the trypsin was mixed with trypsin inhibitor. Concanavalin-A did not inhibit phagocytosis of algae in the presence of α-methyl mannoside. Neither trypsin nor concanavalin-A had any effect on phagocytosis of latex spheres, suggesting that hydra digestive cells possess two recognition systems. One is specific for freshly isolated native symbionts and is sensitive to trypsin and concanavalin-A. The other governs the uptake of latex spheres and is not sensitive to these agents (Pool and Muscatine, 1980).

Fate of Phagocytosed Algae: Migration; Digestion Resistance

Symbiotic algae, as well as other particles described above, are all engulfed by digestive cells from within a few minutes to 2 hr after challenge. Subsequently, over the next 3–24 hr, differences in the spatial and temporal aspects of the fate of these particles can be discerned. On the one hand, particles such as latex spheres, free-living *Chorella,* and heat-treated symbionts remain in the apical portion of the cell after uptake, usually more than one particle per host vacuole, and are egested or digested within the next 24 hr (Fig. 1e, g). On the other hand, freshly isolated symbionts are invariably sequestered in individual vacuoles after phagocytosis (Fig. 1f) and move to the base of the cell (Fig. 1g), maintaining this position thereafter, and effecting reestablishment of the association. From these observations we may further extend our concept of recognition of fully competent symbionts versus other classes of particles, beginning with the various phagocytic morphologies observed by McNeil (1981) and extending through to the dichotomous events leading to either rejection or acceptance of the various challenge particles.

Migration. Migration of algae has been studied by measuring the positions of algae in cells of hydra macerated at intervals after injection of algae. These data show that algae interiorized at the tips of digestive cells 1–2 hr after injection gradually move to the base of the host cells (Fig. 1i). The average rate of movement is 5–6 μm per minute. Migration is completed at about 5 hr after injection of algae. Since all infective algae migrate, and since noninfective algae or other particles do not, we conclude that the process is necessary for establishment of a stable permanent endosymbiosis in these hydra. Weis (1976) has alluded to an analogous situation in *P. bursaria* where algae seem to move from the central region of the cell to the periphery. Coincidental with this is their apparent resistance to or freedom from digestive attack while in the peripheral cytoplasm. Since *Chorella* sp. are nonmotile, and lack locomotory organelles, the movement of each alga-vacuole complex or "symbiosome" from tip to base of the digestive cell must be governed by a transport mechanism intrinsic to the host cell. Because migration can be inhibited by low concentrations of colchicine (Cook, 1980), β-peltatin, and vinblastine (Cooper and Margulis, 1978), all of which bind to tubulin, microtubules are thought to be involved in the migration mechanism. As yet there is no direct evidence to support this hypothesis.

Resistance to digestion. Detailed experimental studies on the digestion of symbiotic *Chlorella* were first carried out on *Paramecium bursaria* (Karakashian and Karakashian, 1973; Weis, 1976). This ciliate normally harbors several hundred intracellular algal symbionts. Digestion or resistance to digestion is apparently a function of the host nutritional state and the location of the algae within the cell. Well-fed hosts seem not to digest algae. Starved hosts, however, will take up homologous algae and digest them in substantial numbers. Digestion apparently takes place in vacuoles in the central cytoplasm. Acid phosphatase activity can be demonstrated in these vacuoles but not in those of the peripheral cytoplasm. In aposymbiotic *P. bursaria,* short periods of feeding on algae results in their digestion but longer feeding periods result in infection of the host cell, possibly the result of cyclosis of many algae from the central to the peripheral cytoplasm. The movement of algae from areas active in digestion to areas which are inactive is reminiscent of migration in *H. viridis.* Heat-killed symbiotic algae are readily digested if located in individual vacuoles but if they occur in a vacuole with a living symbiotic alga the digestive attack is delayed until the heat-killed cells are sequestered into individual vacuoles. The living algae apparently interfere with the normal digestive process, possibly by secretion of enzyme inhibitors or by preventing fusion of lysosomes with the vacuole. A similar picture emerges from studies on *H. viridis.* Hohman (1980) has shown that ferritin-labeled secondary lysosomes are not seen fusing with normal healthy symbiotic algae, but are frequently seen fusing with vacuoles containing heat-killed or damaged symbiotic algae. Acid phosphatase activity can be demonstrated in these vacuoles by electron microscopic cytochemistry. Hohman concludes from these and other observations that healthy symbionts resist digestion and that the mechanism involves the prevention of lysosomal fusion with the vacuole.

Partner Integration

At the time of writing, the level of partner integration (i.e., gene integration) in the *Hydra viridis-Chlorella* association (or any other alga-invertebrate association) has not yet been assessed in any sophisticated way at the molecular level. The state of the art seems to be to list phenotypic traits which describe the extent of evolutionary divergence of the symbiont or host from its free-living counterparts. For *H. viridis* symbionts this list is rather meager. In view of the demonstrated susceptibility of both free-living and symbiotic *Chlorella* to structural modification in response to changing environmental variables, stable evolutionary divergences are difficult to discern (Karakashian, 1970; Pardy, 1976a). Nevertheless, I hypothesize that symbiotic *Chlorella* of *H. viridis* (Fla.) do differ from some free-living *Chlorella* species with respect to at least four traits (Table 1) (a) surface charge; (b) digestion resistance; (c) autospore production; and (d) selective release of soluble photosynthate. Undoubtedly, this list could be modified depending on the species and strains being compared and new information

Table 1. Some Phenotypic Traits Which May Describe the Evolutionary Divergence of Symbiotic *Chlorella* [from *H. viridis* (Fla.)] and Free-living *Chlorella vulgaris*

Trait	Free-living *Chlorella*	Symbiotic *Chlorella*	Selective advantage of the association	Reference
Net surface charge	Weak negative	Strong negative	Initial phase of reinfection	McNeil, 1980
Digestion resistance	Absent	Present	Persistence after phagocytosis	Hohman, 1980
Autospore production	Large numbers	No more than four	Maintenance of growth synchrony	Various
Selective release of metabolites	None	ph-Dependent release of maltose *in vitro* and translocation *in vivo*	Supplement to host nutrition	Muscatine, 1965

on the behavior of symbiotic and free-living *Chlorella* in vitro and in vivo. Surface charge and digestion resistance have been discussed above. Limitation of autospore production might be a facultative trait of *H. viridis* (Fla.) algae in symbiosis. Unfortunately this cannot be tested until symbionts from this strain are cultured in vitro and conditions are found which enable free-living *Chlorella* to persist in *H. viridis* (Fla.) long enough to evaluate their reproductive capacities in vivo.

Translocation is a trait most often correlated with the acquisition of the symbiotic habit, and it deserves some elaboration here. It is now generally known and accepted that a wide range of algae endosymbiotic with aquatic invertebrates exhibit a selective and abundant release of soluble organic material of which carbohydrate in some form is most often the major constituent (Smith, 1974). Symbiotic *Chlorella* generally release maltose along with traces of glycolic acid. Release outside the host is markedly pH-dependent, with acidic media eliciting the greatest level of release (Muscatine, 1965; Cernichiari et al., 1969). Release also occurs within the host. In the latter case as much as 40–50% of the photosynthate may be translocated to animal tissues where it participates in host metabolism and is stored as glycogen (Roffman and Lenhoff, 1969). Translocation, with its nutritional benefit, is usually cited as the selective advantage to the host as a consequence of sustaining a symbiotic algal flora. Translocation may serve other functions. At present, experiments in our laboratory suggest a strong correlation between the ability to translocate and the ability of an alga to resist digestion. For the host hydra the list of divergent phenotypic traits is even shorter. The only observation which unequivocally suggests evolutionary divergence as a result of the symbiotic habit is the fact that symbiotic hydra will accept and establish stable hereditary endosymbiosis with certain algae and virtually all other brown hydra will not (but see section below on specificity). A corollary to this is the ability of the symbiotic hydra to regulate numbers of algae per cell within a range advantageous to the stability of the association. Regulation could be manifested by one or more of the following mechanisms: (a) exocytosis of excess algae; (b) digestion of excess algae; (c) division of host cells to accommodate an increase in numbers of symbionts; (d) inhibition of algal cell division either by withholding some limiting nutrient or by secretion of an inhibitor of algal cell division (Muscatine and Pool, 1979).

Specificity

Early work of Goetsch (1924), later extended by Park et al. (1967), demonstrates that a given green hydra species can take up a range of algal types, many of which initially appear to establish a symbiosis. However, closer scrutiny reveals that the associations are unstable, transient affairs and that the algae lack the morphological disposition of the host's native symbionts. The algae were taken up in relatively small numbers, resided in "packets" in the cells, and neither persisted nor appeared in offspring. Muscatine (1974) noted that the specificity for establishment of a symbiosis should be based, in part, on the features exhibited by the host's own native symbionts. The work of Jolley and Smith (1980) suggests that the degree of host-symbiont specificity may have to be evaluated at the level of each hydra strain and each algal species. They introduced several types of cultured algae, including free-living *Chlorella* species, into aposymbiotic Florida and European hydra and assessed their ability to establish a symbiosis. Although some algal types were taken up and then rejected within 24 hr, others not only persisted to varying degrees, but displayed the morphological disposition characteristic of the host's native symbionts. Since these hydra exhibited a graded response to reinfection, depending on the particular alga, they appear to be more permissive or less selective in algal discrimination processes than has been previously suspected. Jolley and Smith delineate the obstacles that an alga must overcome as it encounters various phases of the establishment process. From these features and from earlier work described here it appears that algae which establish a successful persistent symbiosis with hydra exhibit all of the following properties: (a) surface charge of appropriate sign and strength; (b) ability to evoke a specific phagocytic morphological response; (c) ability to avoid or resist digestion; (d) migration to the base of a host cell; (e) sequestration into individual host vacuoles; (f) cell division, release and exchange of metabolites; and (g) transmission to offspring.

Implications for Evolutionary Hypotheses and Future Research

The criteria for establishment of a contemporary biotrophic endosymbiosis between a eukaryotic alga and a lower metazoan has some relevance to an evaluation of the endosymbiotic theory of the origin of chloroplasts. Without attempting a detailed exposition and rationale, I wish to place emphasis on at least four "events" in the hypothetical evolutionary continuum which gave rise to chloroplasts in eukaryotic cells. The first of these, as noted and elaborated on by Stanier (1970), was the ability of the progenitor of the eukaryotic cell line, the *protoeukaryote* (Margulis, 1970), to perform endocytosis. Engendered by some change in the properties of its plasma membrane, the protoeukaryote acquired this novel means of taking in food, perhaps expressed initially as taking in fluid in small droplets, (i.e., pinocytosis) and later as the taking in of particles (i.e., phagocytosis). With selection favoring more efficient means of predation and its underlying digestive mechanisms, such innovations as Golgi apparatus, lysosomes, and intracellular digestion in vacuoles (phagosomes) were developed. As a consequence of phagocytosis, the anaerobic, photosynthetic *protoplastid* (Margulis, 1970) could then be introduced to the endocellular milieu, probably within a host vacuole. A second event was the avoidance by the protoplastid of digestion by the protoeukaryote. Whereas digestion of protoplastids may have initially served in host nutrition, it may also have resulted in selection for re-

88   L. Muscatine

sistance to digestion. The initial avoidance mechanism may simply have been for protoplastids to multiply slightly faster than they were digested, perhaps in response to the more nutritionally favorable intracellular environment. Alternatively, the protoplastid may have undergone cyclosis within the amoeboidlike protoeukaryote, and thus avoided centers of digestion. Later, more sophisticated means may have been developed, such as a digestion-resistant cell envelope, inhibition of either the lysosome-phagosome fusion event or the enzymes released after fusion, or even escape from the vacuole entirely. Confronted with successful colonization by protoplastids, the host cells may have evolved mechanisms for inhibiting or regulating protoplastid growth. The photosynthetic cells, unable to grow at rapid rates within the host but continuing to carry out photosynthesis, exported surplus organic material to the host. The host then reaped the benefits of biotrophy. Regulation of endosymbiont growth and the subsequent translocation phenomena engendered by it probably fostered the crucial evolutionary shift from phagotrophic predation to phototrophy, simultaneously decreasing selection pressure on the association and giving rise, through events described elsewhere (Margulis, 1970), at many times and in many ways, to what we now know as eukaryotic plant cells.

Can remnants or consequences of these events be detected in contemporary plant cells? Certainly phagotrophy is not a common property of plant cells, and the enigmatic nature of modern plant cell lysosomes might well represent a consequence of the loss of this property (Graham, 1973). The bidirectional exchange of metabolites between chloroplasts and cytoplasm is the sine qua non of plant metabolism. Finally, the nature of chloroplast membranes should be examined not only for knowledge of intrinsic structure and function but also with reference to the intriguing possibility that one or more of these may represent remnants of an ancient host vacuolar membrane (Park, 1971; Schnepf and Brown, 1971; Flavell, 1972; Bisalputra, 1974; Whatley et al., 1979). These features have been and should continue to be fruitful areas of contemporary research.

I wish to thank my graduate students, T. Hohman and P. McNeil, for their contribution to the work, comments, and ideas described in this manuscript, and Drs. E. Jolley and D. C. Smith for permission to cite unpublished observations.

References

Bisalputra, T. 1974. Plastids. *In* Algal Physiology and Biochemistry. W. D. P. Stewart (ed.). Oxford: Blackwells Scientific Publications, pp. 124–160.

Brien, P. and Reniers-Decoen, M. 1950. Etude d'*Hydra viridis* (Linnaeus) (La blastogenèse, la spermatogenèse, l'ovogenèse). Am. Soc. Roy. Zool. Belgique 81:33–110.

Cernichiari, E., Muscatine, L., and Smith, D. C. 1969. Maltose excretion by the symbiotic algae of *Hydra viridis*. Proc. R. Soc. B. 173:557–576.

Cook, C. B. 1980. Infection of invertebrates with algae. *In* Cellular Interactions in

Symbiosis and Parasitism. C. B. Cook, P. W. Pappas, and E. D. Rudolph (eds.). Columbus: Ohio State University Press (in press).

Cook, C. B., D'Elia, C., and Muscatine, L. 1979. Endocytic mechanisms of the digestive cells of *Hydra viridis*. 1. Morphological aspects. Cytobios 23:17–31.

Cooper, C. G., and Margulis L. 1978. Delay in migration of symbiotic algae in *Hydra viridis* by inhibitors of microtubule protein polymerization. Cytobios 19:7–19.

Dubos, R., and Kessler, A. 1964. Integrative and disintegrative factors in symbiotic associations. Symp. Soc. Gen. Microbiol. 13:1–11.

Flavell, R. 1972. Mitochondria and chloroplasts as descendents of prokaryotes. Biochem. Genet. 6:275–291.

Gahan, P. B. 1973. Plant Lysosomes. *In* Lysosomes in Biology and Pathology, Chapter 5. J. T. Dingle (ed.). Amsterdam: North Holland.

Groetsch, W. 1924. Die Symbiose der Susswasser-hydroiden und ihre kunstliche Beeinflussung. Zeit. Morph. Okol. Tiere 1:660–731.

Hohman, T. 1980. Intracellular digestion and symbiosis in *Chlorohydra viridissima*. Ph.D. Dissertation, University of California at Los Angeles 152 pp.

Jolley, E., and Smith, D. C. 1980. The green hydra symbiosis. II. The biology of the establishment of the association. New Phytol. 81:637–645.

Kanaev, I. I. 1952. Hydra. Moscow: Soviet. Academy of Science. Translation by E. T. Burrows and H. M. Lenhoff, 1969. Miami: H. M. Lenhoff.

Karakashian, M. W. 1975. Symbiosis in *Paramecium bursaria*. Symp. Soc. Exp. Biol. 28:145–173.

Karakashian, S. J. 1970. Morphological plasticity and the evolution of algal symbionts. Ann. N.Y. Acad. Sci. 175:474–487.

Karakashian, M. W., and Karakashian, S. J. 1973. Intracellular digestion and symbiosis in *Paramecium bursaria*. Exp. Cell. Res. 81:111–119.

Karakashian, S. J., and Karakashian, M. W. 1965. Evolution and symbiosis in the genus *Chlorella* and related algae. Evolution 19:368–377.

Karakashian, S. J., Karakashian, M. W., and Rudzinska, M. 1968. Electron microscopic observations on the symbiosis of *Paramecium bursaria* and its intracellular algae. J. Protozool. 15:113–128.

McNeil, P. 1980. Mechanisms of nutritive endocytosis. Ph.D. Dissertation, University of California at Los Angeles, 209 pp.

McNeil, P. 1981. Mechanisms of nutritive endocytosis. I. Phagocytic versatility and cellular recognition in *Chlorohydra* digestive cells, a scanning electron microscope study. J. Cell Sci. 49:311–339.

Margulis, L. 1970. Origin of Eukaryotic Cells. New Haven: Yale University Press.

Muscatine, L. 1965. Symbiosis of hydra and algae. III. Extracellular products of the algae. Comp. Biochem. Physiol. 16:77–92.

Muscatine, L. 1974. Endosymbiosis of cnidarians and algae. *In* Coelenterate Biology. L. Muscatine and H. Lenhoff (eds.) New York: Academic Press, pp. 359–395.

Muscatine, L., and Lenhoff, H. M. 1965. Symbiosis of hydra and algae. I. Effects of some environmental cations on growth of symbiotic and aposymbiotic hydra. Biol. Bull. 128:415–424.

Muscatine, L. and Pool, R. 1979. Regulation of numbers of intracellular algae. Proc R. Soc. B. 204:131–137.

Muscatine, L., Cook, C. B., Pardy, R. L., and Pool, R. R. 1975. Uptake, recognition, and maintenance of symbiotic *Chlorella* by *Hydra viridis*. Symp. Soc. Exp. Biol. 29:175–203.

Oschman, J. L. 1967. Structure and reproduction of the algal symbionts of *Hydra viridis*. J. Phycol. 3:221–228.

Pardy, R. L. 1974. Some factors affecting the growth and distribution of the algal endosymbionts of *Hydra viridis*. Biol. Bull. 147:105–118.

Pardy, R. L. 1976a. The morphology of green *Hydra* endosymbionts as influenced by host strain and host environment. J. Cell Sci. 20:655–669.

Pardy, R. L. 1976b. The production of aposymbiotic hydra by the photodestruction of green hydra zoochlorellae. Biol. Bull. 151:255.

Pardy, R. L., and Muscatine, L. 1973. Recognition of symbiotic algae by *Hydra viridis*. A quantitative study of the uptake of living algae by aposymbiotic *H. viridis*. Biol. Bull. 145:565–579.

Park, H., Greenblatt, C. L. Mattern, C. F. T., and Merril, C. R. 1967. Some relationships between *Chlorohydra*, its symbionts and some other chlorophyllous forms. J. Exp. Zool. 164:141–162.

Park, R. B. 1971. The architecture of photosynthesis. *In* Biological Ultrastructure: The origin of cell organelles. P. J. Harris (ed.). Corvallis: Oregon State University Press, pp. 25–40.

Pool, R. R. 1979. The role of algal antigenic determinants in the recognition of potential algal symbionts by cells of *Chlorohydra*. J. Cell Sci. 35:367.

Pool, R. R., and Muscatine, L. 1980. Phagocytic recognition and the establishment of the *Hydra viridis-Chlorella* symbiosis. *In* Endosymbiosis and Cell Biology. W. Schwemmler and H. E. A. Schenk (eds.). Walter de Gruyter & Co. Berlin: pp. 223–238.

Raven, P. H. 1970. A multiple origin for plastids and mitochondria. Science 169:641–646.

Roffman, B., and Lenhoff, H. M. 1969. Formation of polysaccharides by *Hydra* from substrates produced by their endosymbiotic algae. Nature 221:381–382.

Schnepf, E., and Brown R. M. 1971. On relationship between endosymbiosis and the origin of plastids and mitochondria. *In* Origin and Continuity of Cell Organelles. J. Reinert and H. Ursprung (eds.). New York: Springer-Verlag, pp. 229–322.

Smith, D. C. 1974. Transport from symbiotic algae and symbiotic chloroplasts to host cells. Symp. Soc. Exp. Biol. 28:485–520.

Stagni, A. 1974. Some aspects of sexuality in fresh-water hydras. Boll. Zool. 41:340–358.

Stanier, R. Y. 1970. Some aspects of the biology of cells and their possible evolutionary significance. Symp. Soc. Gen. Microbiol. 20:1–37.

Thorington, G., Berger, B. and Margulis, L. 1979. Transmission of symbionts through the sexual cycle of *Hydra viridis*. I. Observations on living organisms. Trans. Amer. Microsc. Soc. 98:401–413.

Trench, R. K. 1979. The cell biology of plant-animal symbiosis. Annu. Rev. Plant Physiol. 30:485–531.

Trench, R. K. 1980. Integrative mechanisms in mutualistic symbiosis. *In* Cellular Interactions in Symbiosis and Parasitism. C. B. Cook, P. W. Pappas, and E. D. Rudolph (eds.). Columbus: Ohio State University Press pp. 275–297.

Valkanov, A. 1950. Untersuchungen uber die Infektion der Eier von *Chlorohydra viridissima* mit Zoochlorellen. Sofia: Universitat Godishnik. Annuaire 46:111–122 (in Russian, German summary).

Vivier, E., Petitprez, A., Chivé, A. F. 1967. Observations ultrastructurales sur les Chlorelles symbiotes de *Paramecium bursaria*. Protistol. 3:325–334.

Weis, D. S. 1976. Digestion of added homologous algae by *Chlorella*-bearing *Paramecium bursaria*. J. Protozool. 23:527–529.
Whatley, J. M., John, P., and Whatley, F. R. 1979. From extracellular to intracellular: The establishment of mitochondria and chloroplasts. Proc. R. Soc. B. 204:165–187.

Discussion of Dr. Muscatine's Presentation

COHEN: Is anything known of the process by which algae are ingested and processed to yield surviving chloroplasts, as in Saccoglossans? Couldn't the enzymes involved in the processing be useful in preparing functional chloroplasts?

MUSCATINE: Phagocytosis of algal chloroplasts by digestive cells of the saccoglossan mollusc *Placida dendritica* has been described by McLean (1976). I suspect that the enzymes that you are referring to are part of a suite of digestive enzymes that come into play during the ingestion phase. These might certainly be useful in preparing functional chloroplasts.

VIGIL: Is there an anchoring of alga and vacuole at the base of host cells? It seems that a distinct coordination of cellular dynamics must exist between host and algae in vacuoles when division of the alga occurs. What is known about the cooperative process(es) involved in formation of new membranes (if necessary) or modification of the existing membranes?

MUSCATINE: We have only circumstantial evidence for "anchoring" based on the observation that treatment of green hydra with 10^{-7} M vincristine results in a more random distribution of algae towards the middle and tip of the digestive cells. Naturally, one is led to speculate on the role of microtubules in anchoring the vacuole and the contained alga at the base of the cell. Nothing is known yet about cooperative processes beyond the observation that daughter cells are immediately sequestered into individual vacuoles.

SHILO: While invertebrate hosts are many and diverse only very few algal species are involved in the endosymbiotic interaction. Is anything known about the uniqueness of the algae that are involved?

MUSCATINE: Probably the most common trait shared by virtually all symbiotic algae is their capacity to selectively release soluble carbohydrate in one or more forms. To this I might add resistance to digestion as a common trait although it has not been studied in any systematic way among the various types of algal symbionts.

RAVIN: I understand that a small proportion of free-living *Chlorella* cells fed to *Hydra* are taken up and are maintained in the host cell's cytoplasm. If these

Chlorella cells are reisolated and then refed to fresh *Chlorella*-free Hydras, do they show their original poor capacity to infect or have they gained an enhanced capacity for infection?

MUSCATINE: The experiment you describe has not yet been carried out. One of the anticipated difficulties is that few free-living *Chlorella* cells are taken up initially, and none persist beyond 24 hr, so that recovery (i.e., reisolation) may yield insufficient algae for reinfection experiments. On the other hand, a strain of English *Hydra* will take up and maintain *Chlorella prototothecoides*, a free-living species, in small numbers and for at least several weeks after the initial infection (Jolley and Smith, 1980). Your experiment could be carried out with this association.

SCHIFF: Something on your first slide relates to our previous discussions on the site of protein synthesis. It has always struck me as curious that although organelles like the chloroplast have only a little less DNA than free-living prokaryotes they need genetic information from the nucleus. The way I explain this to myself is that if the chloroplast had an endosymbiotic origin, some means would have to exist to insure that the cell and the plastid were coordinated in development and replication. This probably occurred by one partner ceding to the other certain synthetic prerogatives. Thus the material supplied to the chloroplast by the rest of the cell allows the cell to regulate the development and replication of the plastid. Similarly, the plastid can regulate the cell's activity by supplying its needs for photosynthate and other molecules. By metering these molecules that flow from one to the other, the cell and its organelles form a coordinated entity. Probably only those endosymbiotic relationships in which the members had mutual needs of this sort could form stable cellular systems in which one member did not outgrow the other.

References to Discussion

Jolley, E., and Smith, D. C. 1980. The green hydra symbiosis. II. The biology of the establishment of the association. New Phytol. 81:637–645.

McLean, N. 1976. Phagocytosis of chloroplasts in *Placida dendritida* (Gastropoda: Sacoglossa). J. Exp. Zool. 197:321–330.

On the Endosymbiotic Origins
of Chloroplasts:
Still Another Approach to the Problem

Seymour S. Cohen

It is very kind of the organizers of this symposium to ask me to introduce the discussions on "The Evolution of Chloroplasts," and in so doing to draw upon some impressions of the recent but relatively early development of this problem. Although Joseph Priestley had detected oxygen evolution by plants in 1771, this is the bicentennial year of the clarification of that discovery by Jan Ingen-Housz. In 1779 the latter demonstrated photosynthetic oxygen production by only the green parts of the plant in specific response to visible light. Obviously this symposium and anniversary present a splendid opportunity for a discussion of the historical aspects of photosynthesis. Nevertheless I have chosen instead to make my remarks in the context of some experiences in teaching and working in seemingly distant fields, and particularly because I have interacted with Roger Stanier, whose work in the field of the cyanobacteria and in microbial physiology is being honored at this symposium. Also, after about 30 years of thinking about and sorting the bits and pieces of the accumulating biochemical and biological information on the evolution of prokaryotic and eukaryotic cells and their organelles, I have begun some experiments myself on the possible relations between blue-green algae or cyanobacteria and higher plant chloroplasts. I would like to communicate some of our early results to you, with the notion that both these and my approach may be relevant to our common problem.

My work began at intersect with that of Roger Stanier in 1947. At that time he was a microbial physiologist interested in the phenomenon of enzymatic adaptation, as it was then called, and I had just observed some interesting

Address reprint requests to Dr. Seymour S. Cohen, Distinguished Professor, Department of Pharmacological Sciences, State University of New York, Stony Brook, New York 11794.

phenomena of nucleic acid biosynthesis in phage-infected *E. coli.* I thought the further clarification of my observations required solutions of the then unsolved problems of the origin of ribose and deoxyribose. To prepare for such work, I undertook to learn something at the Pasteur Institute about the growth and metabolism of *E. coli.* I tested the hypothetical relationship of the uronic acids to the biosynthesis of certain pentoses by the method of simultaneous adaptation, which Stanier (1947) had formulated in several simple rules. The result was unequivocal in *E. coli;* D-xylose and L-arabinose did not arise in the simple decarboxylation of glucuronate and galacturonate respectively (Cohen, 1949). Among the major products of my stay in Paris were discussions of biochemical evolution with André Lwoff, whose book on the subject (Lwoff, 1943) has been praised by Stanier (Stanier, 1971).

We found eventually that ribose in *E. coli* was synthesized via an oxidative decarboxylation of phosphogluconate and that the organism contained alternative paths for metabolizing glucose (Cohen, 1951). At that time the existence of alternative pathways of metabolism was an unexpected complication in biochemical thought. Our observations and problem led us to think about the variety of paths of glucose metabolism and of many other metabolic systems in many organisms and we began to wonder about manipulating the balance of the pathways. I was thinking then about comparative biochemistry as the study of biochemical diversity and the applicability of the concept of diversity to the development of a chemotherapy of infection. I now believe that such an approach has become scientifically and technologically feasible in developing a chemotherapy for any microbial infection of mammals. Of course such a chemotherapy had been found empirically as early as in the late 1930s and early 1940s, with the discovery of the sulfa drugs and antibiotics. Recognizing the significance of such practical findings, Stanier had been among the first to note that these results demonstrated major chemical differences between eukaryotic cells and prokaryotic microbes. I recall our joint participation in a symposium in 1952 in which we both developed these themes; I was delighted to find that we were thinking along similar lines. And we came to know each other much better in summers at Woods Hole, Massachusetts, and elsewhere.

Stanier's training drew him quite naturally to problems of photosynthesis and at least two aspects of these problems would have led him to the much neglected cyanobacteria. These microbes are apparently the first prokaryotic organisms to have discovered photosynthetic O_2 evolution, and are in fact commonly believed to be responsible for creating an O_2-containing atmosphere. Second, the discovery of DNA in chloroplasts in 1963 and 1964 put new life into the endosymbiotic hypothesis, that is, that these plant organelles might have evolved from cyanobacteria. At about this time Stanier was clearly determined to learn more about these neglected organisms. His systematic studies of these O_2-evolving bacteria have been a guidepost in our developing knowledge of the postulated relation of the cyanobacteria to plant chloroplasts.

My own involvement in these questions comes from different paths, from work in bacterial metabolism and viral multiplication, from teaching the problems of biochemical evolution, and via a working association with M. Nass, one of the discoverers of mitochondrial DNA. An examination of the main relevant data until 1973 (Cohen, 1970, 1973) led to my understanding that the initial hypothesis of a symbiotic incorporation of a bacterium into some cellular precursor to form the organelle-containing eukaryotic cell was viable but was much too simple to explain the rapidly emerging data. Mitochondria and chloroplasts did have many features possibly ascribable to once-independent microorganisms. The organelles did contain DNA, which did not hybridize with nuclear DNA; various classes of RNA were derived from the organellar DNA; and the organelles contained a bacterial-type ribosome and possessed the ability to synthesize organellar DNA, RNA, and protein. However, many proteins essential to organelle function were genetically determined by nuclear genes and were synthesized in the eukaryotic cytoplasm under nuclear control. Thus three levels of analysis had become essential in defining the following experimental problems:

1. Is the component being studied a component of the organelle?
2. Is the component synthesized in the organelle?
3. Where is the synthesis determined?

As examples of such complexity, we know now that mitochondrial ribosomal proteins are synthesized in the cytoplasm, and proteins involved crucially in the various steps of photosynthesis are determined in the nucleus. Facts such as these tell us that in addition to problems of biosynthesis and regulation we also have serious problems of cell traffic and transport. The sorting is still going on, and it is evident that the problems of the origins of the organelles and their components are far from having clear solutions. Nevertheless, it is possible to formulate in a far more sophisticated way questions regarding the biosyntheses in cellular compartments and the interactions of various components. Furthermore, it has become possible to make and test new hypotheses, some of which prove to be relevant. For example, when it was observed that many plants make lysine via the diaminopimelate pathway, which is known to operate in bacteria, I asked if the chloroplast contained the essential prokaryotic enzymes (Cohen, 1973). An English group picked this up and demonstrated the presence of diaminopimelate decarboxylase in chloroplasts (Maxelis et al., 1976).

Chloroplast Evolution and Studies of Turnip Yellow Mosaic Virus

I was intrigued by data on the unusual mode of multiplication of the plant virus *turnip yellow mosaic virus* in intraorganellar pockets formed by aggregated chloroplasts (Mouches et al., 1974; Matthews and Sarkar, 1976). This virus is known to contain large amounts of the polyamine *spermidine* and only small amounts of the other polyamines.

The tetramine *spermine* exists in the plant host also, and because spermine binds more tightly to RNA than does spermidine (Sakai et al., 1975), one can ask if the tetramine has been essentially excluded from the chloroplast compartment. Another possibility is that only spermidine is synthesized in the chloroplast; if this were true, the chloroplast would be behaving like other prokaryotic cells (Cohen, 1971).

Because we are interested in polyamines both as organic cations which may neutralize nucleic acids as well as organize their structure (Cohen, 1971) and in problems of biochemical evolution, we have begun to determine whether chloroplasts are indeed able to synthesize polyamines. As a parallel investigation we wish to examine this activity in cyanobacteria, and to prepare ourselves for similar studies on these organisms, I spent some months in Stanier's laboratory in Paris and in Moshe Shilo's laboratory in Jerusalem. The main portion of this chapter will present data on some surprising effects of polyamines on the cyanobacteria.

As a result of our work with turnip yellow mosaic virus (TYMV), we believe that the spermidine of the virion is associated with the viral RNA. We believe this to be true because we have isolated capsids which are devoid of RNA, and these capsids are essentially free of spermidine. We have also isolated viral RNA containing spermidine (Cohen and Fukuma, 1977). It has been demonstrated more recently that exogenous spermidine is not exchangeable with viral spermidine, suggesting that the content of viral spermidine reflects the spermidine content of viral RNA at the time of encapsulation (Torget et al., 1979).

In our earlier studies with the RNA phage R17, we had found that the synthesis of spermidine parallels that of viral RNA and that this triamine neutralizes much of the viral RNA to which it is tightly bound (Fukuma and Cohen, 1975). Spermidine also facilitates the condensation of free viral RNA and markedly increases its infectivity on male E. coli (Leipold, 1977). Similar questions can be posed for the role of spermidine in TYMV multiplication.

TYMV-Infected Chinese Cabbage

When rapidly growing plants are infected on the external (cotyledon) leaves and newly emerging leaves are examined, the infected plant is found to grow (in weight) at a rate only 10-15% less than that of the healthy plant. Virus content in juice expressed from the newly emerged leaves is low but detectable at 6 days and approximately constant at 1.0-1.5 mg/ml at days 11-18. After 18 days, the plant begins to yellow severely and wilt.

Examination of the polyamines (Torget et al., 1979) reveals a markedly higher content of all of these cations in the infected plant, as shown in Table 1. Just prior to the yellowing of the leaves, putrescine accumulation is very considerable. About 25-40% of the spermidine content of the juice is present in the virions and is specifically precipitated by antiviral serum. Such spermidine is not exchangeable with radioactive spermidine added to the initially expressed juice.

Disks of normal infected tissue take up methionine into protein, S-adenosyl-

Table 1. Polyamine Content of Healthy and TYMV-Infected Chinese Cabbage Leaves (in nmol/gm)

Healthy leaves[a] (days)	Pu	Spd	SP	SAM	mg Chlorophyll/g	mg TYMV/g
6	36	250	63	11	1.0	0.0
11	26	270	31	5.6	0.91	0.0
14	23	86	23	7.0	1.1	0.0
18	21	77	42	6.0	1.0	0.0
Infected leaves[a] (days)						
6	41	400	98	14	0.95	≤0.15
11	66	480	110	8.3	0.92	1.5
14	120	490	71	9.1	0.97	0.8
18	200	370	67	10	0.71	0.9

[a]Three plants were used for each time point.
Source: From Torget et al., 1979.

methionine, spermidine, and spermine. However, the two types of tissue incorporate the amino acid into these components at similar rates, suggesting that infection may not increase the capability of the system to synthesize polyamines.

It is of course evident that a study of the kinetics of polyamine biosynthesis and its relation to other functions requires suspensions of simultaneously infected cells. Although such suspensions can be obtained, this has not yet been done for studies of polyamine synthesis. In any case it is evident that virus infection induces significant effects in production of the polyamines. One surprising result was the very high level in putrescine in the tissue just prior to the yellowing of the plant.

Polyamine Metabolism in Cyanobacteria

We have found that essentially all of the laboratory strains of the cyanobacteria, both unicellular and filamentous species, are fairly similar in their patterns of polyamine content during stages of active growth. They lack spermine, contain 1-2 mM spermidine, and about one-tenth this concentration of putrescine (Ramakrishna et al., 1978). We have therefore concentrated on the unicellular *Anacystis nidulans* as our test organism.

In studies of the effects of exogenous amines on this organism during growth in a mineral medium containing bicarbonate, two new types of results were obtained. Spermidine was slowly metabolized to diaminopropane and pyrroline; at millimolar levels the exogenous triamine was relatively nontoxic (Ramakrishna et al., 1978). Spermine at similar concentrations had a significantly higher toxicity. However putrescine was highly toxic when added at the low concentration (0.15 mM) present within the cell, as given in Figure 1. We have looked at this

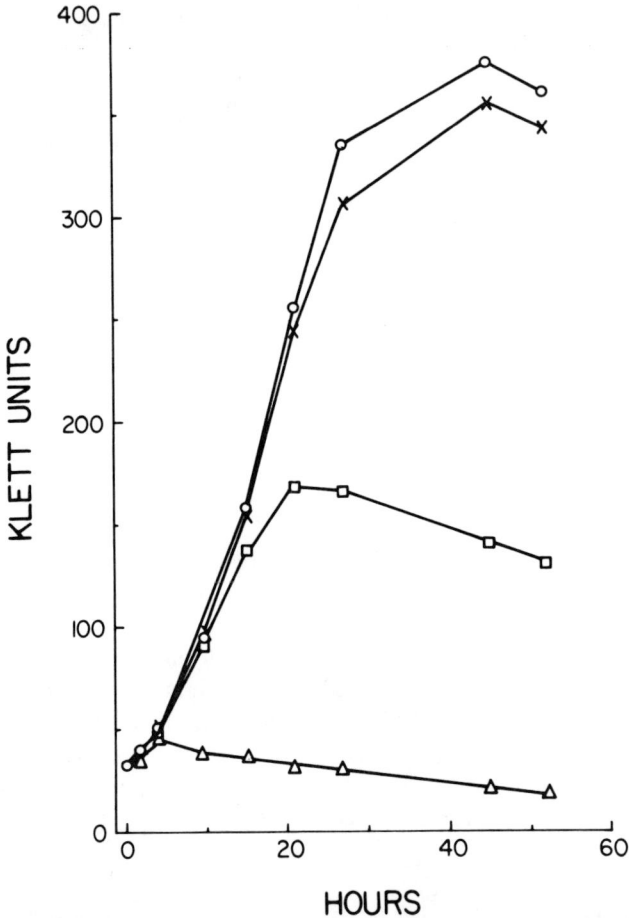

Figure 1. Growth of *A. nidulans* in the presence of polyamines 0.3 m*M* spermine (□), 0.3 m*M* spermidine (X), and 0.15 m*M* putrescine (△), or in the absence of amine (O). *Source:* From Ramakrishna et al., 1978.

phenomenon because it may relate to the phytopathology of TYMV infection as noted above, as well as to the well-known phenomenon of toxicity and putrescine accumulation in plants in acidic or low-sulfur soils (Smith, 1975).

The Toxicity of Putrescine in a Cyanobacterium

It was found that putrescine is rapidly concentrated within *Anacystis*. The rate of putrescine uptake increases exponentially with increasing pH, and is maximal at pH 10.5 (Guarino and Cohen, 1979b). The pK_1 and pK_2 of the amino groups of putrescine are 9.0 and 10.5, respectively. When the cyanobacteria grow in an unbuffered medium, the medium may attain a pH as high as 11. In an alkaline medium it can be imagined that the rate of uptake is a function of the concentration of the unprotonated amine; the data are in agreement with this prediction.

It was also found that the final concentration attained in the cell is a function of the concentration of exogenous amine. The final concentration of intracellular amine at a defined pH—at nontoxic levels of putrescine—can be predicted exactly to be the intracellular concentration of protonated amine trapped by ionization of the diffusible nonprotonated compound at the pH existing within *Anacystis*. The intracellular pH was 7.5-7.6, as estimated by the methylamine method. No evidence has been found for a saturable carrier. All the data are consistent with the hypothesis that the 500- to 3000-fold increase of concentration of intracellular over extracellular putrescine is the result of passive diffusion of nonprotonated amine and trapping of the ionized cation within the cell.

At an intracellular concentration of 0.10 mM putrescine at pH 9.5, the cells are killed rapidly, as determined by plating on agar. Survivors, which numbered 10^{-5} at 2 hr, were not found to be resistant to putrescine, nor have we been able to isolate resistant strains. This result also argues against the existence of a protein carrier of putrescine. The killing effect results in some leakage of putrescine from the cell, that is, toxic conditions do not give a predictable final concentration of the diamine. Light was found to be essential for putrescine uptake, which was also inhibited by inhibitors of photosynthesis such as DCMU and CCCP.

Under toxic conditions, that is, the presence of exogenous putrescine at or above 0.1 mM, the inhibition of protein synthesis (Figure 2), which roughly paralleled the lethal effect, was the first detectable physiological event (Guarino and Cohen, 1979b). This inhibition significantly preceded the inhibition of CO_2 fixation. Thus the accumulation of high intracellular concentrations of putrescine did not affect energy production, a result consistent with the relatively small effects on nucleic acid synthesis.

It was found that 98% of the isotope taken up as putrescine was present as acid-soluble putrescine within the cell (Figure 3). Spermidine was synthesized briefly but this synthesis was inhibited after 1 hr and, as the internal concentration of putrescine exceeded 50 mM, spermidine was lost from the cells, presumably having been displaced from the cell structure. In this system a covalent addition of putrescine to cell structure was also detected. Although this is a small percentage of the total uptake of putrescine in the experiment presented in Figure 3, at 1 hr the quantity of trichloroacetic acid-precipitable putrescine amounted to 70,000 molecules per cell.

In parallel experiments it has been found that acid-precipitable putrescine is not associated with the extracted lipids or nucleic acids. Hydrolysis with 6 M HCl liberated the bulk of the isotope as putrescine from the residual protein fraction. The highest proportion of the putrescine was found in a combined cell-wall-plus-membrane fraction, but putrescine was also found on both the ribosomal and soluble protein fractions.

100

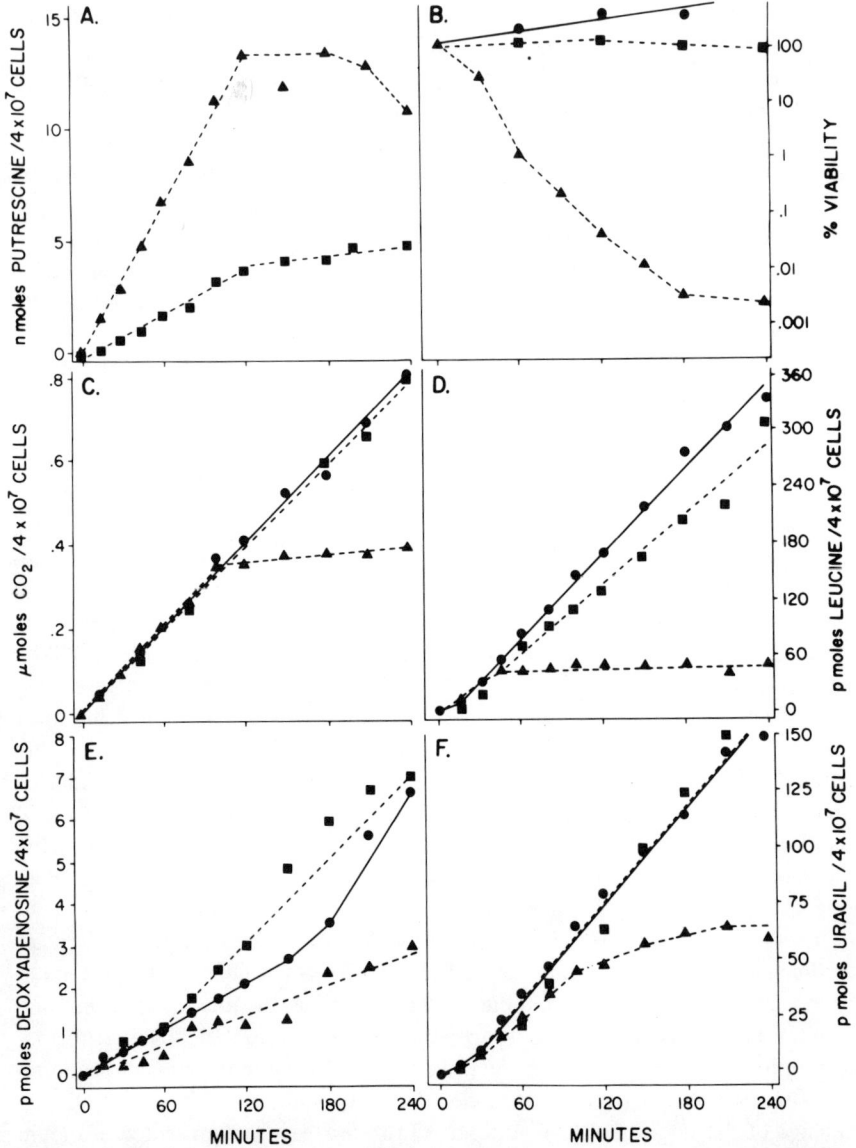

Figure 2. Effects of putrescine on viability (B), CO_2 fixation (C), and synthesis of protein (D), DNA (E), and RNA (F) in *Anacystis nidulans*. Cells in exponential phase were washed and resuspended in Allen's medium (pH 9.5) in 40 μM (■) or 150 μM (▲) or in the absence (●) of [³H] putrescine (16 μCi/μmole). The uptake of putrescine into whole cells is shown in (A). *Source:* From Guarino and Cohen, 1979a.

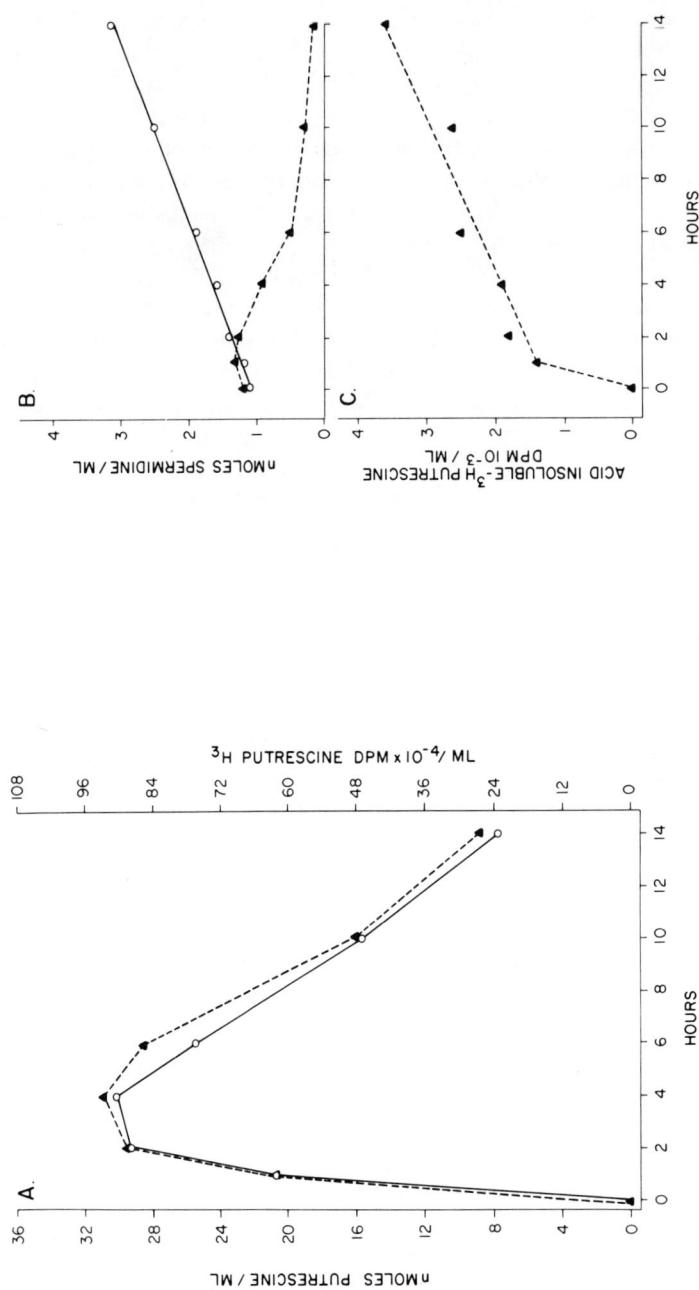

Figure 3. Polyamine analyses in putrescine-treated cells. (A) compares the uptake of 150 μM [^{3}H] putrescine (▲) (13.3 μCi/ μmole) at pH 9.5 and with putrescine recovered from the cells and analyzed as dansyl putrescine (○). (B) shows the intracellular concentration of spermidine in the presence (▲) and absence (○) of 150 μM putrescine. (C) shows the appearance of TCA-precipitable putrescine within the cells. *Source:* From Guarino and Cohen, 1979b.

Putrescine Toxicity and an Irreversible Dissociation of Ribosomes

In order to understand the inhibitory effect on protein synthesis, we have examined the ribosomal fraction. As shown in Figure 4, the ribosomes from putrescine-killed *Anacystis* were found to be irreversibly dissociated and could not be reassociated to 70S monosomes with 5 mM Mg^{2+} and 5 mM spermidine. The degree of dissociation at different putrescine exposures roughly paralleled the inhibition of protein synthesis and lethality. In various experiments the ribosomal subunits of the killed cells have been found to contain 1-3 molecules of covalently bound putrescine per subunit.

Covalently bound putrescine in various cell fractions, including the ribosomes, has also been observed under nontoxic conditions, for example, at 40 mM exogenous putrescine. At this concentration, which results in 15 mM internal putrescine after 2 hr, protein synthesis is inhibited only 10% at most and ribosomal dissociation is only 15% higher than the 9% found in the control culture. However, spermidine is not lost from the cell in such cultures, and we may ask if the specificity of putrescine addition and/or the irreversibility of ribosomal dissociation do not relate to the loss of ribosomal spermidine.

Conclusion

It is easy to see that we are far from answering the questions we posed earlier on the functions and origins of chloroplasts. However, we have made a start by setting up the plant, viral, and algal systems. In that sense this chapter is a progress report to Roger Stanier and Moshe Shilo, who have instructed us on the nurture of *Anacystis nidulans,* the organism on which we have done most of our work. It appears that we may have explored a metabolic and structural parameter that will be of interest in studying the cyanobacteria; hopefully, students of chloroplast synthesis, function, and stability will also find our work relevant.

However, our work on the chloroplasts themselves has just begun, and to make this work more meaningful we will try to improve our experimental systems that employ virus-infected plant cells or virus-infected chloroplast aggregates. In the latter sense this chapter is also our progress report to Dr. J. M. Kaper of the Department of Agriculture at the Virus Laboratory at Beltsville, Maryland, and to Dr. J. Bové at the Agronomy Institute at Bordeaux, who introduced us to numerous aspects of the virus system. We wish to acknowledge the warm hospitality and friendship of key members of the "green belt" who have introduced some color into our laboratory, as well as some sense of reality about the difficulties of the complex problem of chloroplast origins with which all of us are now concerned. It may be that there is no rigorous solution nor a single clean answer to this complex problem, despite our simpleminded notion that we can solve it. Nevertheless, the mere posing of the problem has led to much new

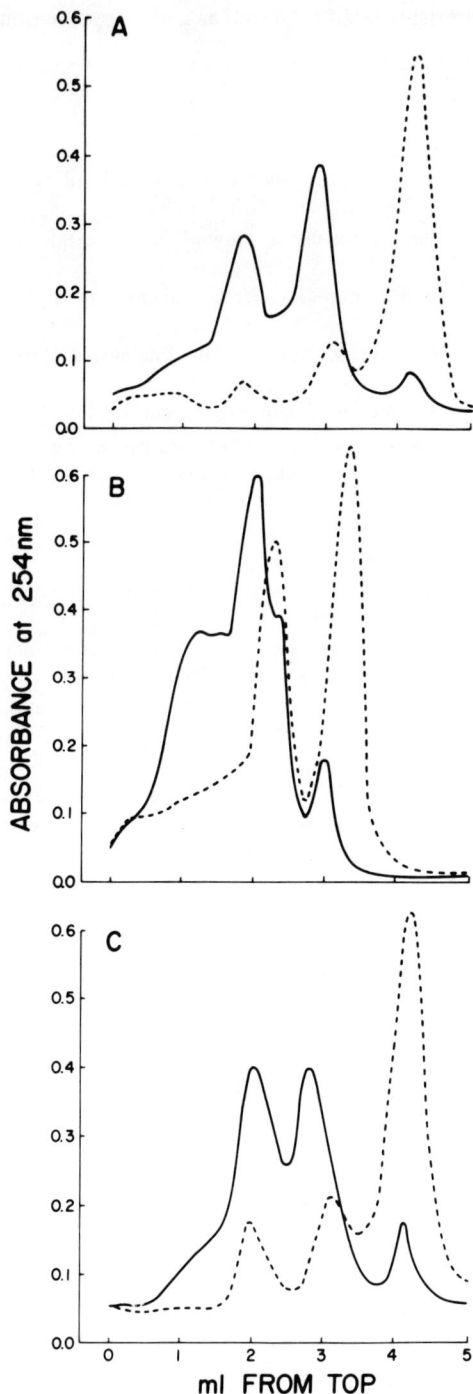

Figure 4. Irreversible dissociation of ribosomes. Cells in exponential phase were washed and resuspended in growth medium pH 9.5 and were incubated at 30°C in the presence (solid line) or absence (dashed line) of 150 μM putrescine for 2 hr. (A) shows the uv scan of ribosomes isolated and centrifuged in buffer A-1 containing 10 mM Mg^{2+}. In (B) the ribosomes were isolated and centrifuged in 1 mM Mg^{2+}. In (C) the ribosomes were isolated in 1 mM Mg^{2+}, reassociated in 5 mM Mg^{2+} and spermidine, and centrifuged in 5 mM Mg^{2+} and 5 mM spermidine. *Source:* From Guarino and Cohen, 1979b.

thinking, many experiments, and surprising results, as well as a new appreciation of our biological world.

References

Cohen, S. S. 1949. Adaptive enzyme formation in the study of uronic acid utilization by the K-12 strain of *Escherichia coli*. J. Biol. Chem. 177:607–619.

Cohen, S. S. 1951. Utilization of gluconate and glucose in growing and virus-infected *Escherichia coli*. Nature 168:746–747.

Cohen, S. S. 1970. Are/were mitochondria and chloroplasts microorganisms? Am. Scient. 58:281–289.

Cohen, S. S. 1971. Introduction to the Polyamines. Prentice Hall: Englewood Cliffs, New Jersey.

Cohen, S. S. 1973. Mitochondria and chloroplasts revisited. Am. Scient. 61:437–445.

Cohen, S. S., and Fukuma, I. 1977. Polyamines and some RNA viruses. *In* The Biochemistry of Adenosylmethionine. F. Salvatore, E. Borek, V. Zappia, H. G. Williams-Ashman, and F. Schlenk (eds.). New York: Columbia University Press.

Fukuma, I., and Cohen, S. S. 1975. Polyamines in bacteriophage R 17 and its RNA. J. Virol. 16:222–227.

Guarino, L. A., and Cohen, S. S. 1979a. Mechanism of toxicity of putrescine in *Anacystis nidulans*. Proc. Nat. Acad. Sci. USA 76:3660–3664.

Guarino, L. A., and Cohen, S. S. 1979b. Uptake and accumulation of putrescine and its lethality in *Anacystis nidulans*. Proc. Nat. Acad. Sci. USA 76:3184–3188.

Leipold, B. 1977. Effect of spermidine on the RNA-A protein complex isolated from the RNA bacteriophage MS2. J. Virol. 21:445–450.

Lwoff, A. 1943. L'Evolution Physiologique. Paris: Hermann et Cie.

Matthews, R. E. F., and Sarkar, S. 1976. A light-induced structural change in chloroplasts of Chinese cabbage cells infected with turnip yellow mosaic virus. J. Gen. Virol. 33:435–446.

Maxelis, M., Milfin, B. J., and Pratt, H. M. 1976. A chloroplast-localized diaminopimelate decarboxylase in higher plants. FEBS Lett. 64:197–200.

Mouches, C., Bové, C., and Bové, J. M. 1974. Turnip yellow mosaic virus-RNA replicase: Partial purification from the solubilized enzyme-template complex. Virology 58:409–423.

Ramakrishna, S., Guarino, L., and Cohen, S. S. 1978. Polyamines of *Anacystis nidulans* and metabolism of exogenous spermidine and spermine. J. Bact. 134:744–750.

Sakai, T. T., Torget, R., I, Josephine, Freda, C. E., and Cohen, S. S. 1975. The binding of polyamines and of ethidium bromide to tRNA. Nucl. Acid. Res. 2:1005–1022.

Smith, T. A. 1975. Recent advances in the biochemistry of plant amines. Phytochemistry 14:865–890.

Stanier, R. Y. 1947. Simultaneous adaptation: A new technique for the study of metabolic pathways. J. Bact. 54:339–348.

Stanier, R. Y. 1971. L'évolution physiologique: a retrospective appreciation. *In* Of Microbes and Life. J. Monod and E. Borek (eds.). New York: Columbia University Press.

Torget, R., Lapi, L., and Cohen, S. S. 1979. Synthesis and accumulation of polyamines and S-adenosylmethionine in Chinese cabbage infected by turnip yellow mosaic virus. Biochem. Biophys. Res. Commun. 87:1132–1139.

Discussion of Presentation by Dr. Cohen

MARGULIES: Are the vesicles formed during infection by turnip yellow mosaic virus actually closed off from the cytoplasm or open to it?

COHEN: During accumulation of virus in the fused vesicles in the chloroplast aggregate there appears to be very little virus present in the cytoplasm until the disaggregation of the chloroplasts late in the infection. However, it is difficult to believe that the fused breakdown vesicle is tightly sealed for the reason that the coat protein of the virus is thought to be synthesized in the cytoplasm, that is, the synthesis of virion protein is inhibited by cycloheximide (Renaudin et al., 1975).

WILDMAN: Tobacco mosaic virus stops chloroplast DNA transcription, as evidenced by cessation of chloroplast rRNA synthesis and fraction I protein synthesis (Hirai and Wildman, 1969). Viral RNA and protein synthesis occur in the cytoplasm. Viral rods which appear in chloroplasts are a late event and represent passive transport from cytoplasm to chloroplasts.

COHEN: No comment. I did not know that tobacco mosaic virus infection affects chloroplast syntheses as an early event.

STREERE: Would it be useful to locate a virus which would behave to spinach chloroplasts as turnip yellow mosaic does with respect to Chinese cabbage chloroplasts?

COHEN: Someone will eventually figure out how to prepare good protoplasts and chloroplasts from Chinese cabbage. *Note added in press:* Our studies on such protoplasts are in press in *Plant Physiology.*

SCHIFF: In view of your remarks it is evident that comparisons between free-living procaryotes and the prokaryotic properties of the organelles (chloroplasts and mitochondria) will be important in talking about origins and evolution. What is your opinion of this similarity and its consequences for accounting for the origins of the organelles?

COHEN: As I indicated in this chapter, I think that the hypothesis is still viable that mitochondria and chloroplasts might have originated as a symbiotic association of a bacterium with some cellular precursor. However, investigation of the real world has told us that if this hypothesis is correct, a further evolution has resulted in the transfer of many genes from the symbiotic microorganism to the new cellular nucleus. Most discussions have not yet confronted the problem of this postulated transfer. If this transfer has in fact occurred it should be taken note of very seriously by the groups working with recombinant DNA. It seems to me that the molecular biologists have not yet discussed this particular possibility of gene transfer in terms of the question of security and caution in their experimentation.

MURRAY: Is there a limited number of putrescine-binding proteins characteristic of *Cyanobacteria?*

COHEN: We don't know the answer for the microbiological world. There are many putrescine-binding proteins known in animals. An enzyme known as *transglutaminase* catalyzes this reaction with putrescine and other polyamines; it was discovered some time ago (Clarke et al., 1959). In recent years protein complexes containing covalently-bound polyamines have been found in animal tissues, cells, and fluids.

MARGULIES: Is covalent binding of putrescine to ribosomal subunits an enzymatic process?

COHEN: Putrescine is bound to ribosomal subunits (one to three molecules per subunit) but the diamine is also bound to soluble proteins and to cell wall components (Guarino and Cohen, 1979). In fact most of the covalently bound putrescine is associated with the latter, and Dr. Guarino in our lab has observed that the purified cell wall fraction provides the best substrate for covalent addition in a cell-free reaction. However, very little has been done with this system and we would prefer to be cautious at this time before describing these covalent additions as occurring via an enzymatic reaction.

References to Discussion

Clarke, D. D., Mycek, M. J., Neidle, A., and Waelsch, H. 1959. The incorporation of amines into protein. Arch. Biochem. Biophys. 79:338–354.

Guarino, L. A., and Cohen, S. S. 1979. Mechanism of toxicity of putrescine in *Anacystis nidulans*. Proc. Nat. Acad. Sci. USA 76:3660–3664.

Hirai, A., and Wildman, S. G. 1969. Effect of TMV multiplication on RNA and protein synthesis in tobacco chloroplasts. Virology 38:73–82.

Renaudin, J., Bové, J. M., Otsuki, Y., and Takebe, I. 1975. Infection of *Brassica* leaf protoplasts by turnip yellow mosaic virus. Mol. Gen. Genet. 141:59–68.

Rapporteur's Summary:
Plastids and Their Precursors

Elisabeth Gantt

There has been a resurgence of interest in the origin of chloroplasts. Discovery of DNA in chloroplasts revived the hypothesis that chloroplasts as well as mitochondria had endosymbiotic origins. A polyphyletic origin of chloroplasts resulting from the endosymbiotic uptake of prokaryotic photosynthetic organisms into eukaryotic hosts is often postulated (Raven, 1970). The topic of the conference "On the Origins of Chloroplasts" in addition to being timely is especially appropriate to honor Dr. Roger Stanier. He has been a moving force in extolling the advantages of the cyanobacteria as experimental organisms, and it is these organisms which show the most convincing path for an endosymbiotic chloroplast evolution. These photosynthetic prokaryotes had previously been considered to be blue-green algae, mainly because of their oxygenic type of photosynthesis. By stressing such features as their bacterialike cell wall layers, absence of organelles, and ability to carry out oxygenic and anoxygenic photosynthesis (Padan, 1979), Stanier has made a convincing case for the linkage of the blue-green algae to the bacteria. Accordingly, Dr. Stanier and his colleagues have proposed that they belong with the bacteria and should, therefore, be called cyanobacteria (Stanier and Cohen-Bazire, 1977; Stanier et al., 1978).

The assumption that chloroplasts arose by endosymbiosis between a prokaryote and a eukaryote is best supported by a comparison between the red algae and the cyanobacteria. This is partly because of the similarity in their photosynthetic apparatuses; they both have only chlorophyll a, their phycobiliproteins are nearly identical, and both form phycobilisomes.

Address reprint requests to Dr. Elisabeth Gantt, Senior Research Biologist, Radiation Biology Laboratory, Smithsonian Institution, 12441 Parklawn Drive, Rockville, Maryland 20852.

One may ask whether there are extant prechloroplasts in transition? Trench (1981) surveyed a number of endosymbiotic associations between photosynthetic and heterotrophic hosts, as well as between photosynthetic eurkaryotes and heterotrophic hosts (Trench, 1979). He pointed out that at this point in our knowledge it is difficult to assess which of these represent a true transition from the free-living state to an integrated state between the chloroplast and the cytoplasm. However, there is one possible exception: that of *Cyanophora* in regard to which more is known about the morphological, physiological, and biochemical characteristics than in the case of any other endosymbiotic relationship. *Cyanophora paradoxa* is an association between *Cyanocyta korschikoffiana*, a blue-green organelle, and a colorless host.

Cyanophora, largely on the basis of its genome size and the calculated redundancy of the number of its genome copies, has been called a chloroplast (Herdman and Stanier, 1977). Although it does not yet appear as a fully integrated chloroplast, it does in fact perform that function for the cell. Trench stressed the necessity for establishing criteria to be used in distinguishing chloroplasts from endosymbionts. Of major importance are the relative autonomy of the cyanelle, and where the control over major biosynthetic pathways resides. This work has been greatly hampered by the lack of adequate conditions for separate culturing of the host and the cyanelle. Trench pointed out that at the time of writing there are conflicting reports of the buoyant density of the cyanelle DNA; one report is consistent with values characteristic of free-living organisms, while the other is closer to those measured in plastids. The maintenance of a peptidoglycan layer around the cyanelle suggests that its association with the host may be relatively recent.

Convincing results as to the relative status of the cyanelle were obtained by Trench and his colleagues from studies of phycocyanin synthesis, ribosomal-RNA degradation patterns, and chlorophyll synthesis. In studying the synthesis of the large and small subunits of C-phycocyanin, they found that synthesis of the small subunit is under cytoplasmic control, whereas the synthesis of the large subunit is under organellar control. Furthermore, the degradation pattern of ribosomal RNA resembles that characteristic of chloroplasts. These are the properties of an integrated plastid. On the contrary, cyanellar ribosomal RNA appeared to be independent of the host cytoplasmic protein synthesis, which would not be expected in an organelle with a limited genome. Perhaps such apparently contradictory results are to be expected in a system that is evolving into a mutually obligate symbiosis.

In prokaryotes, polyamines may play an essential role in neutralizing the nucleic acids and in stabilizing their organized structure. Dr. Cohen (1981) and his colleagues are exploring polyamine metabolism in various systems in order to get an indication of endosymbiotic relationships. By infecting Chinese cabbage with turnip yellow mosaic virus, they found that polyamine production is significantly affected. The viruses seem to accumulate in pockets between chloroplasts, and spermidine appears to be associated with the viral RNA. Since exogenous spermidine is not freely exchangeable, and since empty capsids have been found without spermidine, it is expected that the association of RNA and spermidine occurs prior to encapsulation.

Anacystis nidulans contains about 1-2 mM spermidine and about one-tenth this concentration of putrescine. In studying the polyamine metabolism in this photosynthetic prokaryote Cohen and his colleagues found a ready uptake of putrescine, but—unlike the situation in *E. coli*—it was toxic at concentrations as low as 0.15 mM. They suggested that the toxicity may be a result of the irreversible dissociation of 70S ribosomes which, by inhibiting protein synthesis, results in cell death.

The exploration of this metabolic parameter in cyanobacteria may become significant in understanding chloroplast origins. When questioned about the possibility of a chloroplast origin from prokaryotes, Dr. Cohen suggested that it may have occurred once, and that in the process some of the DNA from the microorganism may have become incorporated into the nuclear DNA.

Dynamic cellular processes are involved in the establishment of symbioses. Endosymbiosis may be in the making between the metazoan *Hydra viridis* and *Chlorella*, as described by Muscatine (1981). In *H. viridis*, uptake of the photosynthetic alga is not as specific as the maintenance of the alga in the host, because although many strains of *Chlorella* could be introduced only one strain formed a stable association.

Muscatine elaborated on the criteria for establishing an endosymbiotic system. Cell uptake was very rapid once contact was made, and it seemed to require a difference in the surface charge of the membranes. Recognition of the symbiont by the host appeared to be specific and could be blocked by antisera. Digestion of the algae had to be avoided and they must have been able to divide and grow. Dead algal cells were digested by the host, but live ones were maintained. Throughout the association, however, the algal cells remained in a vacuole, but this did not interfere with the nutritional exchange between the host and the alga.

A major selective advantage in establishing a symbiotic relationship is the translocation of soluble organic material from the endosymbiont to the host. Living symbiotic *Chlorella* cells release maltose with traces of glycolic acid as a result of active photosynthesis. This also appears to be involved in the ability to resist digestion. Temporary phagocytic associations between a photosynthetic and a nonphotosynthetic eukaryote is not uncommon, and although several of these associations may have led to the establishment of a permanent endosymbiosis, it is doubtful whether many chloroplasts resulted from the reduction of a photosynthetic cell through loss of mitochondria and nucleus and secondary integration into another cell. Incorporation of cells similar to *Prochloron* (Lewin, 1977) would provide the same benefits to a host, and would entail considerably less reduction. Because it has been so well characterized by Dr. Muscatine and his colleagues, the *Hydra-Chlorella* system promises to provide further insight into the process of endosymbiosis.

Is there a common developmental pathway in plastid morphogenesis which may suggest a similar origin? This question was addressed by Klein (1981). In reviewing the developmental sequence he also summarized old and new evidence on the role of the prolamellar body, chloroplast division, and the role of the inner chloroplast membrane. He clarified several misconceptions. In the past, the difference between chloroplast development in higher plants and algae has probably been overemphasized. This was largely due to the importance placed on the role of the prolamellar body. Much of the previous work on this involved etiolated tissues. In cereal plants, as well as in beans, where etiolated plants were generally used for developmental studies, the prolamellar body had been postulated to give rise to the chloroplast thylakoids, and was considered to be the site of the protochlorophyll(ide) to chlorophyll(ide) conversion. Some aspects of the etioplast to chloroplast transformation are now questionable because the protochlorophyll(ide) to chlorophyll(ide) conversion can be accomplished without changing the prolamellar body; the prolamellar bodies have a significantly higher lipid/protein ratio than prothylakoids (Lütz, 1978); and most of the protochlorophyll(ide) is recoverable in the prothylakoid fractions instead of in the prolamellar body fraction.

The distinction formerly made between chloroplast development in algae and higher plants on the basis of the role of the prolamellar body seems now to be of secondary significance. Furthermore, the distinction made on the basis of chloroplast division—according to which in algae mature chloroplasts tend to divide more frequently, whereas in higher plants the division of proplastids is more common—is also now considered to be less significant than was formerly presumed.

Involvement of the double chloroplast envelope in thylakoid production is also in question because direct connections between the inner membrane and the thylakoids remain unsubstantiated. If chloroplasts originated from prokaryotes, one could expect an involvement of the inner membrane similar to the role of the plasma membrane in photosynthetic membrane development in photosynthetic bacteria (Kaplan, 1978). Indeed in the cyanobacterium *Gloeobacter* (Rippka et al., 1974) the plasma membrane serves as the photosynthetic membrane. Of particular interest in this regard would be to know if the plasma membranes of other cyanobacteria, or of *Prochloron,* are involved in thylakoid differentiation.

The phylogenetic relationships and possible insight into polyphyletic origins of chloroplasts was considered by Shilo (1981). He pointed out that between 1970 and 1980 several exciting and important discoveries have been made which require consideration in evaluating the interrelationships of photosynthetic prokaryotes. The most basic type of photosynthesis occurs in the purple membrane of halobacteria. They carry out direct light/energy conversion across the membrane and produce ATP. Because these membranes contain bacteriorhodopsin instead of chlorophyll and have a primitive membrane system, they appear not to be related to either the cyanobacteria or the chloroplasts.

One of the most important discoveries to date is the ability of cyanobacteria to carry on anoxygenic as well as oxygenic photosynthesis (Padan, 1979). The fact that many cyanobacteria (16 of 30 strains tested) can photoreduce CO_2 using only photosystem I with sulfide as the electron donor, brings them functionally close to the photosynthetic bacteria. Since the cyanobacteria are capable of oxygenic and anoxygenic photosynthesis, they have an adaptive advantage, and are probably more highly evolved than the photosynthetic bacteria. However, they can be suggested as a bridging link between oxygen-evolving plants with two photosystems and those growing in anaerobic conditions with one photosystem.

The discovery by Lewin (1977) of *Prochloron* provides another example of a missing link. It is a prokaryote which contains chlorophyll *b* in addition to chlorophyll *a*, lacks phycobiliproteins, and has bacterial-type wall layers. Previous to this discovery, the blue-green algal line had been proposed as the most direct path to the origin of the chloroplasts of green plants (Raven, 1970). *Prochloron* may be an organism in transition, for its photosynthetic unit size was found to be smaller than that of chloroplasts, although somewhat larger than that of cyanobacteria (Withers et al., 1978). Furthermore, it is not a free-living prokaryote, but is obligated to exist in the extracellular cavities of its hosts. Knowledge of its nutritional requirements, and of its genome size will be critical in evaluating its role as a transition form to a chloroplast.

The phycobiliproteins occurring in the red algae and cyanobacteria have often been regarded as one of the more convincing examples of a direct phylogenetic relationship between these two groups. Glazer (1981), in showing the amino acid sequences of the α- and β-polypeptides of C-phycocyanin, emphasized that the structure is highly conserved in three algal species (*Cyanidium, Mastigoclaclus,* and *Synechococcus*). He stressed that since the sequence is virtually identical in these three examples, as well as in cryptophyte phycocyanin, one can assume that they came from the same ancestral gene. Any elaboration manifested in specific pigments, such as B- and R-phycoerythrin and R-phycocyanin, are the results of later divergent elaborations. The absence of phycobilisomes in the cryptophytes appears to set them apart from the red and blue-green lines.

Examination of photosynthetic catalysis by Krogmann (1981) revealed an interesting point of evolutionary divergence in the cyanobacteria. Plastocyanin, which in higher plants and many cyanobacteria is involved in the reduction of P_{700}, can be functionally replaced in some cyanobacteria by cytochrome c_{553}. It is similar to cytochrome c_{552} found in some photosynthetic bacteria *(Chromatium)*. The isoelectric points of cytochrome c_{553} were higher in the apparently more primitive cyanobacteria (*Anabaena* and *Aphanizomenon*) than in the more advanced ones (*Microcystis* and *Spirulina*) which had lower isoelectric points similar to those of related cytochromes from red algae. Furthermore, a similar relationship is suggested from examination of their amino acid sequences. Along with the changes in the net charge of these catalysts, evidence was also presented suggesting other site alterations in membrane components to accommodate these changes during evolutionary development.

References to Rapporteur's Summary

Cohen, S. 1981. On the endosymbiotic origins of chloroplasts: Still another approach to the problem. *In* On the Origins of Chloroplasts. J. A. Schiff (ed.). New York: Elsevier North Holland.

Glazer, A. 1981. Amino acid sequences of biliproteins and the endosymbiotic origins of chloroplasts. *In* On the Origins of Chloroplasts. J. A. Schiff (ed.). New York: Elsevier North Holland.

Herdman, M., and Stanier, R. Y. 1977. The cyanelle, chloroplast or endosymbiotic prokaryote? FEMS Lett. 1:7–12.

Kaplan, S. 1978. Control and kinetics of photosynthetic membrane development. *In* The Photosynthetic Bacteria. R. C. Clayton and W. R. Sistrom (eds.). New York: Plenum Press, pp. 809–840.

Klein, S. 1981. The diversity of chloroplast structure. *In* On the Origins of Chloroplasts. J. A. Schiff (ed.). New York: Elsevier North Holland.

Krogmann, D. W. 1981. The evolution of photosynthetic catalysts in cyanobacteria. *In* On the Origins of Chloroplasts. J. A. Schiff (ed.). New York: Elsevier North Holland.

Lewin, R. A. 1977. *Prochloron*, type genus of the prochlorophyta. Phycologia 16:217.

Lütz, C. 1978. Separation and comparison of prolamellar bodies and prothylakoids of etioplasts from *Avena sativa* L. *In* Chloroplast Development. G. Akoyunoglou and J. H. Argyroudi-Akoyunoglou (eds.). Amsterdam: Elsevier North-Holland, pp. 481–488.

Muscatine, L. 1981. The establishment of photosynthetic eukaryotes as endosymbionts in animal cells. *In* On the Origins of Chloroplasts. J. A. Schiff (ed.). New York: Elsevier North Holland, pp.

Padan, E. 1979. Facultative anoxygenic photosynthesis in cyanobacteria. Annu. Rev. Plant Physiol. 30:27–40.

Raven, P. 1970. A multiple origin of plastids and mitochondria. Science 169:641–646.

Rippka, R., Waterbury, J., and Cohen-Bazire, G. 1974. A cyanobacterium which lacks thylakoids. Arch. Microbiol. 100:419–436.

Shilo, M. 1981. The diversity of the photosynthetic prokaryotes. *In* On the Origins of Chloroplasts. J. A. Schiff (ed.). New York: Elsevier North Holland.

Stanier, R. Y., and Cohen-Bazire, G. 1977. Phototrophic prokaryotes: The cyanobacteria. Annu Rev. Microbiol. 31:225–274.

Stanier, R. Y., Sistrom, W. R., Hansen, T. A., Whitton, B. A., Castenholz, R. W., Pfennig, N., Gorlenko, V. N., Krondratieva, E. N., Eimhjellen, K. E., Whittenbury, R., Gherna, R. Z., and Trüper, H. G. 1978. Proposal to place the nomenclature of the Cyanobacteria (blue-green algae) under the rules of the International Code of Nomenclature of Bacteria. J. System. Bact. 28:335–366.

Trench, R. K. 1979. The cell biology of plant-animal symbiosis. Annu. Rev. Plant Physiol. 30:485–531.

Trench, R. K. 1981. Cyanelles. *In* On the Origins of Chloroplasts. J. A. Schiff (ed.). New York: Elsevier North Holland.

Withers, W. N., Alberte, R. S., Lewin, R. A., Thornber, J. P., Britton G., and Goodwin, T. W. 1978. Photosynthetic unit size, carotenoids, and chlorophyll-protein composition of *Prochloron* sp. a prokaryotic green alga. Proc. Nat. Acad. Sci. USA 75:2301–2305.

PART II
ORIGIN AND EVOLUTION
OF PLASTID METABOLISM

I do not think it at all degrading to the business of experimental philosophy to compare it, as I often do, to the diversion of *hunting*, where it sometimes happens that those who have beat the ground the most, and are consequently the best acquainted with it, weary themselves without starting any game; when it may fall in the way of a mere passenger, so that there is but little room for boasting in the most successful termination of the chase.

—Joseph Priestley

Paths of Carbon and Their Regulation

James A. Bassham

Two questions arise when one considers the paths of carbon and their regulation with respect to the objectives of the conference on which this book is based. First, one might hope to gain some clues regarding the possible evolutionary relationships between bacteria carrying-out oxygenic photosynthesis and chloroplasts from eukaryotic cells, from a comparison of carbon pathways and their regulation. Second, one might attempt to write a scenario for the evolution of the carbon pathways from the most primitive of these organisms onward.

The Carbon Pathways of Chloroplasts

The reductive pentose phosphate (RPP) cycle, the Calvin cycle, is ubiquitous and occurs in all eukaryotic photosynthetic plants, in the cyanobacteria, and in many photosynthetic bacteria (Norris et al., 1955; Fuller, 1978). This pathway (Figure 1) is required for the synthesis of sugar phosphates from CO_2, utilizing reduced pyridine nucleotides and ATP derived from the light reactions in photosynthetic organisms.

Additional pathways of CO_2 fixation supplementary to the RPP cycle exist in certain higher plants. These are the pathways of the C-4 plants, which include certain tropical grasses and other species from diverse families, and the plants that exhibit Crassulacean acid metabolism (CAM) such as the succulents, in-

This work was supported by the Division of Biomedical and Environmental Research of the U.S. Department of Energy under contract No. W-7405-ENG-48.

Address reprint requests to Dr. James A. Bassham, Senior Staff Scientist, Building 3—Chemical Biodynamics, Lawrence Berkeley Laboratory, 1 Cyclotron Road, Berkeley, California 94720.

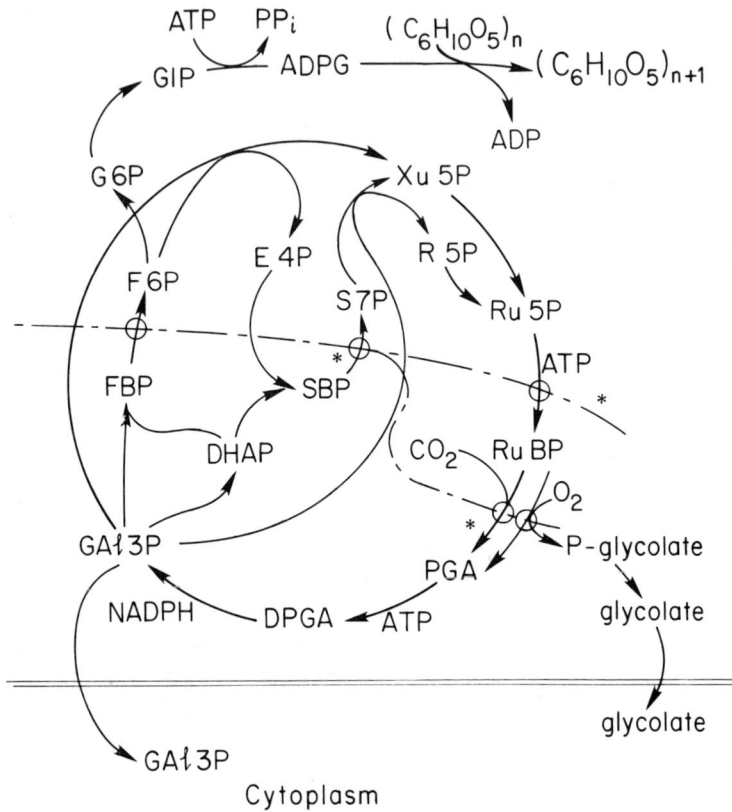

Figure 1. Reductive pentose phosphate (RPP) cycle. Regulated steps are indicated by
⊖, unique steps by asterisks. Abbreviations: G1P, glucose-1-phosphate; G6P, glucose-6-phosphate; F6P, fructose-6-phosphate; E4P, erythrose-4-phosphate; Ru5P, ribulose-5-phosphate; Xu5P, xylulose-5-phosphate; DHAP, dihydroxyacetone phosphate; Gal3P, glyceraldehyde-3-phosphate; FBP, fructose-1,6-bisphosphate; SBP, sedoheptulose-1,7-bisphosphate; RuBP, ribulose-1,5-bisphosphate; ADPG, adenosine diphosphoglucose; PGA, 3-phosphoglycerate.

cluding the cactuses. Neither of these photosynthetic carbon fixing pathways is relevant to the purposes of our present discussion since the evolution of such pathways has occurred much later than the evolution of the chloroplasts.

The RPP, or Calvin, cycle (Bassham et al., 1954) consists of three stages. The first stage consists of the phosphorylation with ATP of ribulose 5-phosphate (Ru5P) to give ribulose 1,5-bisphosphate (RuBP), followed by the carboxylation and hydrolytic split of the resulting 6-carbon enzyme-bound compound to give

2 molecules of 3-phosphoglyceric acid (3-PGA). This stage is unique to the RPP cycle.

The second stage of the cycle begins with the phosphorylation of PGA to give phosphoryl PGA and ADP. The reaction is mediated by the enzyme phosphoglyceric acid kinase. The resulting acyl phosphate is then reduced with NADPH in the presence of triose phosphate dehydrogenase yielding oxidized $NADP^+$ and 3-phosphoglyceraldehyde (GAl3P) plus inorganic phosphate (P_i). These reactions are essentially the reverse of steps of the glycolytic pathway.

The third stage of the cycle consists of all of the remaining reactions which work together to convert five molecules of the triose phosphate (Gal3P) to three molecules of Ru5P. The reactions are mediated by aldolase, transketolase, and isomerases, epimerases, and bisphosphatases. Sedoheptulose-1,7,-bisphosphatase (SBPase) is unique to the RPP cycle. The transketolase, aldolase, and isomerization reactions are common to the oxidative pentose phosphate cycle.

The oxidative pentose phosphate (OPP) cycle includes the hexose monophosphate shunt (Figure 2). This cycle, beginning with glucose-6-phosphate (G6P), includes the oxidation of G6P to 6-phosphogluconic acid (6PGluA), mediated by glucose-6-phosphate dehydrogenase and transferring electrons to $NADP^+$ producing NADPH. The second step is a further oxidation of 6PGluA to CO_2 and Ru5P, with two more electrons being transferred to $NADP^+$. The resulting NADPH is widely used in plant cells for the biosynthesis of fatty acids from acetyl CoA. Subsequent steps of the OPP cycle result in the conversion of three molecules of Ru5P to one molecule of fructose-6-phosphate and one molecule of Gal3P which in turn can be converted back to F6P and G6P, thus completing the cycle (when and if it functions as a complete closed cycle) (Kaiser and Bassham, 1979a).

Regulation of RPP and OPP Cycles in Chloroplasts

The regulation of the RPP cycle in chloroplasts has been worked out over the past 12 years or so (for reviews see Bassham, 1971, 1979). During photosynthesis the rate-limiting steps (Figure 1) are those mediated by phosphoribulokinase, ribulose-1,5-bisphosphate carboxylase, and the two phosphatases involved in the conversion of SBP and fructose-1,6-bisphosphate (FBP) to their respective monophosphates (Bassham and Krause, 1969). The principal sites at which carbon is removed from the cycle during photosynthesis are as follows: (a) Gal3P is exported from the chloroplast to the cytoplasm where it can be used for subsequent biosynthetic reactions; (b) F6P is converted to glucose-6-phosphate (G6P), the starting point for the synthesis of starch; and (c) the oxygenase reaction mediated by RuBP carboxylase/oxygenase (in which oxygen binds competitively at the CO_2 binding site) oxidizes RuBP to PGA and phosphoglycolate. Free glycolate is exported from the chloroplast after which its metabolic fate is different in algae and higher plants. In higher plants the glycolate is converted in peroxisomes and mitochondria via the glycolate pathway, producing one

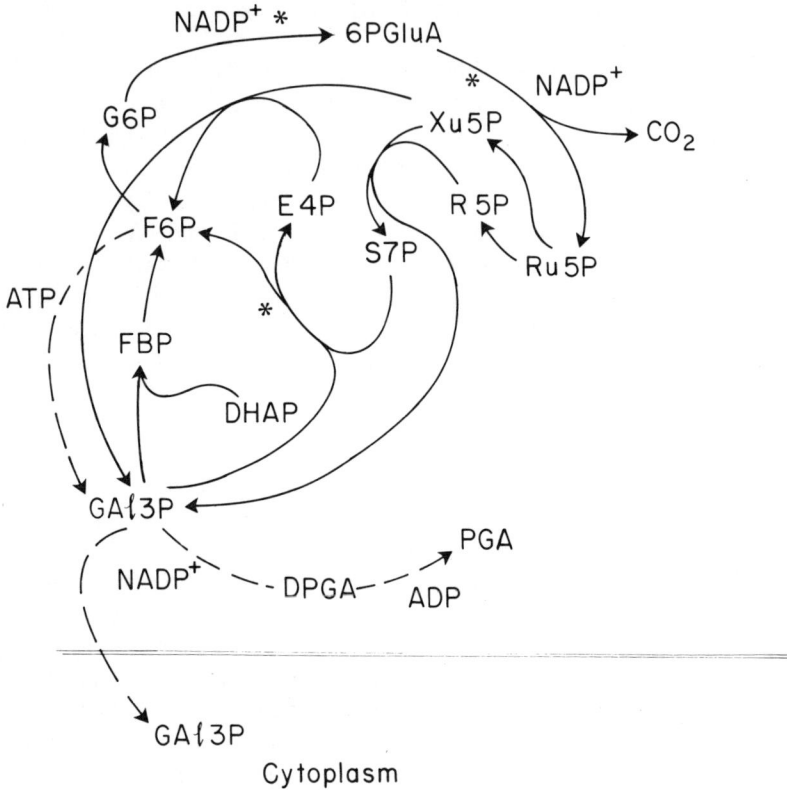

Figure 2. Oxidative pentose phosphate (OPP) cycle. Regulated steps are indicated by ⊖→, unique steps by asterisks. Glycolytic steps are indicated by dashed lines. Abbreviations (see Figure 1), plus 6PG1uA, 6-phosphogluconate.

molecule of CO_2 and one molecule of glycerate from each two molecules of glycolate. The glycerate is reincorporated into the chloroplast (Tolbert, 1971). The CO_2 evolved is considered to be the product of photorespiration. This production of glycolate (and the whole pathway of photorespiration) can be considered as a late evolutionary phenomenon, occurring only after the level of CO_2 had declined to its present low level, and it is not relevant to the question of the evolution of the chloroplast or its metabolic cycles.

Regulation of the rate-limiting steps of the RPP cycle in the light is required to balance the levels of cycle intermediates when there is a change in the relative proportions of cycle metabolites exported or converted to starch. The export of Gal3P, controlled in part by cytoplasmic P_i (Heldt, 1976; Walker, 1976), should vary inversely with the conversion of F6P to starch when there is a constant rate of carboxylation. In order for physiological concentrations of each sugar phosphate to be maintained within the chloroplast; therefore, the activities of the FBPase and SBPase must be finely adjusted in comparison with the activity of phosphoribulokinase and RuBP carboxylase.

It appears that the internal P_i level in the chloroplasts serves to provide the inverse relation between the rate of export of Gal3P to the cytoplasm and conversion of F6P to starch. Inasmuch as Gal3P export is balanced against P_i import, chloroplast internal P_i concentration increases with increased rate of triose phosphate export. This increased P_i concentration in the chloroplast slows the rate of conversion of glucose-1-phosphate (G1P) with ATP to adenosine diphosphoglucose (ADPG) and inorganic pyrophosphate (PP_i).

If PGA should accumulate in the chloroplast, as would be expected when the rate of triose phosphate export declines, it would be advantageous for the rate of starch formation to accelerate. The stimulation of the ADPG pyrophosphorylase reaction by PGA and its inhibition by P_i were reported for the isolated enzyme (Priess, 1967; Sanwal et al., 1968). When concentrations of controlling factors (pH, P_i, PGA and ATP) observed in whole chloroplasts in the light and dark (Kaiser and Bassham, 1979b) were administered to reconstituted chloroplasts, strong regulation of ADPG formation was observed (Kaiser and Bassham, 1979c).

For regulation of the light-dark transition from the RPP cycle to glycolysis and the OPP cycle, those enzymes of the RPP cycle which were rate-limiting in the light become inactivated in the dark (Bassham and Kirk, 1968; Pedersen et al., 1966). This inactivation is the result of changes in Mg^{2+} ion concentration, pH, and the level of reduced to oxidized cofactors (Anderson, 1973; Schürmann and Buchanan, 1975; Schürmann et al., 1976; Breazeale et al., 1978; Buchanan et al., 1979). Besides the dark-induced decline in the activities of the four rate-limiting steps from the light, there appears to be some evidence for a decrease in the activity of triose phosphate dehydrogenase in the dark (Buchanan et al., 1979).

In the OPP cycle, the key regulated step is the oxidation of glucose-6-phosphate to 6-phosphogluconate mediated by glucose-6-phosphate dehydrogenase. The principal mechanism of regulation appears to be by the ratio of $NADPH/NADP^+$ (Lendzian and Bassham, 1975, 1976). High ratios of $NADPH/NADP^+$ decrease the activity of the enzyme and this effect is more pronounced at pH 8.2 than at pH 7.6 (the light and dark pH values of the chloroplast stroma, respectively). Considerable additional inactivation of glucose-6-phosphate dehydrogenase at the physiological levels of $NADPH/NADP^+$ in the light occurs in the presence of physiological levels of ribulose-1,5-bisphosphate. With such high ratios of $NADPH/NADP^+$ and in the presence of 0.5 mM RuBP, the activity of the enzyme is reduced to less than 1% of its level under conditions of full activity.

Regulation of RPP and OPP Cycles in Cyanobacteria

Regulation of carbon metabolism in the cyanobacteria (see Stanier and Cohen-Bazire, 1975) exhibits many similarities to the regulation of the RPP and OPP cycles in chloroplasts of green algae and higher plants. Except for glucose-6-phosphate dehydrogenase, which increases in specific activity somewhat in the dark (Pelroy et al., 1972), all enzymes of the OPP and RPP cycles are synthesized under both light and dark conditions. In *Aphanocapsa* 6714 the rate-limiting steps of the RPP cycle in the light appear to be those mediated by FBPase and SBPase and by phosphofructokinase and perhaps the carboxylase (Pelroy et al., 1976). Even the time required to reactivate FBPase and SBPase on going from dark to light (20-30 sec) is reminiscent of that observed with *Chlorella pyrenoidosa* (Bassham and Kirk, 1968). *Aphanocapsa*, when metabolizing glucose in the dark, however, maintains a high level of NADPH, a condition not likely to be found in chloroplasts of eukaryotic cells. Phosphoribulokinase activity was dependent on the presence of sulfhydryl reducing agents such as dithiothreitol, a response similar to that reported for the enzyme from eukaryotic cells (Anderson, 1973). Perhaps cyanobacteria, like chloroplasts of eukaryotic cells, regulate SBPase, FBPase, phosphoribulokinase, and even triose phosphate dehydrogenase through the ratios of reduced to oxidized ferridoxin, acting through the intermediacy of thioredoxin, as has been shown in the case of eukaryotic cells (Breazeale et al., 1978). The presence of the thioredoxin system in bacterial systems including cyanobacteria has been reported (Buchanan and Wolosiuk, 1976).

It appears that the OPP cycle can operate to some extent in *Aphanocapsa* even in the presence of high ratios of NADPH to $NADP^+$. For cyanobacteria such as *Anacstis nidulans* which are only poorly permeable to glucose (Pelroy et al., 1972) such high ratios of NADPH to $NADP^+$ would not be expected. In view of the specific inhibition of the eukaryotic chloroplast glucose 6-phosphate dehydrogenase by RuBP mentioned earlier, it is interesting that one study of glucose-6-phosphate dehydrogenase activity in crude, cell-free extracts of a variety of cyanobacteria showed this enzyme to be inhibited by RuBP (Pelroy et al., 1972). Other studies of the same enzyme partly purified from *Anabaena* found NADPH, but not RuBP, to be an inhibitor (Grossman and McGowan, 1975). Considering the reported complex nature of the *Anabaena* enzyme (Stanier and Cohen-Bazire, 1977), it seems possible that just as in the case of chloroplasts, the bacterial enzyme may be regulated both by the $NADPH/NADP^+$ ratio and by RuBP concentration, perhaps with requirements for regulatory behavior that are lost during the purification of the enzyme.

Although the cyanobacteria synthesize glycogen as a carbohydrate storage product rather than starch, there appear to be similarities in the regulation of the synthetic pathway. The rate-limiting enzyme, ADPG pyrophosphorylase, is activated 8- to 25 fold by 3-phosphoglycerate and is inhibited by inorganic phosphate (Levi and Preiss, 1976). The enzyme from other prokaryotic cells is not activated by phosphoglycerate. It is significant that the cyanobacterial enzyme exhibits this regulatory property characteristic of eukaryotic photosynthetic organisms.

Endosymbiotic Origin of Chloroplasts from Oxygenic Photosynthetic Bacteria

What are the implications of the similarities between carbon pathways and their regulation in chloroplasts of higher plants and in the oxygenic photosynthetic bacteria? As pointed out by Stanier and Cohen-Bazire (1977), the similarities in carbon metabolism between the cyanobacteria and chloroplasts of higher plants are striking, but the pigment systems are quite different. All photosynthetic oxygenic organisms contain chlorophyll *a*, but whereas higher plants have chlorophyll *b* as the principal second pigment, the cyanobacteria contain phycobilins. The red algae contain a light-harvesting pigment system very similar to that of the cyanobacteria, making the origin of the rhodophytan chloroplast from cyanobacteria an attractive evolutionary hypothesis. For the origin of chloroplasts of higher plants and green algae, one must look for other precursors, possibly the oxygenic photosynthetic bacteria isolated by Lewin and his colleagues from the marine organism *Didemnum* (Lewin and Withers, 1975; Lewin, 1976; Withers et al., 1978). These organisms, for which the name *Prochloron* sp. has been proposed, have Chl *a*/Chl *b* ratios in the range of 4.4 to 6.9, contain carotene and zeaxanthin, and generally resemble green algae in the arrangement of paired or stacked thylakoids. Moreover, a chlorophyll protein complex isolated from these prokaryotes is indistinguishable from that obtained from the chloroplasts of higher plants.

From the foregoing it seems clear that although similarities in carbon metabolism and its regulation point to a common evolutionary origin for all photosynthetic oxygenic organisms, they do not provide a basis for establishing a sequential relationship between the cyanobacteria and green chloroplasts of eukaryotic cells.

Origin of Carbon Metabolism in Chloroplasts

What can we deduce about the origin of chloroplast metabolism in the first primitive organisms and its subsequent evolution? Much has been written on chemical evolution and subsequent biochemical evolution. Broda (1975) has provided an excellent comprehensive description of the evolution of bioenergetic processes which brings together a great many of the excellent reviews and original papers which have appeared on this subject over the years. The following speculations are based in part on information reviewed by Broda. No claim is made either for originality or for agreement with all the existing literature which in at least some cases is still controversial.

It is widely believed that after the earth had reached its approximate present size and structure, most of the primary atmosphere had been lost, and the new atmosphere was formed by outgassing of the earth's rocks. There is considerable controversy over how reducing this primitive earth's atmosphere was, but it is generally conceded that it was basically free of oxygen. Various simple organic molecules are thought to have been formed from hydrides (methane, water, ammonia, hydrogen sulfide) and perhaps from CO, CO_2, N_2, and H_2 under the influence of electrical discharges from storms, ultraviolet radiation, heat and volcanism, and perhaps catalysis by certain mineral clays. Under a variety of assumed conditions, amino acids, sugars, and more complex biological building blocks have been formed in laboratory experiments employing these various agents. From concentrations of such organic chemicals, primitive life could have evolved in the form of very simple cells capable of utilizing the wealth of organic substrates in their environments and of reproducing themselves.

All life requires energy, and the earliest organisms must have derived their vital supply of energy by carrying out various oxidation-reduction reactions on the abiologically formed materials available to them. Thus we can suppose that the fermentative pathway of glycolysis from hexoses to pyruvate followed by either lactogenesis or formation of alcohol and CO_2 may have evolved in very primitive organisms as a means for generating the energy currency of living cells, ATP.

Although a variety of sugars, including pentoses and hexoses, would be expected to be formed from the condensation of formaldehyde in the ponds of the primitive earth, it is apparent that pentoses became the sugar of choice for nucleotides and hence for genetic material. Since thermodynamics somewhat favors the formation of hexoses over pentoses, it seems possible that the supply of pentoses may have become exhausted, necessitating the evolution of biochemical pathways to convert hexoses to pentoses. The most primitive of such pathways may well have been the conversion of hexose phosphates to pentose phosphates via the reactions mediated by transketolase, transaldolase, and isomerases.

Further depletion of the reduced compounds in the primitive environment, particularly as the atmosphere became less reducing (due to H_2 escape from the earth) could have led to a shortage of the supply of endogenous long-chain hydrocarbons required for the synthesis of fatty acids needed for cellular membranes. Assuming that at this point carbohydrates were still available, a way had to be found for the conversion of carbohydrate to fatty acids. If acetyl CoA became the building block for the synthesis of fatty acids, these two-carbon fragments could have been generated by pyruvate formation via glycolysis and oxidation of pyruvate with the electrons being used for fatty acid synthesis. Since this supply of electrons would be stoichiometrically inadequate, the oxidative stages of the OPP cycle may have evolved in order to provide a supply of reduced NADPH. The oxidative pentose phosphate cycle is used primarily for that purpose in many types of cells to this day.

The evolution of the reductive pentose phosphate cycle would have occurred very much later, at a time when the endogenous supply of carbon in forms other than CO_2 had become depleted. Long before this happened, we may suppose that photosynthetic organisms analogous to modern day photosynthetic bacteria had developed which were capable of utilizing light energy to drive the process of photophosphorylation. We can surmise that the first photophosphorylation was of the cyclic type in which electrons are cycled from the product of the photochemical reaction (electron acceptor) to the resulting oxidized pigment via an electron donor, with the energy of electron flow being used to form a proton gradient across membranes in order to drive the conversion of ADP and inorganic phosphate to ATP. A later stage of evolution would lead to noncyclic photophosphorylation, that is, a net flow of electrons from donor to acceptor. With the establishment of photophosphorylation and photochemical electron flow, the stage was set for the development of a great diversity of types of photochemical metabolism, many of which are seen in present day photosynthetic bacteria, both aerobic and anaerobic.

Depending upon the oxidation level of available substrate and the oxidation level of cellular material, photosynthetic bacteria may absorb CO_2 and reduce it to various organic molecules required for cell synthesis. Long before supplies of endogenous reduced carbon were exhausted, it seems likely that the reductive pentose phosphate cycle may have evolved in photosynthetic bacteria. The evolution of the RPP cycle from the oxidative pentose phosphate pathway requires the evolution of the reaction converting R5P to RuBP, utilizing one molecule of ATP, and the subsequent carboxylation reaction which produces two molecules of PGA. A third unique reaction is involved in the RPP cycle, namely the conversion of SBP to sedoheptulose-7-phosphate. The enzyme for this reaction, SBPase, is closely related to FBPase and could easily have evolved from it. Even though phosphoribulokinase (PRK) is similar in function to phosphofructokinase, it should be noted that PRK activity is turned on in the light and off in the dark, whereas, PFK, an enzyme of glycolysis, has to be regulated in exactly the opposite way. Since RuBP has no other known function than to serve as substrate for the photosynthetic carboxylation reaction of the RPP cycle, it would appear that both the PRK and RuBP carboxylase enzymes had to evolve simultaneously. The evolution of these two enzymes has to be considered as one of the most significant events in evolutionary history, given the present day importance of the carbon-fixing reactions in the biosphere. The evolution of the genetic information, both chloroplastic and nuclear, for the various subunits of RuBPCase is an extremely interesting problem which has been discussed elsewhere at this conference.

Although it has been suggested that the formation and carboxylation of RuBP were evolved by certain photosynthetic bacteria, another possibility might be that these steps could have evolved in anaerobic chemolithotrophs living in proximity to anaerobic fermenters producing hydrogen and CO_2.

Presumably, oxygenic prokaryotic photosynthetic bacteria, the cyanobacteria and others, evolved in time from some primitive photosynthetic bacteria and

126 J. A. Bassham

retained the capability of carrying out the reductive pentose phosphate cycle. With the evolution of O_2 formation by oxidation of water in oxygenic photosynthetic bacteria, the resulting electrons and ATP could be used to drive the RPP cycle. In addition, a smaller portion of the electron flow could be used for the reduction of nitrite and the reduction of sulfate.

Finally, with the incorporation of these oxygenic prokaryotes into eukaryotic cells, the last stages of the evolution of chloroplastic carbon metabolism could proceed with the development of the special regulatory mechanism required in these organelles.

References

Anderson, L. E. 1973. Regulation of pea leaf ribulose 5-phosphate kinase activity. Biochim. Biophys. Acta 321:484–488.

Bassham, J. A. 1971. Control of photosynthetic carbon metabolism. Science 172:526–534.

Bassham, J. A. 1979. The Reductive Pentose Phosphate Cycle and its Regulation. *In* Encyclopedia of Plant Physiology (new series). Vol. 6, Photosynthesis II: "Regulation of Photosynthetic Carbon Metabolism and Related Processes." M. Gibbs and E. Latzko (eds.). Berlin and New York: Springer-Verlag.

Bassham, J. A., and Kirk, M. 1968. Dynamic metabolism regulation of the photosynthetic carbon reduction cycle. *In* Comparative Biochemistry and Biophysics of Photosynthesis. K. Shibata, A. Tamamiya, A. T. Jagendorf, and R. C. Fuller (eds.). Tokyo: University of Tokyo Press, pp. 365–378.

Bassham, J. A., and Krause, G. H. 1969. Free energy changes and metabolic regulations in steady-state photosynthetic carbon reduction. Biochim. Biophys. Acta 189:207–221.

Bassham, J. A., Benson, A. A., Kay, L. D., Harris, A. Z., Wilson, A. T., and Calvin, M. 1954. The path of carbon in photosynthesis. XXI. The cyclic regeneration of CO_2 acceptor. J. Am. Chem. Soc. 76:1760–1770.

Breazeale, V. D., Buchanan, B. B., and Wolosiuk, R. A. 1978. Chloroplast sedoheptulose 1,7-biphosphatase: Evidence for regulation by the ferredoxin/thioredoxin system. Z. Naturforsch 33c:521–528.

Broda, E. 1975. The Evolution of the Bioenergetic Processes. Oxford: Pergamon Press.

Buchanan, B. B., and Wolosiuk, R. A. 1976. Photosynthetic regulatory protein found in animal and bacterial cells. Nature 264:669–670.

Buchanan, B. B., Wolosiuk, R. A., and Schürmann, P. 1979. Thioredoxin and enzyme regulation. Trends Biochem. Sci. 4:93–94.

Buchanan, B. B., Schürmann, P. and Kalberer, P. T. 1971. Ferredoxin-activated fructose diphosphatase of spinach chloroplasts. J. Biol. Chem. 246:5952–5959.

Fliege, R., Flugge, U., Werdan, K., and Heldt, H. W. 1968. Specific transport of inorganic phosphate, 3-phosphoglycerate and triosephosphates across the inner membrane of the envelope in spinach chloroplasts. Biochim. Biophys. Acta. 502:232–247.

Fuller, R. C. 1978. Photosynthetic carbon metabolism in the green and purple bacteria. *In* The Photosynthetic Bacteria. K. Clayton and W. C. Systrom (eds.). New York and London: Plenum Press, pp. 691–705.

Grossman, A. and McGowan, R. E. 1975. Regulation of glucose 6-phosphate dehydrogenase in blue-green algae. Plant Physiol. 55:658–662.

Heldt, H. W. 1976. Metabolite transport in intact spinach chloroplasts. *In* The Intact Chloroplast. J. Barber (ed.). Amsterdam: Elsevier-North Holland Biomedical Press.

Kaiser, W. M., and Bassham, J. A. 1979a. Carbon metabolism of chloroplasts in the dark: Oxidation pentose phosphate cycle versus glycolytic pathways. Planta 144:193–200.

Kaiser, W. M., and Bassham, J. A. 1979b. Light-dark regulation of starch metabolism in chloroplasts. I. Levels of metabolites in chloroplasts and medium during light and dark. Plant Physiol. 63:105–108.

Kaiser, W. M. and Bassham, J. A. 1979c. Light-dark regulation of starch metabolism in chloroplasts. II. Effect of chloroplastic metabolite levels on the formation of ADP-glucose by chloroplast extracts. Plant Physiol. 63:109–113.

Lendzian, K., and Bassham, J. A. 1975. Regulation of glucose 6-phosphate dehydrogenase in spinach chloroplasts by ribulose -1,5- diphosphate and NADPH/NADP$^+$ ratios. Biochim. Biophys. Acta 396:260–275.

Lendzian, K. and Bassham, J. A. 1976. NADPH/NAD$^+$ ratios in photosynthesizing reconstituted chloroplasts. Biochim. Biophys. Acta 430:478–489.

Levi, C., and Preiss, J. 1976. Regulatory properties of the ADP-glucose pyrophosphorylase of the blue-gree bacterium *Synechococcus* 6301. Plant Physiol. 58:753–756.

Lewin, R. A. 1976. Prochlorophyta as a proposed new division of algae. Nature 261:697–698.

Lewin, R. A., and Withers, N. W. 1975. Extraordinary pigment composition of a prokaryotic alga. Nature 256:735–737.

Norris, L., Norris, R. E., and Calvin, M. 1955. A survey of the rates and products of short-term photosynthesis in plants of nine phyla. J. Exp. Bot. 6:64–74.

Pedersen, T. A., Kirk, M., and Bassham, J. A. 1966. Light-dark transients in levels of intermediate compounds during photosynthesis in air-adapted *Chlorella*. Physiol. Plant. 19:219–231.

Pelroy, R. A., Levine, G. A., and Bassham, J. A. 1976. Kinetics of light-dark CO$_2$ fixation and glucose assimilation by *Aphanocapsa* 6714. J. Bacteriol. 128:633–644.

Preiss, J. Ghosh, H. P., and Wittkop, J. 1967. Regulation of the biosynthesis of starch in spinach leaf chloroplasts. *In* The Biochemistry of Chloroplasts, Vol II. T. W. Goodwin (ed.). London: Academic Press.

Sanwal, G. G., Greenberg, E., Hardie, J., Cameron, E. C., and Preiss, J. 1968. Regulation of starch biosynthesis in plant leaves: Activation and inhibition of ADP-glucose pyrophosphorylase. Plant Physiol. 43:417–427.

Schürmann, P., and Buchanan, B. B. 1975. Role of feredoxin in the activation of sedoheptulose diphosphatase in isolated chloroplasts. Biochim. Biophys. Acta 376:189–192.

Schürmann, P., Wolosiuk, R. A., Breazeale, V. D., and Buchanan, B. B. 1976. Two proteins function in the regulation of photosynthetic CO$_2$ assimilation in chloroplasts. Nature 263:257–258.

Stanier, R. Y., and Cohen-Bazier, G. 1977. Prokaryotes: The cyanobacteria. Annu. Rev. Microbiol. 31:225–274.

Tolbert, N. W. 1971. Microbodies, perioxisomes and glyoxysomes. Annu. Rev. Plant Physiol. 22:45–74.

Walker, D. A. 1976. Plastids and intracellular transport. *In* Encyclopedia of Plant Physiology. Vol. 3, Transport in Plants. C. R. Stocking and U. Heber (eds.). Berlin and New York: Springer-Verlag.

Withers, N. W., Alberti, R. S., Lewin, R. A., Thornber, J. D., Britton, G., and Goodwin, R. W. 1978. Photosynthetic unit size carotenoids, and chlorophyll-protein composition of *Prochloron* *sp.*, a procaryotic green algae. Proc. Nat. Acad. Sci. USA 75:2301–2305.

Discussion of Presentation by Dr. Bassham

SCHMIDT: There are two thioredoxins in spinach chloroplasts (Crawford et al., 1979) and two thioredoxins in cyanobacteria (Schmidt, 1979a). Fructose bisphophatase in cyanobacteria is stimulated by thioredoxin (Schmidt, 1979b).

HALLICK: You spoke of the oxygenase activity of ribulose bisphosphate carboxylase-oxygenase as a "latter day" phenomenon. Do you view this reaction as gratuitous to the chloroplast?

BASSHAM: It is often thought to be. We known of no metabolic requirement for this reaction unless it is to handle "excess" reducing power. Glycolate can provide carbon for glycine synthesis, but glycine can also be made by other pathways, and is, when the CO_2 level is increased and glycolate is no longer formed.

COHEN: When we think of the structures of the sugar phosphates during the oxidative and reductive pentose cycles it is clear that there are sharp changes from cyclic to straight-chain structures. The key role of ketopentose phosphates must relate to the straight acyclic aspect of these structures (Lim and Cohen, 1966). It might be asked if acyclic hexoses might affect these cycles and reactions in interesting ways. Thus 5-0 methyl ketohexoses and their phosphates, for example, 5-0 methylfructose-6-phosphate, might affect reactions of the ketopentose-5-phosphates. Since 2-deoxyhexoses and 2-deoxypentoses exist in considerable measure as the acyclic sugars it would be interesting to test the effects of their 6- and 5-phosphates respectively on many of the reactions of the cycles.

BASSHAM: I agree.

STANIER: These so-called "wasteful" reactions (i.e., the role of ribulose biphosphate carboxylase as oxygenase and the role of nitrogenase as hydrogenase) are perhaps *chemically* inevitable. Thus no possible device of natural selection could overcome this disability.

COHEN: The inhibition of glucose-6-phosphate dehydrogenase by ribulose bisphosphate suggests that the product of the phosphogluconate lactonase might also be an inhibitor and controlling element for glucose-6-phosphate dehydrogenase. There has been little discussion of the lactonase and its possible role in regulating the oxidative cycle and in producing a possible inhibitor and controlling element for the cycle. Can you tell us anything about this step in the dark reactions of photosynthetic organisms?

BASSHAM: The conversion of glucose-6-phosphate and $NADP^+$ to 6-phosphogluconolactone and NADPH is reversible, the hydrolysis of the lactone is the rate-limiting step, and therefore the expected site of regulation. We don't see the lactone by our paper chromatographic analysis, but it is reasonable, as you suggest, that similarities of structure between ribulose bisphosphate and 6-phosphogluconolactone could be responsible for the inhibition of glucose-6-phosphate dehydrogenase by ribulose bisphosphate as I discussed.

WILDMAN: As an extension of your statement that phosphoribulose kinase appears only in company with ribulose bisphosphate carboxylase: In contrast to the latter, where coding information is divided between chloroplast and nuclear DNA, Dr. Takashi Kagawa has obtained evidence that phosphoribulose kinase is coded exclusively by nuclear DNA.

BASSHAM: Thank you.

VON WETTSTEIN: Regarding the carboxylase and oxygenase activities of ribulose bisphosphate carboxylase, it is of interest that in the evolution of higher plants restriction of photorespiration has been accomplished by banning the enzyme from the mesophyll cells and placing it in the bundle sheath cells, and by evolving mechanisms to insure a high CO_2 concentration around the enzyme. On the other hand, evolution of the enzyme to segregate the two activities in separate molecules is not known. Do you know if the oxygenase and carboxylase activities are carried out by the same catalytic sites, that is, by the same amino acid sequence?

BASSHAM: Most evidence suggests that it is the same catalytic site although there is one report to the contrary.

SCHIFF: The Calvin cycle is so widespread and takes the same form even in nonphotosynthetic organisms such as the chemosynthetic bacteria. This suggests that this cycle is very primitive or, perhaps, that the ways of effectively reducing CO_2 are very few.

BASSHAM: Yes. I neglected to mention that the evolution of the Calvin cycle could have first occurred either in the photosynthetic bacteria, or in chemosynthetic bacteria. The Calvin cycle is thermodynamically efficient; of the energy supplied as ATP and NADPH, 85% is stored in the reduction of CO_2 to sugar phosphates.

SCHIFF: People have suggested that cyanobacteria lack metabolic regulation. However, since most of these organisms are obligate phototrophs there would be little evolutionary demand for end product regulation in response to substrates entering the cells. Thus, it is heartening that you find a regulation in the CO_2 reduction cycle of cyanobacteria similar to that in higher plant chloroplasts. It's not that the cyanobacteria lack metabolic regulation, as some have suggested, but rather that as obligate phototrophs there was no selective advantage for them to have that part of regulation associated with utilization of external substrates.

BASSHAM: I agree.

SCHIFF: Pigiet and Conley, (1978) have found that a large proportion of the thioredoxin of *E. coli* is phosphorylated at one of the sulfur atoms. If this is true for the chloroplast, phosphate may control the availability of thioredoxin for reductive regulation, that is, there may be dual regulation by ATP/phosphate and reducing power through thioredoxin.

BASSHAM: The regulation of fructose bisphosphatase and sedoheptulose bisphosphatase by phosphate, via the mechanism you suggest, would be useful in providing a balance between rates of carboxylation and rates of conversion of hexose/haptose bisphosphates to monophosphates, when rates of triose phosphate export compared to starch synthesis change.

MUDD: Is there a role for reduced glutathione in the regulation of enzymes of CO_2 metabolism?

SCHIFF: Glutathione, NADPH, and glutathione reductase can serve to reduce thioredoxin in place of NADPH and thioredoxin reductase in *E. coli* (Tsang and Schiff, 1978); in the chloroplast thioredoxin is reduced by system I. Perhaps glutathione serves as an alternative means of reducing thioredoxin in chloroplasts.

References to Discussion

Crawford, N. A., Yee, B. C., Hishizawa, A. N., and Buchanan, B. B. 1979. Occurrence of cytoplasmatic *f*- and m-type thioredoxins in leaves. FEBS Lett. 104:141–145.

Lim, R., and Cohen, S. S. 1966. D-Phosphoarabinoisomerase and D-ribulokinase in *Escherichia coli*. J. Biol. Chem. 241:4304–4315.

Pigiet, V., and Conley, R. R. 1978. Isolation and characterization of phosphothioredoxin from *Escherichia coli*. J. Biol. Chem. 253:1910–1920.

Schmidt, A. 1979a. Different types of sulfotransferases in cyanobacteria. Third International Symposium on Photosynthetic Prokaryotes (abs.). Oxford, B36.

Schmidt, A. 1979b. A thioredoxin-activated fructose bisphosphatase in the cyanobacterium *Synechococcus* 6301. Plant Physiol. 63:S-11.

Tsang, M. L. S., and Schiff, J. A. 1978. Assimilatory sulfate reduction in a mutant of *Escherichia coli* lacking thioredoxin activity. J. Bact. 134:131–138.

Lipid Metabolism

J. B. Mudd

Glycerolipid Composition

The main glycerolipids of the chloroplast thylakoids are monogalactosyldiacylglycerol (MGDG), digalactosyldiacylglycerol (DGDG), phosphatidylglycerol (PG), and sulfoquinovosyldiacyglycerol (SL) (Table 1). These lipids are also found in the chloroplast envelope, but the envelope contains a higher proportion of phospholipids, particularly phosphatidycholine (PC), and a notably high content of DG (diacylglycerol). The overall lipid composition of the chloroplast reflects the composition of the thylakoid, since the envelope contributes relatively little mass.

The glycerolipid composition of cyanobacteria (blue-green algae) and photosynthetic bacteria shows the former to be very similar to the thylakoid of the higher plant chloroplast (Nichols et al., 1965; Allen et al., 1966). Thin-layer chromatograms of total lipids from cyanobacteria reveal negligible amounts of phospholipids other than PG. Even when the heterocyst envelope lipids are examined there is no indication of the presence of phospholipids characteristic of the extrachloroplastic compartment in higher plants such as phosphatidylcholine (PC) or phosphatidylethanolamine (PE) (Winkenbach et al., 1972).

Abbreviations: ACP, acyl carrier protein; BCCP, biotin carboxyl carrier protein; CDP-DG, cytidinediphosphodiacylglycerol; DG, diacylglycerol; DGDG, digalactosyldiacylglycerol; G3P, sn-glycerol-3-phosphate; MGDG, monogalactosyldiacylglycerol; PC, phosphatidylcholine; PE, phosphatidylethanolamine; PG, phosphatidylglycerol; PGP, phosphatidylglycerolphosphate; PI, phosphatidylinositol; SL, sulfoquinovosyldiacylglycerol; TGDG, trigalactosyldiacylglycerol; UDP-gal, uridinediphosphogalactose.

Address reprint requests to Dr. J. B. Mudd, Professor, Department of Biochemistry, University of California, Riverside, California 92521.

Table 1. Glycerolipid Composition

Lipid Composition of Spinach Chloroplasts

		Lipid (μg/mg) protein in	
		Envelope	Thylakoid
Monogalactosyldiacylglycerol	(MGDG)	300	210
Digalactosyldiacylglycerol	(DGDG)	380	110
Trigalactosyldiacylglycerol	(TGDG)	60	0
Sulfoquinovosyldiacylglycerol	(SL)	80	30
Phosphatidylcholine	(PC)	240	20
Phosphatidylglycerol	(PG)	100	40
Diacylglycerol	(DG)	200	0
Chlorophyll		0	147
Carotenoids		8.2	27

Comparison of Lipid Composition of Spinach Chloroplast Thylakoids and *Anacystis nidulans*

		Molar ratio relative to SL = 10	
		Thylakoids	*A. nidulans*
Monogalactosyldiacylglycerol	(MGDG)	36	32
Digalactosyldiacylglycerol	(DGDG)	20	12
Sulfoquinovosyldiacylglycerol	(SL)	10	10
Phosphatidylglycerol	(PG)	14	9
Phosphatidylcholine	(PC)	5	—
Phosphatidylinositol	(PI)	3	—

Source: Data from Joyard and Douce (1976); and from Allen et al. (1966).

The photosynthetic bacteria sometimes have glycerolipids in common with cyanobacteria and higher plant chloroplasts. *Chloropseudomonas ethylicum* contains MGDG as well as other glycolipids (Constantopoulos and Bloch, 1967). Although glycosyldiacylglycerols are widely distributed in nonphotosynthetic bacteria, their structures do not include the galactosyl diacylglycerols found in cyanobacteria and higher plant chloroplasts (Shaw, 1970). The presence of SL in *Rhodopseudomonas spheroides* has been reported by Wood et al. (1965) and by Radunz (1969). In the case of *Rhodospirillum rubrum,* Benson et al. (1959) reported SL to be present, but Wood et al. (1965) could not find SL in this organism. PG is widely distributed in bacteria and in higher and lower plants: it is always present in photosynthetic bacteria, cyanobacteria, and chloroplasts.

On the basis of glycerolipid composition, the chloroplast is very much more comparable to the cyanobacteria than to the photosynthetic bacteria.

Fatty Acid Composition

The fatty acid composition of glycerolipids from chloroplasts is given in Table 2. It has been found that the galactosyldiacylglycerols are more unsaturated than PG and SL, that PG contains an unusual fatty acid 16:1[3t] at the sn 2 position, and that the fatty acid distribution of SL is unusual in that saturated fatty acids are predominant at the sn 2 position.

The fatty acids of *Anabaena variabilis* (Table 2) show some similarities with those from higher plant chloroplasts; the two galactosyl diacylglycerols are similar and are relatively unsaturated, whereas PG and SL have somewhat similar fatty acid compositions.

The unicellular cyanobacteria have relatively low amounts of polyunsaturated fatty acids (Kenyon, 1972), whereas the filamentous forms have large amounts of both 18:2 and 18:3 (Kenyon et al., 1972). In both cases, however, two types of 18:3 are found: α-linolenic acid (18:3[9,12,15]) and γ-linolenic acid (18:3[6,9,12]). The latter acid is not usually found in higher plants. It is found in *Euglena,* and it is the usual product of 18:2[9,12] desaturation in animal tissues. The 18:4 fatty acid also found in some filamentous cyanobacteria has unusual double-bond positions: 6,9,12,15.

A strong contrast is seen in the fatty acid composition of photosynthetic bacteria (Wood et al., 1965) since they do not contain any polyunsaturated fatty acids.

Table 2. Comparison of Chloroplast Thylakoid and Cyanobacterial Fatty Acid Composition

		\multicolumn{8}{c}{Fatty acid}
		14:0	16:0	16:1	16:3	18:0	18:1	18:2	18:3
Chloroplast thylakoids (Allen et al. 1966)									
monogalactosyldiacylglycerol	(MGDG)	—	—	—	25	—	—	2	72
digalactosyldiacylglycerol	(DGDG)	—	3	—	5	—	2	2	87
phosphatidylglycerol	(PG)	1	11	32	2	—	2	4	47
sulfoquinovosyldiacylglycerol	(SL)	—	30	—	—	—	—	6	52
Anacystis nidulans (Nichols et al. 1965)									
monogalactosyldiacylglycerol	(MGDG)	1.2	42.6	33.6	—	4.0	20.4	—	—
digalactosyldiacylglycerol	(DGDG)	1.7	51.8	28.3	—	3.6	16.2	—	—
phosphatidylglycerol	(PG)	8.0	35.2	17.3	—	7.5	22.0	—	—
sulfoquinovosyldiacylglycerol	(SL)	4.5	43.1	28.2	—	4.2	16.2	—	—
Anabaena variabilis (Nichols et al. 1965)									
monogalactosyldiacylglycerol	(MGDG)	—	27.1	27.5	—	—	12.0	18.7	14.7
digalactosyldiacylglycerol	(DGDG)	—	26.4	26.0	—	2.1	9.0	20.4	17.0
phosphatidylglycerol	(PG)	7.5	31.0	8.9	—	5.9	21.1	11.5	6.3
sulfoquinovosyldiacylglycerol	(SL)	—	52.7	5.3	—	2.3	18.7	13.7	7.1

Fatty Acid Biosynthesis

Aspects of fatty acid biosynthesis in the chloroplast are summarized in Figure 1.

Sources of Carbon for Fatty Acid Synthesis

The most commonly used substrate for the study of fatty acid synthesis by isolated chloroplasts is acetate. If acetate is the physiological substrate, its origin is not obvious. Alternatives such as pyruvate have been considered. It has been

Figure 1. Schematic representation of fatty acid synthesis in plastids.

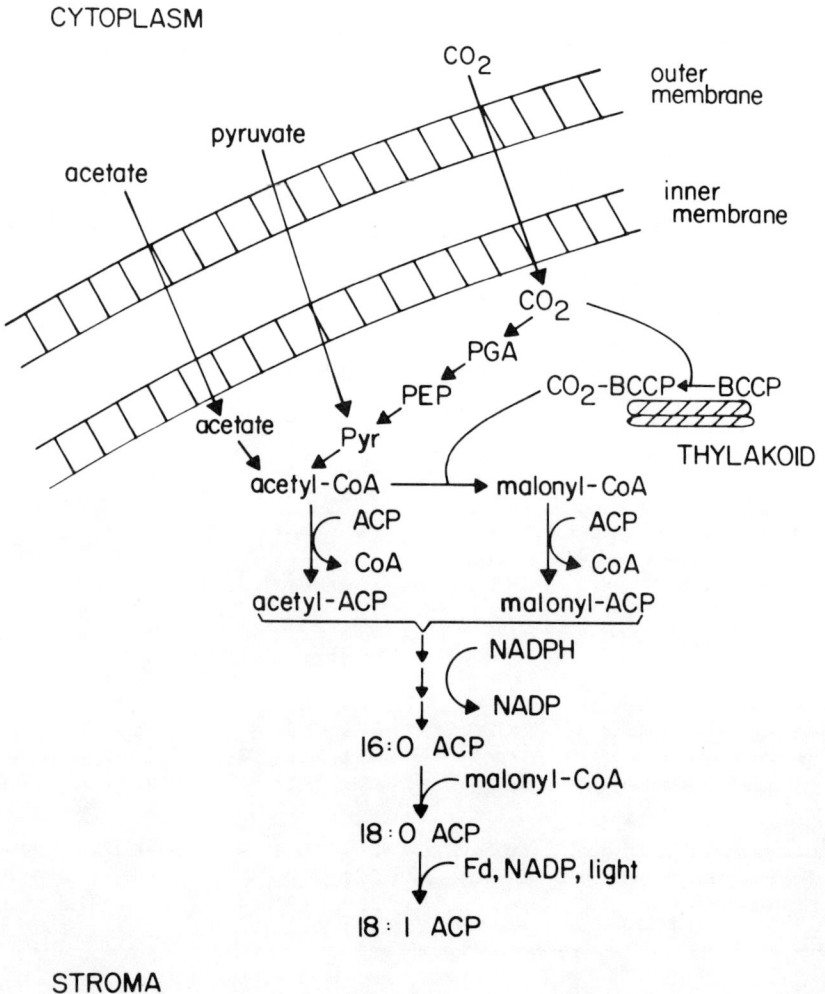

found that fatty acid synthesis from ^{14}C-pyruvate must be carefully checked for ^{14}C-acetate contamination; direct comparison of acetate, pyruvate, and malonate showed the former to be the most efficient substrate (Roughan et al., 1980). Murphy and Leech (1977, 1978) and McKee and Hawke (1978) have reported the incorporation of $H^{14}CO_3^-$ into fatty acids of isolated chloroplasts. The pathway for CO_2 fixation is suggested in Figure 1, although the evidence for the steps to acetyl CoA is not absolutely convincing. (See chapter by Lea et al. in this volume, Assimilation of Nitrogen and Synthesis of Amino Acids in Chloroplasts and Cyanobacteria (Blue-Green Algae), for the requirement of chloroplasts for pyruvate for amino acid synthesis).

This subject has not been well studied in cyanobacteria. Exogenous acetate is incorporated into fatty acids. While the chloroplast may depend on the cytoplasmic compartment for the supply of precursors for fatty acid synthesis, cyanobacteria cannot function analogously; photosynthetically fixed CO_2 must be converted into fatty acid precursors in the compartment surrounding the photosynthetically active membranes.

Carboxylation of Acetyl CoA

The formation of malonyl CoA from acetyl CoA in chloroplasts is comparable to the process in prokaryotes. Three proteins are necessary: a biotin carboxyl carrier protein (BCCP), a BCCP carboxylase, and a transcarboxylase which transfers CO_2 from BCCP to acetyl CoA. The BCCP is bound to the thylakoids whereas the other two proteins are present in the stroma. Acetyl CoA and malonyl CoA are the starting materials for fatty acid synthesis.

Kannangara and Stumpf (1973) have examined chloroplasts from higher plants, cyanobacteria, and photosynthetic bacteria for the distribution of the BCCP. They found the biotin of *Rhodospirillium rubrum* to be soluble as it is in *E. coli,* whereas the biotin in chloroplasts from spinach or *Chlamydomonas reinhardi* was bound to thylakoids. The filamentous cyanobacterium *Anabaena flos-aquae* was similar to the chloroplast, whereas in unicellular *Anacystis nidulans (Synechococcus)* the biotin was about equally distributed between soluble and membrane-bound forms. These observations again suggest that chloroplasts are closer to cyanobacteria than to photosynthetic bacteria.

Acyl Carrier Protein Reactions

Fatty acid synthesis by plastid preparations can be greatly stimulated by the addition of acyl carrier protein (ACP). During the preparation of plastids it seems to be easy to lose ACP and lower the concentration in the plastid (Brooks and Stumpf, 1965). It is now apparent that all of the ACP of a plant leaf is in the plastid (Ohlrogge et al., 1979). It is most probable that acetyl ACP and malonyl ACP are used as precursors and all steps of condenstion, reduction and dehydration take place using ACP derivatives. The major product appears to be 16:0

(palmitoyl)-ACP. Enzyme preparations from plastids are capable of elongating 16:0-ACP to 18:0-ACP.

It is notable that organotrophically grown *Euglena* contains a fatty acid synthetase of the multienzyme type, comparable to that found in animals and fungi, but when grown phototrophically the plastid develops an ACP-dependent fatty acid synthetase comparable to that in bacteria or chloroplasts of higher plants. (Delo et al., 1971; Ernst-Fonberg and Bloch, 1971). There is no evidence of a multienzyme fatty acid synthetase in the cytoplasmic compartment of higher plants.

The type of fatty acid synthetase in cyanobacteria has not been described. One can confidently predict that it will be an ACP-dependent system with separable, soluble enzymes catalyzing independent steps of the synthesis.

Desaturation of 18:0 Acyl Carrier Protein

Plastids contain the machinery for converting 18:0 ACP to 18:1 ACP (Jaworski and Stumpf, 1974). Ferredoxin is required and NADPH can serve as the reductant for the soluble enzyme system. However, the most efficient reducing system is ferredoxin, chloroplast lamellae, ascorbic acid, dichlorophenolindophenol, and light. This system is comparable to that reported in *Euglena* grown phototrophically (Nagai and Bloch, 1965). When *Euglena* is grown organotrophically it contains a desaturase system which uses acyl CoA as substrate (Nagai and Bloch, 1965).

Cyanobacteria incorporate ^{14}C-acetate into fatty acids, including the monounsaturated 16:1 and 18:1 molecules. They also have the capability of desaturating 16:0 and 18:0 fatty acids supplied exogenously (Nichols et al., 1965). The latter activity distinguishes them from the photosynthetic bacteria, which form monounsaturated fatty acids by introducing the double bond during the elongation procedure; the final position of the double bond depends on the degree of elongation (Wood et al., 1965). The substrate for monodesaturation in cyanobacteria is not known. The exogenously supplied fatty acid is presumably activated (via CoA and ACP) but it is not known whether desaturation takes place using one of these activated forms as substrate.

Formation of Polyunsaturated Fatty Acids

Desaturation of 18:1 → 18:2 → 18:3 is still not elucidated. The pathway 12:3 → 14:3 → 16:3 → 18:3 proposed earlier (Jacobson et al., 1973) probably accounts for a very small proportion of 18:3 synthesized (Murphy and Stumpf, 1979). The conversion of 18:1 to 18:2 has been proposed to occur while the fatty acids are attached to the glycerol backbone of phosphatidylcholine (Gurr et al., 1969), but this is a property of the microsomes. We have found that fatty acids synthesized by chloroplasts can be desaturated to 18:2 and 18:3 provided that galactolipids are being synthesized (Roughan et al., 1979). It appears possible that formation of polyunsaturated fatty acids takes place, at least to some extent, when the acyl chains are attached to the glycerol backbone of galactosyl diacylglycerol.

The fatty acid composition of cyanobacteria is consistent with the scheme of desaturation proposed by Kenyon et al. (1972).

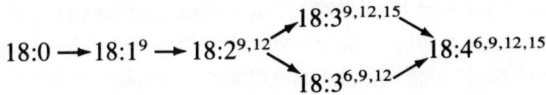

$$18:0 \longrightarrow 18:1^9 \longrightarrow 18:2^{9,12} \underset{18:3^{6,9,12}}{\overset{18:3^{9,12,15}}{\diagdown}} 18:4^{6,9,12,15}$$

The $\Delta 6$-desaturase is a feature unfamiliar to investigators of desaturation in higher plants. The substrate for formation of polyunsaturated fatty acids is not known. Clearly PC cannot be a substrate as it is in some cases in higher plants. When ^{14}C-acetate is supplied to cyanobacteria the fatty acids formed are rapidly found in the galactolipids. It seems likely that galactolipids are the substrates for at least some desaturation reactions in cyanobacteria as they are in green algae (Nichols et al., 1967) and in chloroplasts of higher plants (Roughan et al., 1979).

Supply of Chloroplast Fatty Acids to the Cytoplasm

It has been suggested that the sole site of fatty acid synthesis in the plant cell is the plastid. Certainly there can be no ACP-dependent fatty acid synthetase in the cytoplasm (Ohlrogge et al., 1979), and there is so far no good evidence of a multifunctional enzyme such as that found in yeast, rat liver, or pigeon liver. If fatty acids are transported to the cytoplasm, the question arises as to how this is accomplished. The chloroplast membrane(s) has (have) a very active acyl CoA synthetase whose function is not known (Roughan and Slack, 1977). Perhaps this enzyme is involved in fatty acid transport. The chloroplast membrane(s) is (are) rich in diacylglycerol (Joyard and Douce, 1976b). Possibly this is the form in which acyl groups are transferred to the cytoplasm. The apparent dependence of the extra chloroplastic compartment on the plastid for fatty acids is an unusual event since we are accustomed to the dependence of the chloroplast on the cytoplasm for much in the way of polypeptides; in only a few authenticated cases does it supply its own needs.

Acylation of Glycerol-3-phosphate

The acylation of glycerol-3-phosphate (G3P) is readily measured in plant tissue. The product, phosphatidic acid, is presumed to be the precursor of the chloroplast glycerolipids.

Sources of Glycerol-3-phosphate

In animal tissues, dihydroxyacetonephosphate (DHAP) is converted to G3P by an NADH-dependent reductase. In some cases DHAP is acylated before reduction and the second acylation. There is no evidence that DHAP is acylated in plant tissue. There is difficulty in measuring activity of G3P dehydrogenase—so much so that Heinz (1977) suggests that G3P is formed in plants by phosphorylation of glycerol. The G3P dehydrogenase has recently been purified from plant tissue (Santora et al., 1979). The enzyme was found in the cytoplasm and was highly specific for NAD(H) and DHAP/G3P. It seems certain that photosynthetically fixed CO_2 is incorporated into the glycerol backbone of chloroplast glycerolipids by way of triosephosphates, but the exact mechanism has to be shown. Formation of G3P is a potentially useful regulation point for lipid synthesis.

Characteristics of Glycerol-3-phosphate Acylation

The acylation of G3P is actively catalyzed by the microsomal fraction of plant cells (Cheniae, 1965), and to a significant extent by both inner and outer mitochondrial membranes (Sparace and Moore, 1979). These systems utilize CoA derivatives of fatty acids, and the characteristics of the enzyme systems are closely comparable to those from animals.

The acylation of G3P by enzyme preparations from purified chloroplasts has been studied (Joyard and Douce, 1977). Acyl-CoA was used to acylate G3P, and this does not seem an ideal choice since the newly formed fatty acids exist as the acyl-ACP compounds. It is conceivable that acyl-CoA is formed for the acylation of G3P; in bacteria where acyl-ACP is also the product of fatty acid synthesis, ACP and CoA derivatives are used equally well for the acylation of G3P. Shine et al. (1976) have compared acyl-CoA and acyl-ACP as precursors for lipid synthesis. Avocado fruit microsomes used acyl-CoA more efficiently than acyl-ACP. Grana from spinach chloroplasts also used acyl-CoA more efficiently than acyl-ACP. It should be noted that these experiments probably did not measure synthesis de novo, but rather the acylation of endogenous lysoglycerolipids.

Subchloroplastic Localization

Joyard and Douce (1977) reported that the first acylation of G3P by chloroplast preparations depended on enzymes in the stroma. The second acylation depended on enzyme activities associated with the chloroplast envelope. The fatty acid specificity for the two steps and the positional specificity of fatty acid insertion was not described. It is known that diacylglycerols made by the chloroplast from acetate have 16:0 predominately at position 2 and 18:0 or 18:1 predominately at position 1 (McKee and Hawke, 1979; Roughan et al., 1980). The existence of a soluble G3P acyltransferase is without precedent. In both mitochondria and microsomes this activity is membrane bound. Since the G3P acyltransferase from microsomes can be solubilized, and since chloroplast membranes are rich in diacylglycerol (possibly indicating lipid degradation during isolation), one can still entertain the possibility that the first acylation step is a function of an enzyme attached to the chloroplast envelope.

Galactolipid Biosynthesis

Neufeld and Hall (1964) first demonstrated that chloroplasts formed galactosyl-diacylglycerol from uridinediphosphogalactose (UDP-gal). These reactions have since become one of the best studied aspects of plant lipid metabolism. The cooperation between chloroplast and cytoplasm required for galactolipid syn-thesis is shown in Figure 2.

Membrane Localization

Studies of the subcellular location of galactolipid biosynthesis showed high activity in the chloroplast fraction, but there was always considerable activity in the 100,000g supernatant fraction. In retrospect, this activity could be attrib-uted to fragments of the chloroplast envelope. Douce (1974) has shown that the

Figure 2. Schematic representation of the biosynthesis of galactolipids in plastids.

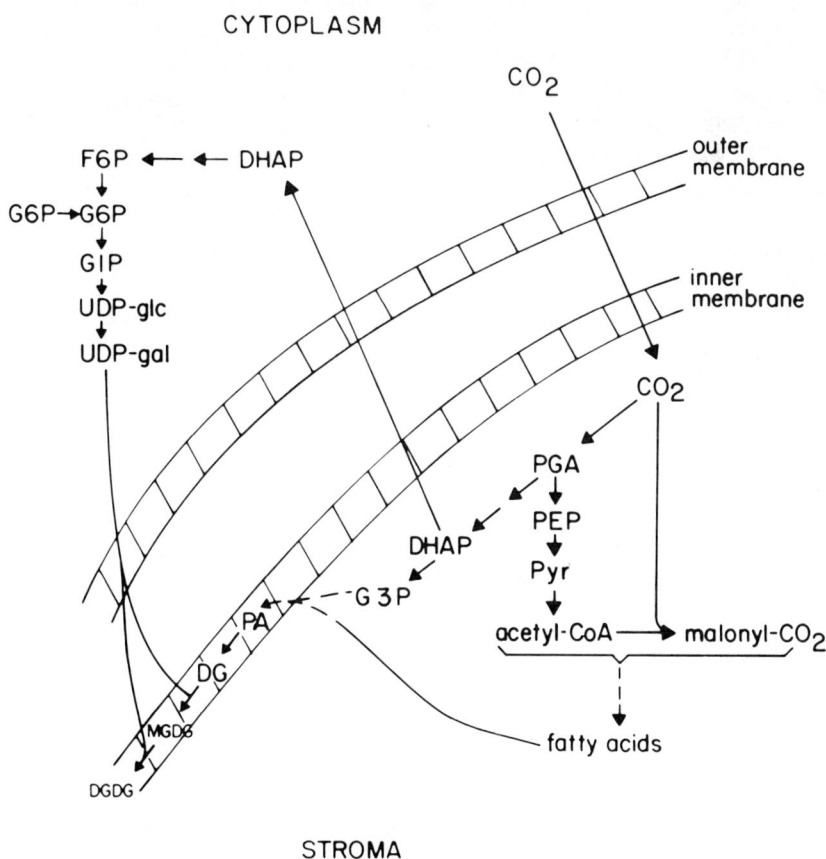

chloroplast envelope has a very active UDP-galactose:diacylglycerol galactose transferase. It is notable that the chloroplast envelope is well endowed with the diacylglycerol substrate and this may be oriented in a fashion favorable for reaction. The chloroplast fraction catalyzes the synthesis of galactosyldiacylglycerol independent of added diacylglycerol, and can only be made dependent on this substrate by extraction with acetone (Mudd et al., 1969). At this point, vast excesses of DG must be added; there clearly are problems in efficient presentation of lipid substrates to the enzyme. One may reserve final judgment on the localization of galactolipid synthesis until DG has been presented to all subcellular (subchloroplastic) fractions in an efficient manner.

Source of Diacylglycerol

What is the biosynthetic origin of DG used in MGDG and DGDG synthesis? DG synthesized by chloroplasts from ^{14}C-acetate is quite saturated whereas endogenous galactolipids are highly unsaturated. It has been suggested that the fatty acids synthesized in the chloroplast are processed in the extrachloroplastic compartment (i.e., to the level of 18:2) before being returned to the chloroplast. This suggestion stemmed from the lack of demonstration of fatty acid desaturation in the chloroplast. Fatty acid desaturation in the chloroplast can take place provided that conditions are appropriate for galactolipid biosynthesis (Roughan et al., 1979). It is therefore conceivable that the chloroplast is autonomous as far as the diacylglycerol moiety is concerned. However, labeling in vivo from $^{14}CO_2$ shows fatty acids of phosphatidylcholine (PC) and PG becoming radioactive much sooner than MGDG and digalactosyldiacylglycerol (DGDG), suggesting a DG pool which is not readily available for galactosylation (Williams et al., 1976). It is impossible to decide at the moment whether the DG moiety used in galactolipid synthesis originates in the chloroplast or in the cytosol compartment.

The high degree of unsaturation in MGDG could arise in a number of ways: (a) MGDG is an excellent substrate for fatty acid desaturation; (b) galactosylation is specific for highly unsaturated DG; or (c) galactosylation is not specific, but the DG pool is highly unsaturated. Evidence for (a) is presented by Roughan et al. (1979). Evidence for (b) has been presented (Mudd et al., 1969; Joyard and Douce, 1976a); however, one report found no specificity for the DG moiety (Eccleshall and Hawke, 1971).

Source of Uridine Diphosphate Galactose

Heinz and co-workers have studied the reactions leading to UDP-gal (Königs and Heinz, 1974). The most likely pathway is the conversion of glucose-1-phosphate to UDP-glc and conversion to UDP-gal by action of UDP-glucose-4-epimerase. Both of these enzymes are localized in the cytoplasmic compartment.

The UDP-gal must move to the chloroplast membrane where the second substrate, DG, is present, having arrived by an unknown pathway. Thus, an important precursor for the predominant lipids of the chloroplast is synthesized outside of the chloroplast.

The Second Galactosylation

It is generally concluded that MGDG is converted to DGDG by reaction with UDP-gal. Siebertz and Heinz (1977) have studied this reaction with respect to the specificity of the fatty acid composition of MGDG (galactosylation is not affected by the fatty acid composition). On the other hand van Besouw and Wintermans (1978) suggest that two molecules of MGDG react to form DGDG and DG.

Formation of Galactolipids in Cyanobacteria

Nichols et al. (1965) have noted the incorporation of long-chain fatty acids, synthesized from ^{14}C-acetate, into MGDG of cyanobacteria. The labeling of MGDG was more rapid than in *Chlorella vulgaris* or in plant leaves. Further examination of galactolipid biosynthesis in cyanobacteria awaits further research. One may expect the presence of the reaction of UDP-gal with DG, and it is clear that the cooperation between chloroplast and cytoplasm found in higher plants cannot be operative in the cyanobacteria.

The fatty acid composition and distribution at the 1- and 2-positions of sn-glycerol may provide ideas about the biosynthetic pathway of galactosyldiacyl-glycerols. Safford and Nichols (1970) made such analyses for cyanobacteria, green algae, and higher plants. The striking features of the cyanobacteria were the predominance of 16C acids at position 2 and 18C acids at position 1. The same was true of *Chlorella vulgaris* grown organotrophically, less so for those grown phototrophically, and not true for higher plants. The predominance of 16C acids at position 2 is characteristic of PG and SL in higher plants. There may be specificity for chain length in the first acylation reactions of sn G3P both in higher plants and in cyanobacteria. In the case of higher plants the galactosyldiacylglycerols can be further modified by deacylation-reacylation at position 2.

Sulfoquinovosyldiacylglycerol and Phosphatidylglycerol

There is very little known about the biosynthesis of these two compounds in the chloroplast and really nothing known about their biosynthesis in cyanobacteria. Possible biosynthetic relationships are shown in Figure 3.

CYTOPLASM

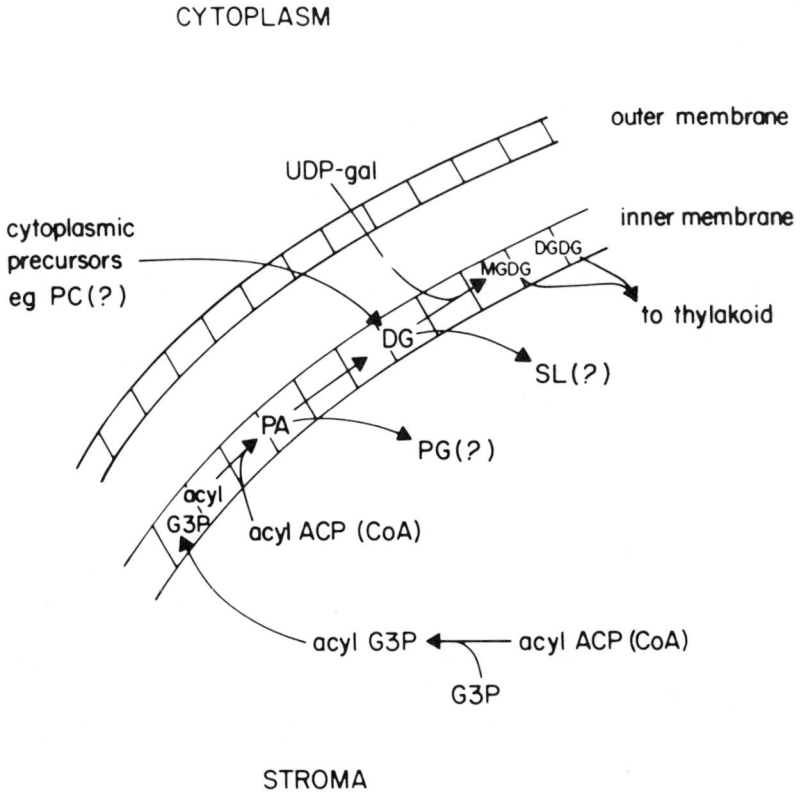

Figure 3. Schematic representation of possible routes of PG and SL in plastids.

Phosphatidylglycerol Biosynthesis

The pathway of biosynthesis of this compound in higher plants is:

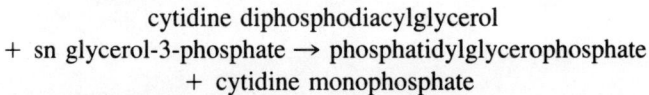

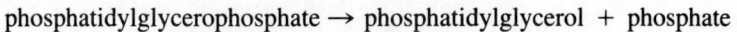

cytidine diphosphodiacylglycerol
+ sn glycerol-3-phosphate → phosphatidylglycerophosphate
+ cytidine monophosphate

phosphatidylglycerophosphate → phosphatidylglycerol + phosphate

The reaction is well authenticated in endoplasmic reticulum (Marshall and Kates, 1972) and in mitochondria (Douce and Dupont, 1969). However, attempts to measure such a reaction in chloroplasts have failed. It is conceivable that a sequence of reactions other than that above operates in the chloroplast. In any case it would be interesting to see if the sequence operates in the cyanobacteria.

Sulfoquinovosyldiacylglycerol

The most sophisticated precursor used for the study of SL biosynthesis is cysteic acid. In *Euglena* the label from both ^{35}S- and 3-^{14}C-cysteic acid was found in the sulfoquinovose moiety of SL, and it was proposed that the carbon-sulfur skeleton was intact after incorporation (Davies et al., 1966). Evidence obtained in higher plants is in agreement with this conclusion (Harwood, 1975). The hypothetical intermediates would be analogous to those of the gluconeogenic pathway, finally forming UDP-sulfoquinovose. Although the presence of UDP-sulfoquinovose has been reported in *Chlorella ellipsoidea* (Shibuya et al., 1963), there is no proof that it is the precursor of SL.

In both cyanobacteria and in chloroplasts of higher plants, the fatty acid compositions of PG and SL are similar, while those of MGDG and DGDG are similar. These results suggest that the diacylglycerol moiety of the galactolipids might have a different origin or different metabolic treatment from that of PG and SL. It seems possible that the study of these problems in cyanobacteria may provide useful insights for the further study of this process in chloroplasts.

References

Allen, C. F., Hirayama, O., and Good, P. 1966. Lipid composition of photosynthetic systems. In The Biochemistry of Chloroplasts, Vol. 1. p 195–200 T. W. Goodwin (ed.). London, New York: Academic Press.

Benson, A. A., Daniel, H., and Wiser, R. 1959. A sulfolipid in plants. Proc. Nat. Acad. Sci. USA 45:1582–1587.

Brooks, J. L., and Stumpf, P. K. 1965. A soluble fatty acid synthesizing system from lettuce chloroplasts. Biochim. Biophys. Acta 98:213–216.

Cheniae, G. M. 1965. Phosphatidic acid and glyceride synthesis by particles from spinach leaves. Plant Physiol. 40:235–263.

Constantopoulos, G., and Bloch, K. 1967. Isolation and characterization of glycolipids from some photosynthetic bacteria. J. Bact. 93:1788–1793.

Davies, W. H., Mercer, E. I., and Goodwin, T. W. 1966. Some observations on the biosynthesis of the plant sulpholipid by *Euglena gracilis*. Biochem. J. 98:369–373.

Delo, J., Ernst-Fonberg, M. L. and Bloch, K. 1971. Fatty acid synthetases from *Euglena gracilis*. Arch. Biochem. Biophys. 143:384–391.

Douce, R. 1974. Site of biosynthesis of galactolipids in spinach chloroplasts. Science 183:852–853.

Douce, R., and Dupont, J. 1969. Biosynthèse du phosphatidylglycérol dans les mito-chondries végétales isolées: mise en évidence du phosphatidylglycérol-phosphate. C. R. Acad. Sci. (Paris) 268D:1657–1660.

Eccleshall, T. R., and Hawke, J. C. 1971. Biosynthesis of monogalactosyl diglyceride by chloroplasts from *Spinacia oleracea* and from some *Gramineae*. Phytochemistry 10:3035–3045.

Ernst-Fonberg, M. L., and Bloch, K. 1971. A chloroplast-associated fatty acid synthetase system in *Euglena*. Arch. Biochem. Biophys. 143:392–400.

144   J. B. Mudd

Gurr, M. I., Robinson, M. P., and James, A. T. 1969. The mechanism of formation of polyunsaturated fatty acids by photosynthetic tissue. The tight coupling of oleate desaturation with phospholipid synthesis in *Chlorella pyrenoidosa*. Eur. J. Biochem. 9:70–78.

Harwood, J. L. 1975. Synthesis of sulfoquinovosyl diacylglycerol by higher plants. Biochim. Biophys. Acta. 398:224–230.

Heinz, E. 1977. Enzymatic reactions in galactolipid biosynthesis. In Lipids and Lipid Polymers in Higher Plants. M. Tevini and H. K. Lichtenthaler (eds.). Berlin, Heidelberg: Springer-Verlag.

Jacobson, B. S., Kannangara, C. G., and Stumpf, P. K. 1973. The elongation of medium chain trienoic acids to α-linolenic acid by a spinach chloroplast stroma system. Biochem. Biophys. Res. Commun. 52:1190–1198.

Jaworski, J. G., and Stumpf, P. K. 1974. Properties of a soluble stearyl-acyl carrier protein desaturase from maturing *Carthamus tinctorius*. Arch. Biochem. Biophys. 162:158–165.

Joyard, J., and Douce, R. 1976a. Mise en évidence et rôle des diacylglycerols de l'enveloppe des chloroplastes d'épinard. Biochim. Biophys. Acta. 424:125–131.

Joyard, J., and Douce, R. 1976b. Préparation et activités enzymatiques de l'enveloppe des chloroplastes d'épinard. Physiol. Vég. 14:31–48.

Joyard, J., and Douce, R. 1977. Site of synthesis of phosphatidic acid and diacylglycerol in spinach chloroplasts. Biochim. Biophys. Acta 486:273–285.

Kannangara, C. G., and Stumpf, P. K. 1973. Distribution and nature of biotin in chloroplasts of different plant species. Arch. Biochem. Biophys. 155:391–399.

Kenyon, C. N. 1972. Fatty acid composition of unicellular strains of blue-green algae. J. Bact. 109:827–834.

Kenyon, C. N., Rippka, R., and Stanier, R. Y. 1972. Fatty acid composition and physiological properties of some filamentous blue-green algae. Arch. Mikrobiol. 83:216–236.

Königs, B., and Heinz, E. 1974. Investigation of some enzymatic activities contributing to the biosynthesis of galactolipid precursors in *Vicia faba*. Planta 118:159–169.

Marshall, M. O., and Kates, M. 1972. Biosynthesis of phosphatidylglycerol by cell-free preparations from spinach leaves. Biochim. Biophys. Acta. 260:558–570.

McKee, J. W. A., and Hawke, J. C. 1978. The incorporation of [14C]bicarbonate and 14CO2 into the constituent fatty acids of monogalactosyl diacylglycerol by spinach chloroplasts and leaves. FEBS Lett. 94:273–276.

McKee, J. W. A., and Hawke, J. C. 1979. Incorporation of [14C]acetate into the constituent fatty acids of monogalactosyldiglyceride by isolated spinach chloroplasts. Arch. Biochem. Biophys. 197:322–332.

Mudd, J. B., van Vliet, H. H. D. M., and van Deenen, L. L. M. 1969. Biosynthesis of galactolipids by enzyme preparations from spinach leaves. J. Lipid Res. 10:623–630.

Murphy, D. J., and Leech, R. M. 1977. Lipid biosynthesis from [14C]bicarbonate, [2-14C]pyruvate and [1-14C]acetate during photosynthesis by isolated spinach chloroplasts. FEBS Lett. 77:164–168.

Murphy, D. J., and Leech, R. M. 1978. The pathway of [14C]bicarbonate incorporation into lipids in isolated photosynthesizing spinach chloroplasts. FEBS Lett. 88:192–196.

Murphy, D. J., and Stumpf, P. K. 1979. Elongation pathway for α-linolenic acid synthesis in spinach leaves. A reexamination. Plant Physiol. 64:428–430.

Nagai, J., and Bloch, K. 1965. Synthesis of oleic acid by *Euglena gracilis*. J. Biol. Chem. 240:PC3702–3703.

Neufeld, E. F., and Hall, C. W. 1964. Formation of galactolipids by chloroplasts. Biochem. Biophys. Res. Commun. 14:503–508.

Nichols, B. W., Harris, R. V., and James, A. T. 1965. The lipid metabolism of blue-green algae. Biochem. Biophys. Res. Commun. 20:256–262.

Nichols, B. W., James, A. T., and Breuer, J. 1967. Interrelationships between fatty acid biosynthesis and acyl-lipid synthesis in *Chlorella vulgaris*. Biochem. J. 104:486–496.

Ohlrogge, J. B., Kuhn, D. N., and Stumpf, P. K. 1979. Subcellular localization of acyl carrier protein in leaf protoplasts of *Spinacia oleracea*. Proc. Nat. Acad. Sci. USA 76:1194–1198.

Radunz, A. 1969. Uber das Sulfochinovosyl-diacylglycerin aus höheren Pflanzen, Algen und Purpurbakterien. Z. Physiol. Chem. 350:411–417.

Roughan, P. G., and Slack, C. R. 1977. Long-chain acyl-coenzyme A synthetase activity of spinach chloroplasts is concentrated in the envelope. Biochem. J. 162:457–459.

Roughan, P. G., Holland, R., Slack, C. R., and Mudd, J. B. 1979. Acetate, pyruvate and bicarbonate as precursors of fatty acids in isolated chloroplasts. Biochem J. 184:565–569.

Roughan, P. G., Holland, R., and Slack, C. R. 1980. Role of chloroplasts and microsomes in polar lipid synthesis from [1-^{14}C]acetate by cell-free preparations from *Spinacia oleracea* leaves. Biochem J. 188:17–24.

Roughan, P. G., Mudd, J. B., McManus, T. T., and Slack, C. R. 1979. Linoleate and linolenate synthesis by isolated chloroplasts. Biochem. J. 184:571–574.

Safford, R., and Nichols, B. W. 1970. Positional distribution of fatty acids in mono-galactosyl diglyceride fractions from leaves and algae. Biochim. Biophys. Acta 210:57–64.

Santora, G. T., Gee, R., and Tolbert, N. E. 1979. Isolation of a sn-glycerol-3-phosphate: NAD oxidoreductase from spinach leaves. Arch. Biochem. Biophys. 196:403–411.

Shaw, N. 1970. Bacterial glycolipids. Bact. Rev. 34:365–377.

Shine, W. E., Mancha, M., and Stumpf, P. K. 1976. Differential incorporation of acyl-coenzymes A and acyl-acyl carrier proteins into plant microsomal lipids. Arch. Biochem. Biophys. 173:472–479.

Shibuya, I., Yagi, T., and Benson, A. A. 1963. Sulfonic acids in algae. In: Studies on Microalgae and Photosynthetic Bacteria. Tokyo: University of Tokyo Press, pp. 627–636.

Siebertz, M., and Heinz, E. 1977. Galactosylation of different monogalactosyldiacyl-glycerols by cell-free preparations from pea leaves. Z. Physiol. Chem. 358:27–34.

Sparace, S. A., and Moore, T. S. 1979. Phospholipid metabolism in plant mitochondria. Submitochondrial sites of synthesis. Plant Physiol. 63:963–972.

van Besouw, A., and Wintermans, J. F. G. M. 1978. Galactolipid formation in chloroplast envelopes. Evidence for two mechanisms in galactosylation. Biochim. Biophys. Acta 529:44–53.

Williams, J. P., Watson, G. R., and Leung, S. P. K. 1976. Galactolipid synthesis in *Vicia faba* leaves. Formation and desaturation of long chain fatty acids in phosphati-dylcholine, phosphatidylglycerol and the galactolipids. Plant Physiol. 57:179–184.

Winkenbach, F., Wolk, C. P., and Jost, M. 1972. Lipids of membranes and of the cell envelope in heterocysts of a blue-green alga. Planta 107:69–80.

Wood, B. J. B., Nichols, B. W., and James, A. T. 1965. The lipids and fatty acid metabolism of photosynthetic bacteria. Biochim. Biophys. Acta 106:261–273.

Discussion of Presentation by Dr. Mudd

SCHIFF: Since the sulfolipid is found in *Euglena* W₃BUL in which plastid DNA
is undetectable, the enzymes which make it must be coded elsewhere, probably
in nuclear DNA. I believe other plastid lipids are also found in mutants lacking
plastid DNA; perhaps Dr. Hendren can recall these.

HENDREN: Dr. Ernst-Fonberg (Ernst-Fonberg et al., 1974) has demonstrated the
presence of substantial acyl carrier protein (ACP)-dependent fatty acid synthetase activity in the *Euglena* W₃BUL mutant. I have also observed 50% of
the wild-type ACP-dependent fatty acid synthetase activity in the W₁₅ZHB
mutant of *Euglena* which also lacks chloroplast DNA. However, most of the
Euglena mutants lacking functional chloroplasts which have been examined
completely lack the ACP-dependent fatty acid synthetase.

VON WETTSTEIN: Do you think that the very long-chain lipids like, for example,
the β-diketones (which are eventually excreted to the surface of the plant) are
also synthesized in the chloroplast or could they be elongated from short-chain
fatty acids in the cytoplasm?

MUDD: The experiments of Ohlrogge et al. (1979) showed that the antibody to
spinach ACP completely inhibited fatty acid synthesis de novo from malonyl
CoA in a spinach leaf homogenate. It is possible that elongation systems
forming long-chain fatty acids would not be inhibited by the antibody.

WILDMAN: Where CO_2 was fixed by isolated chloroplasts into fatty acids, was
the phenomenon light-dependent and did it occur at high rates ($\sim$100 μmol
hr^{-1} mg^{-1} chlorophyll) of CO_2 fixation into phosphoglyceric acid intermediates?

MUDD: The incorporation of CO_2 into long-chain fatty acids by isolated chloroplasts was light-dependent. Total CO_2 fixation in the papers cited was 60-
75 μmoles hr^{-1} mg^{-1} chlorophyll in one case and 100 μmoles hr^{-1} mg^{-1}
chlorophyll in the other. Of the CO_2 fixed 0.6% was found in the fatty acids.

GANTT: Does the isolated chloroplast envelope in the work of Douce include
both the inner and outer membranes?

VON WETTSTEIN: Concerning the separation of inner and outer chloroplast envelope membranes, Dr. R. Douce reported in June 1979 at a meeting in
Copenhagen that separation of the two envelope membranes has so far not
been achieved. Potato slices have been used for studies of fatty acid synthesis
to compare the capacity of amyloplasts to that of chloroplasts (see Willemot
and Stumpf, 1967a,b).

SCHIFF: The problem of the formation of sulfonic acids in general is still a vexing
one. In animals sulfonic acids can be formed by oxidation of the respective
thiols, avoiding the problem of how to remove the oxygen between the sulfur

and carbon which arises with precursors containing oxidized sulfur. However, sulfonic acids also come from sulfate and in *Chlorella* sulfate is a more efficient precursor than cysteine. *Chlorella* mutants blocked in early sulfate reduction still appear to form sulfolipid. It seems that the sulfonic acid group of the sulfolipid comes either from sulfate directly or from adenosine-5'-phosphosulfate (APS) or adenosine-3'-phosphate-5'-phosphosulfate (PAPS) by as yet unknown reactions.

MUDD: Sulfonic acids such as sulfoacetic acid and sulfolactic acid have been isolated and characterized in green algae and higher plants. There is good evidence that these are degradation products of sulfoquinovose. We have no direct information about sulfonic acids which may be precursors of sulfoquinovose.

BASSHAM: With respect to the questions of whether acetyl CoA is made from CO_2 in the chloroplast, and whether the chloroplast may be the sole site of fatty acid synthesis in green cells, the effects of NH_4^+ administered to cells of *Chlorella* is relevant. Added NH_4^+ (1.0 mM) stimulates pyruvate kinase (raising the pyruvate level and lowering phosphoenolpyruvate) as part of a general shift from sucrose formation to amino acid and fatty acid synthesis. Lipid formation is stimulated. The effects of NH_4^+ are, for several reasons, known to occur in the cytoplasm, not the chloroplast, under these conditions. These results suggest either that fatty acid synthesis is not all in the chloroplasts, or that pyruvate formation occurs outside chloroplasts (i.e., the complete pathway from CO_2 to fatty acids would not be in the chloroplasts, but rather phosphoglyceric acid would come out and pyruvate or acetyl CoA would go back in).

MUDD: In the papers by Murphy and Leech (1977, 1978) and McKee and Hawke (1978), in which ^{14}C-bicarbonate was incubated with purified chloroplasts, there seems to be little possibility that an intermediate leaving the chloroplast could be metabolized and returned to the chloroplast.

CASTELFRANCO: In cyanobacteria do you find 18:1[6] or 18:2[6,9] fatty acids? In one of your slides you showed pyruvate decarboxylase in chloroplasts; has this been shown?

MUDD: The identification of fatty acids in cyanobacteria reported by Kenyon (1972) showed the double bond at the 6 position only when double bonds were also present at positions 9 and 12. One infers that the sequential desaturation is first at position 9, then at 12, and then at either 15 or 6.

In the report of Murphy and Leech (1978), the evidence for conversion of pyruvate to acetyl CoA is cited, and includes the work of Yamada and Nakamura (1975).

SCHMIDT: A comment on sulfolipid formation: Cysteic acid should be treated carefully, since it turns on respiration.

MUDD: When cysteic acid has been used as an efficient precursor for the sulfolipid, the tracer used was ^{35}S or ^{14}C. There is no evidence from double-labeling experiments that the sulfur and the three carbon fragment were incorporated together. It has been shown, however, that sulfolactic acid, a proposed intermediate in the incorporation from cysteic acid, does prevent the incorporation of $^{35}SO_4{}^{2-}$ into sulfolipid, and that the carbon skeleton from cysteic acid is found only in the sugar moiety of the sulfolipid.

TRENCH: When $^{14}CO_2$ is provided to intact *Codium*, synthesis of a glycolipid by the chloroplasts can be readily demonstrated. However, when these same plastids are sequestered by slugs (e.g., *Elysia*) incubation in $^{14}CO_2$ for up to 12 hr shows no further synthesis of monogalactosyldiglyceride, digalactosyl-diglyceride, or sulfoquinovosyldiglyceride. This suggests that there is something very subtle about the plant cytoplasm and its role in plastid glycolipid synthesis.

MUDD: In higher plants, the cytoplasm is required to provide UDP-gal for the galactolipid synthesized in the chloroplast envelope. The evidence you provide suggests a similar cooperation for the synthesis of the sulfolipid.

References to Discussion

Ernst-Fonberg, M. J., Dubinskas, R., and Jonak, Z. L. 1974. Comparison of two fatty acid synthetases from *Euglena gracilis* variety *bacillaris*. Arch. Biochem. Biophys. 165:646-655.

Kenyon, C. N. 1972. Fatty acid composition of unicellular strains of blue-green algae. J. Bact. 109:827-834.

McKee J. W. A., and Hawke, J. C. 1978. The incorporation of [^{14}C]bicarbonate and $^{14}CO_2$ into the constituent fatty acids of monogalactosyldiacylglycerol by spinach chloroplasts and leaves. FEBS Lett. 94:273-276.

Murphy, D. J., and Leech, R. M. 1977. Lipid biosynthesis from [^{14}C] bicarbonate [2-^{14}C]pyruvate and [1-^{14}C]acetate during photosynthesis by isolated spinach chloroplasts. FEBS Lett. 77:164-168.

Murphy, D. J., and Leech, R. M. 1978. The pathway of [^{14}C]bicarbonate incorporation into lipids in isolated photosynthesizing spinach chloroplasts. FEBS Lett. 88:192-196.

Ohlrogge, J. B., Kuhn, D. N., and Stumpf, P. K. 1979. Subcellular localization of acyl carrier protein in leaf protoplasts of *Spinacia oleracea*. Proc Nat. Acad. Sci. USA 76:1194-1198.

Willemot, C., and Stumpf, P. K. 1967a. Fat metabolism in higher plants. XXXIII. Development of fatty acid synthetase during the "aging" of storage tissue slices. Canad. J. Bot. 45:579-584.

Willemot, C., and Stumpf, P. K. 1967b. Fat metabolism in higher plants. XXXIV. Development of fatty acid synthetase as a function of protein synthesis in aging potato tuber slices. Plant Physiol. 42:391-397.

Yamada, M., and Nakamura, Y. 1975. Fatty acid synthesis by spinach chloroplasts. II. The path from PGA to fatty acids. Plant Cell Physiol. 16:151-162.

Assimilation of Nitrogen and Synthesis of Amino Acids in Chloroplasts and Cyanobacteria (Blue-Green Algae)

Peter J. Lea, W. Ronald Mills, Roger M. Wallsgrove, and Benjamin J. Miflin

The aim of this chapter is to compare the ability of chloroplasts and cyanobacteria to carry out the reactions involved in the conversion of nitrate to all of the amino acids involved in protein synthesis. The reduction of sulfate and its conversion to cysteine is discussed in Chapter 10 of the present volume. The formation of the carbon skeleton of methionine will be discussed in this section.

The experimental evidence suggesting that chloroplasts can carry out a number of reactions involved in nitrogen metabolism will be considered under three headings: (a) studies of the ability of isolated intact chloroplasts to metabolize labeled compounds in the light; (b) measurements of oxygen evolution carried out by isolated chloroplasts in the presence of potential substrates; and (c) investigations into the localization of key enzymes within the plant cell.

Nitrate Reduction

$$NO_3^- + 2e^- + 2H^+ \rightarrow NO_2^- + H_2O$$

In cyanobacteria the above reaction is catalyzed by a ferredoxin-dependent enzyme (Vennesland and Guerrero, 1979), but in the green algae (Solomonson et al., 1975) and higher plants (Hewitt et al., 1979) the enzyme is predominantly NADH-dependent, although NADPH activity has been reported (Campbell, 1976; Shen et al., 1976). The enzymes from higher plants (mol wt $\sim$ 200,000) and *Chlorella* (mol wt 356,000) are more complex than the enzyme from blue-

Address reprint requests to Dr. Peter John Lea, Principal Scientific Officer, Department of Biochemistry, Rothamsted Experimental Station, Harpenden, Herts., United Kingdom.

green bacteria (mol wt 75,000) (Losada and Guerrero, 1979). The cyanobacterial enzyme consists of one polypeptide, but the enzyme from barley leaves consists of three separate subunits (containing flavinadenine dinucleotide [FAD] and cytochrome b_{557}), plus one or two small molybdenum-containing units (Wray et al., 1979).

The regulation of nitrate reduction is exceedingly complex, and beyond the scope of this chapter (Vennesland and Guerrero, 1979; Hewitt et al., 1979; Oaks, 1979; Beevers and Hageman, 1980). Nitrate reduction can take place either in the root or leaf; the relative proportion varies between different species of plant (Pate, 1973) and is dependent upon the level of nitrate applied and the physiological age of the plant (Kirkman and Miflin, 1979).

It was originally thought that light was required for the reduction of nitrate under aerobic conditions (Canvin and Atkins, 1974), although the ability of leaves to reduce nitrate in the dark anaerobically has long been used as an assay for in vivo nitrate reductase. It has been suggested that oxygen inhibits nitrate reduction by competing for the reducing power available in the mitochondria (Sawhney et al., 1978; Mann et al., 1979; Canvin and Woo, 1979). However, data has suggested that leaves which have a sufficient carbohydrate store can reduce nitrate aerobically in the dark at rates 50% of that obtained in the light (Jones and Sheard, 1978; Aslam et al., 1979).

There is some evidence that illuminated isolated chloroplasts can reduce nitrate at low rates (Heber and French, 1968; Grant and Canvin, 1970), but this may well be due to cytoplasmic contamination (Swader and Stocking, 1971).

The localization of nitrate reductase within the leaf cell also continues to be a subject of controversy. Chloroplasts produced by aqueous (Del Campo et al., 1963; Rathnam and Das, 1974; Plaut and Littan, 1975) and nonaqueous (Coupe et al., 1967) methods have been shown to contain nitrate reductase activity. Rathnam and Das (1974) concluded that the enzyme was attached to the envelope membrane of the chloroplast, a proposal that is not disproved by the data of Ritenour et al., 1967; Grant et al., 1970; and Eaglesham and Hewitt, 1971. This suggestion is supported by a nitrate transport theory proposed by Butz and Jackson (1977), in which nitrate is reduced when it is transported across a membrane. However, a number of other workers have failed to demonstrate nitrate reductase in mechanically isolated chloroplasts (Swader and Stocking, 1971; Dalling et al., 1972; Harel et al., 1977) or chloroplasts isolated from enzymically isolated protoplasts (Wallsgrove et al., 1979a).

It may be concluded that if nitrate reductase is associated with the chloroplasts, it is readily lost on isolation and purification of the organelles. It is possible, however, that if suitable proteins (e.g., bovine serum albumin) are not included in the isolation medium that there may be nonspecific adsorption of nitrate reductase present in the cytoplasm onto chloroplast membranes.

The source of NADH for nitrate reduction is also still a subject of debate. Glyceraldehyde-3-phosphate transported out of the chloroplast has been suggested as a light-dependent source (Klepper et al., 1971; Rathnam, 1978). Malate oxidation to yield NADH catalyzed by malate dehydrogenase has been suggested as a source of reductant. Malate may be formed by the reduction of oxaloacetate either generated in the Krebs cycle in the mitochondria, by the oxidation of glycine during photorespiration (Wood and Osmond, 1976) or in the chloroplast by photosynthetically derived NADPH (Rathnam, 1978). Mann et al. (1978, 1979) examined the ability of a number of compounds to act as sources of reductant in a freeze-thaw assay for nitrate reductase; fructose 1,6-diphosphate and glyceraldehyde-3-phosphate had the higher activity.

The cyanobacteria *Anabaena* and *Anacystis* contain a nitrate reductase that is tightly bound to a membrane (Hattori and Uesugi, 1968). The enzyme may be solubilized and is able to use ferredoxin and a range of artificial dyes as a source of reductant. The nitrate reductase is unable to utilize reduced pyridine nucleotides directly, although a system involving ferredoxin and ferredoxin-NADP reductase may be used to couple NADPH to nitrate reduction (Ortega et al., 1976). Candau et al. (1976) have been able to isolate chlorophyll-containing particles from *Anacystis nidulans* which, in the light, can use water as the electron donor to reduce nitrate to nitrite and ammonia with a corresponding evolution of oxygen. The reaction is totally dependent upon light and is inhibited by the photosystem II inhibitor DCMU. Further work on the ability of particles from blue-green bacteria to reduce nitrate and nitrite has been discussed by Losada and Guerrero (1979). The data suggest that there is a fundamental difference between cyanobacteria and higher plants in their respective modes of nitrate reduction. In the prokaryote, the enzyme is a single small peptide that is attached to a chlorophyll-containing membrane and receives its electrons directly from photosystem 1, probably via ferredoxin. In the plant, the enzyme has been enlarged to include a transport chain to transfer the electrons from NADH via FAD and cytochrome b_{557} to Mo and subsequent nitrate reduction. The enzyme is probably in the cytoplasm or, at the most, very loosely attached to the outer chloroplast membrane, and the reducing power is generated indirectly from photosynthetic reactions or from reduced carbon substrates.

Nitrite Reduction

$$NO_2^- + 6e^- + 8H^+ \rightarrow NH_4^+ + 2H_2O$$

The reduction of nitrite involves the transfer of six electrons, a reaction which is catalyzed by a single protein, nitrite reductase. Although there is a large negative free energy change in the reaction (103.5 kcal/mol) a low potential electron carrier which is probably ferredoxin in all plants is required for the reduction process. In spinach the enzyme is composed of one polypeptide chain (mol wt 61,000 daltons) containing a siroheme group plus two additional iron and labile sulfur atoms (Vega and Kamin, 1977). The enzyme has not been extensively studied in cyanobacteria (Hattori and Uesugi, 1968; Losada and Guerrero, 1979). The enzyme is, however, tightly bound to a membrane (as opposed to the soluble higher plant enzyme) and is extremely resistant to denaturation by heat or acid, or to inhibition by p-chloromercuribenzoate (Guerrero et al., 1974).

Isolated illuminated chloroplasts have been shown by a number of workers to reduce nitrite (Swader and Stocking, 1971; Dalling et al., 1972; Magalhaes et al., 1974; Miflin, 1974b; Plaut et al., 1977; Anderson and Done, 1978; Wallsgrove et al., 1979b), and to convert it stoichiometrically to α-amino nitrogen. The reduction of nitrite is considerably stimulated by the addition of α-ketoglutarate, which is then converted to glutamate (Table 1). The reaction is only carried out by intact chloroplasts, and is very sensitive to DCMU (Anderson and Done, 1978), suggesting the involvement of the chloroplast electron transport mechanism.

Initial studies by Swader and Stocking (1971) with *Wolffia* chloroplasts suggested that there was a stoichiometric evolution of 1.5 mol of O_2 for each mol of nitrite reduced. In more detailed studies of pea chloroplasts, Anderson and Done (1978) showed that in short-term experiments the ratio was 1.5 (but thereafter decreased), and 80% of the nitrite was recovered as ammonia. In further experiments α-ketoglutarate restored the ratio to 1.35, and the ammonia was converted to amino acids.

Nitrite reductase has been demonstrated in the chloroplast by a number of workers (Swader and Stocking, 1971; Dalling et al., 1972; Magalhaes et al., 1974; Miflin, 1974a; Rathnam and Das, 1974; Harel et al., 1977) although its localization in the cytoplasm (Grant et al., 1970) and in the peroxisome (Lips and Avissar, 1972) has also been suggested. We have recently obtained results which suggest that in pea leaves the enzyme is probably totally located in the chloroplast (Wallsgrove et al., 1979a) (Table 1).

Table 1. Distribution of Enzymes of Nitrogen Assimilation from Lysed Pea Leaf Protoplasts in Different Regions of Sucrose Density Gradients

	Distribution of gradient (percent of recovered activity), Region[a]		
Enzyme	I	II	III
Chlorophyll	2.6	17.4	80.0
Nitrite reductase	16.5	2.5	81.0
Glutamate synthase	19.5	0	80.5
Glutamine synthetase	41.8	12.4	45.8
NAD-glutamate dehydrogenase	13.0	83.0	4.0
Nitrate reductase	93.5	0.5	6.0

Source: From Wallsgrove et al., 1979a.

[a]The regions on the gradients were defined as: I, the volume corresponding to the protoplast lysate applied to the gradient; II, the portion of the gradient from the top to the beginning of the intact chloroplast band, which included most of the mitochondria marker enzymes and the sedimentable catalase (microbody marker); and III, the intact chloroplast band down to the bottom of the gradient.

There would, therefore, appear to be a strong similarity between cyanobacterial nitrite reduction and that of the higher plant chloroplast. The only major difference is that in the cyanobacteria the enzyme is tightly bound to a membrane, whereas in the chloroplast it is soluble and readily liberated on rupture of the external membrane.

Ammonia Assimilation

Pathway

The current hypothesis for the pathway of ammonia assimilation in plants has been reviewed extensively (Lea and Miflin, 1979; Miflin and Lea, 1976, 1977, 1980) (Figure 1). Two reactions are involved; the first reaction is catalyzed by glutamine synthetase (GS).

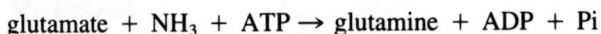

$$\text{glutamate} + NH_3 + ATP \rightarrow \text{glutamine} + ADP + Pi$$

The second reaction is catalyzed by glutamate synthase (GOGAT), which may require either NAD(P)H or reduced ferredoxin as an electron donor.

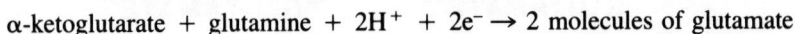

$$\alpha\text{-ketoglutarate} + \text{glutamine} + 2H^+ + 2e^- \rightarrow 2 \text{ molecules of glutamate}$$

The two enzymes work consecutively to yield a net synthesis of glutamate from ammonia and α-ketoglutarate; one glutamate molecule is continuously recycled. Two inhibitors, methionine sulfoximine (MSO) and azaserine, have frequently been used in studies of ammonia assimilation. MSO is a relatively specific inhibitor of GS (Tate and Meister, 1973), but azaserine inhibits a number of amide transfer reactions, including GOGAT. The problems involved in using these inhibitors have been discussed by Miflin and Lea (1976, 1977, 1980). Prior to 1974, it was assumed that amino acids could be synthesized by the direct incorporation of ammonia into the 2-position of an α-keto acid. The most widely favored reaction is that catalyzed by glutamate dehydrogenase (GDH).

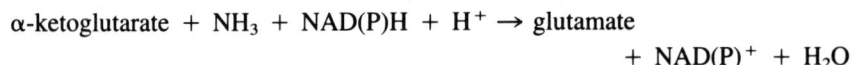

$$\alpha\text{-ketoglutarate} + NH_3 + NAD(P)H + H^+ \rightarrow \text{glutamate}$$
$$+ NAD(P)^+ + H_2O$$

Similar reactions involving the action of aspartate dehydrogenase (AsDH) on oxaloacetate and alanine dehydrogenase (AlDH) on pyruvate have also been suggested.

Cyanobacteria are unique among chlorophyll-containing organisms in that they can convert N_2 gas to ammonia, a reaction which usually takes place in a modified cell known as a *heterocyst*. For this reason it has been possible to study ammonia assimilation using ^{13}N-labeled dinitrogen with a half-life of 10 min, which may be obtained from a cyclotron. In *Anabaena cylindrica* it has been demonstrated that after 15 sec of $^{13}N_2$ fixation almost all of the label has been incorporated into the amide group of glutamine. Subsequently the label was transferred to the amino group of glutamate and other amino acids (Wolk et al., 1976; Meeks et

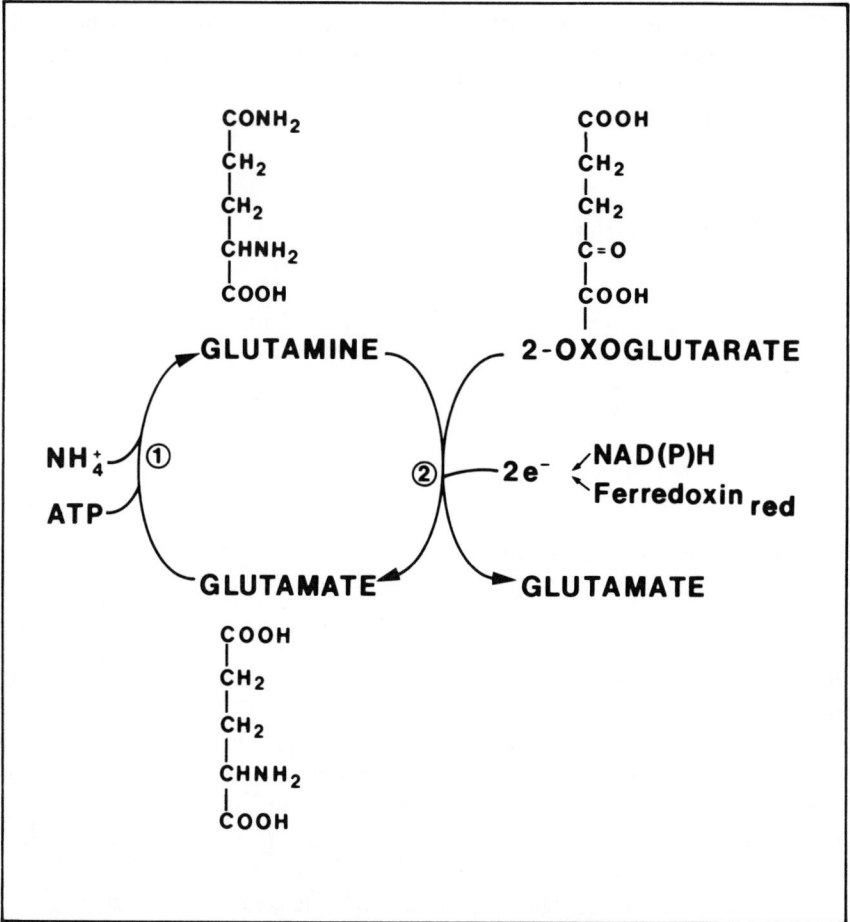

Figure 1. The glutamate synthase cycle. Enzymes: 1, glutamine synthetase; 2, glutamate synthase.

al., 1977; Thomas et al., 1977). The addition of methionine sulfoximine (MSO), caused all the ^{13}N to be trapped in ammonia. The addition of azaserine caused the label to remain in the amide position of glutamine (Wolk et al., 1976). Similar results have been obtained with unlabeled N_2 (Stewart and Rowell, 1975; Ladha et al., 1978) and ^{15}N tracer studies (Stewart et al., 1975). MSO has also been shown to increase the frequency of heterocyst formation in *Anabaena* (Ownby, 1977). Thomas et al. (1977) concluded that after the initial reduction of N_2 to ammonia, glutamine was synthesized in the heterocyst. Glutamine was then transferred to the vegetative cell, where glutamate was formed by the enzyme GOGAT. If this scheme is correct then there is a requirement for glutamate transport back to the heterocyst, by a mechanism so far not determined. There are few data available as to the mechanism of ammonia assimilation when cyanobacteria utilize NO_3^- as a nitrogen source.

The light-dependent conversion of α-ketoglutarate to glutamate has been demonstrated in isolated chloroplasts (Givan et al., 1970; Tsukamoto, 1979), although only in the latter case was the reaction dependent upon the presence of ammonia. The synthesis of glutamine from ^{14}C-glutamate has also been demonstrated in chloroplasts (Santarius and Stocking, 1969; Givan, 1975; Mitchell and Stocking, 1975). Mitchell and Stocking were able to show that a light-dependent synthesis of glutamine took place at 10^{-4} M ammonia but was inhibited by concentrations above 2 mM. Data concerning the ability of isolated pea chloroplasts to synthesize glutamate and glutamine from ammonia are shown in Table 2 (Wallsgrove et al., 1979b; Lea and Miflin, 1979). The data in Table 2 suggest that ammonia is being converted to glutamate via glutamine.

The initial isolation of ferredoxin-dependent GOGAT was from pea chloroplasts, which can convert glutamine and α-ketoglutarate to glutamate in a light-dependent reaction (Lea and Miflin, 1974). The ability to carry out this reaction has been demonstrated in a number of other plant chloroplasts (Wallsgrove et al., 1977). Anderson and Done (1977b) showed that the reaction between glutamine and α-ketoglutarate was coupled to oxygen evolution as predicted by the requirement for reduced ferredoxin. Using chloroplasts in which CO_2 fixation had been inhibited by the addition of DL-glyceraldehyde, they found that light-dependent O_2 evolution was inhibited by azaserine and DCMU, but not by MSO.

Ammonia was able to substitute for glutamine in the oxygen-evolving system providing ADP, Mg^{2+}, and PPi were present in the chloroplast incubation medium (Anderson and Done, 1977a). The reaction was light-dependent, was not stimulated by the addition of glutamine, and was inhibited by MSO and azaserine.

Table 2. Production of Glutamate and Glutamine by Isolated Pea Chloroplasts[a]

Nitrogen source	Net nmoles formed (light-dark)	
	Glutamine	Glutamate
1. None	0	+26
2. KNO_3 (5 mM)	0	+30
3. $NaNO_2$ (1 mM)	0	+337
4. $NaNO_2$ (1 mM) (α-ketoglutarate omitted)	+51	-97
5. $(NH_4)_2SO_4$ (1 mM)	+15	+496
6. $(NH_4)_2SO_4$ (1 mM) + methionine sulfoximine (2.5 mM)	0	+68
7. $(NH_4)_2SO_4$ (1 mM) + Azaserine (1 mM)	+165	-115

Source: From Wallsgrove et al., 1979b.
[a]Assays contained, in a volume of 0.5 ml: ADP (2.25 μmol); α-ketoglutarate (1.25 μmol) (except 4); Na pyrophosphate (2.5 μmol); $MgCl_2$ (5 μmol), and chloroplasts (200 μl), containing 0.123 mg chlorophyll. Incubation was for 30 min at 25°C.

Such results suggest that chloroplasts are able to synthesize glutamine from ammonia in an ATP/Mg^{2+}-dependent reaction. The glutamine formed can then react with α-ketoglutarate as above, thereby leading to O_2 evolution. α-Ketoglutarate could not be replaced by glyoxylate or pyruvate in the above reactions, but oxaloacetate sustained high rates of O_2 evolution on its own, presumably due to reduction to malate.

Reassimilation of Ammonia

It is normally assumed that ammonia is formed either by nitrate or dinitrogen reduction. There are however a number of other mechanisms by which ammonia may be generated: (a) the conversion of glycine into serine during photorespiration; (b) the hydrolysis of urea formed by the catabolism of arginine and the ureides allantoin and allantoic acid; and (c) the breakdown of asparagine. The first mechanism deserves special mention because of its potential magnitude; the latter two have been discussed in detail elsewhere (Lea and Miflin, 1980). In higher plants photorespiration rates have been estimated at 80 mol gm^{-1} fr wt hr^{-1} (Keys et al., 1977; Thomas et al., 1978). If all of the CO_2 released is assumed to have come from the conversion of glycine to serine, then ammonia must be released at rates an order of magnitude higher than that of nitrate reduction.

$$2 \text{ glycine } + H_2O \rightarrow \text{serine} + NH_3 + CO_2 + 2H^+ + 2e^-$$

Even if the rates of photorespiration are considered to be overestimated, or that CO_2 may also be derived from glyoxylate oxidation (Grodzinski, 1978), ammonia reassimilation must still be quantitatively more important than primary assimilation. Keys et al. (1978) have put forward a scheme for higher plants in which the ammonia released is transported out of the mitochondria and is converted to glutamine by GS in the cytoplasm. The glutamine is then converted to glutamate by a light-dependent GOGAT reaction in the chloroplast. If this cycle operates, only one of every 3 ATP molecules formed in the mitochondria is utilized in glutamine synthesis; the other two may be used for other energy-requiring reactions.

Cyanobacteria also photorespire, the rate increasing linearly with O_2 concentration up to at least 0.2 atm with a subsequent decrease in nitrogen fixing capacity (Lex et al., 1972). Cyanobacteria are also able to excrete glycolate, but only 1% of the total C fixed is excreted in this way, as opposed to up to 10% in the green alga *Chorella* (Cheng et al., 1972). Although *Anabaena* has been shown to have the enzymes capable of converting glycine to serine, studies by ^{14}C-1-glycolate feeding (Codd and Stewart, 1973) and the photorespiration inhibitor isonicotinylhydrazide (Codd and Stewart, 1973; Cheng et al., 1971) suggest that this is not the major mechanism of CO_2 evolution during photorespiration in the cyanobacteria.

Location and Properties of the Enzymes of Ammonia Assimilation

Glutamine synthetase. GS has been shown to be present in the chloroplasts of many species of plants (Haystead, 1973; O'Neal and Joy, 1973; Miflin, 1974a; Rathnam and Edwards, 1976; Harel et al., 1977). Miflin (1974a) suggested in his earlier paper that GS may also be present in the cytoplasm. This result was later confirmed when chloroplasts were isolated from pea leaf protoplasts (Wallsgrove et al., 1979a) and only 50% of the total GS activity could be attributed to the organelles (Table 1). Two types of GS have now been separated from barley (Mann et al., 1979) and rice leaves (Guis et al., 1979). In barley the two types of GS are not interconvertible and appear to be located separately in the cytoplasm and in the chloroplasts.

It has been suggested that mitochondria also contain GS, at levels capable of assimilating the ammonia released during photorespiration (Jackson et al., 1979). At Rothamsted we have been unable to detect any significant levels of GS activity in the mitochondria isolated from pea leaf protoplasts. Neither have we any evidence that isolated mitochondria are capable of reassimilating the ammonia released from glycine oxidation occurring within them (Wallsgrove et al., 1980).

Although the properties of GS from higher plants have been studied, it is difficult to make direct comparisons with the cyanobacteria, as the localization of the purified GS was not ascertained. It remains to be seen whether the enzyme so far investigated is localized in the chloroplasts. The properties of relatively crude preparations of cyanobacterial GS have been described (Dharmawardene et al., 1973; Rowell et al., 1977; Sawhney and Nicholas, 1978), and two groups of workers have examined the characteristics of the purified enzyme (Stacey et al., 1977; Sampaio et al., 1979).

The enzyme from *Anabaena* and *Nostoc* has a mol wt of the order of 600,000 and is composed of 12 identical subunits of 50,000 arranged in two superimposed hexagonal rings (Sampaio et al., 1979). The structure of the cyanobacterial enzyme thus resembles very closely that of other bacterial enzymes (Valentine et al., 1968; Wedler and Hoffman, 1974). The plant enzyme has been studied in pea leaves (O'Neal and Joy, 1973), food yeast (Sims et al., 1974b), *Lemna* (Rhodes et al., 1979), and the nonbacteroid portion of soybean root nodules (McParland et al., 1976). The mol wt of the plant enzyme ranges from 330,000 to 390,000 and has eight subunits of 45,000 arranged in two sets of planartetramers (McParland et al., 1976).

Glutamine synthetase has been shown to be regulated by a number of different mechanisms.

Adenylation. The mechanism by which ammonia or a high glutamine: α-ketoglutarate ratio can regulate bacterial GS is extremely complex and has been discussed in detail by Magasanik (1976). The GS enzyme molecule is able to bind 12 molecules of ATP in a reaction catalyzed by adenylyl transferase; in the fully adenylated state the enzyme is unable to synthesize glutamine. The adenylation state of the enzyme is regulated indirectly by the ammonia level via uridyl transferase. There is no evidence of an adenylation mechanism in the regulation of GS in the cyanobacteria, despite a number of attempts to demonstrate it (Dharmawardene et al., 1973; Stewart et al., 1975; Rowell et al., 1977; Sampaio et al., 1979), even with the use of cetyltrimethylammonium bromide, a known stabilizer of the adenylated state (Johansson and Gest, 1977). Attempts to demonstrate the adenylation of GS from higher plants have also been unsuccessful (Kingdom, 1974).

Energy charge. There is considerable evidence that GS from all sources is inhibited by AMP and ADP, and this may be regulated by energy charge (Atkinson, 1968; Weissman, 1976). In *E. coli* 5′-AMP is a noncompetitive inhibitor with respect to ATP (Woolfolk and Stadtman, 1967), and in *Bacillus subtilis* the enzyme is inhibited by AMP and CTP (Duel and Stadtman, 1970). The higher plant enzyme is inhibited by AMP and ADP (Kanomari and Matsumoto, 1972; O'Neal and Joy, 1975; Weissman, 1976; McParland et al., 1976; Stewart and Rhodes, 1977), the inhibition being competitive with respect to ATP. Miflin (1977) proposed that GS activity could be controlled in chloroplasts in the light by magnesium concentration, energy charge, and pH, which have been shown to increase upon illumination. The enzyme from cyanobacteria is also inhibited by AMP and ADP (Stewart et al., 1975; Rowell et al., 1977).

Feedback inhibition. A frequently encountered property of GS is its inhibition by a number of protein amino acids, in particular alanine and glycine; only the enzyme isolated from rice roots does not appear to show this property (Kanamori and Matsumoto, 1972). The effect is more apparant when Mn^{2+} rather than Mg^{2+} is used in the assay. In bacteria (Hubbard and Stadtman, 1967), pea leaves (O'Neal and Joy, 1975) and *Lemna* (Stewart and Rhodes, 1977) the action is apparently cumulative. In the cyanobacterium *Anabaena,* alanine and glycine are able to almost totally inhibit GS when added separately (Rowell et al., 1977).

Reversible deactivation. If *Lemna* plants are placed in the dark for 6 hr in the presence of ammonia there is an 80% loss of extractable GS activity (Stewart and Rhodes, 1977). If the plants are reilluminated there is a rapid increase in activity. The reactivation process can be carried out in vitro by the addition of Mg^{2+}, glutamate, ATP, and a thiol reagent to the isolated enzyme. A similar type of deactivation of GS occurs when *Anabaena* cultures are placed *either* in ammonia medium or in the dark (Rowell et al., 1979). GS levels were restored upon reillumination. Dark or ammonia deactivation of GS did not take place, however, in two other genera of the cyanobacteria tested. The reactivation of the *Anabaena* enzyme could also be carried out in vitro in the presence of the substrates of the biosynthetic reaction and 40 mM 2-mercaptoethanol (Sampaio et al., 1979). The yeast enzyme, which exists as an octamer in the active form, is able to dissociate into an inactive tetramer. The dissociation is stimulated by NAD, NADPH, ATP, and phosphoenolpyruvate and the active enzyme is stabilized by α-ketoglutarate and glutamate (Sims et al., 1974b). It must be remembered that GS in the fungi is not involved in the assimilation of ammonia into the 2-amino position of amino acids.

In summary, GS from the cyanobacteria appears to exhibit properties that are found in both the bacterial and higher plant enzymes. The evidence of Rowell et al. (1977) that antibodies raised to *E. coli* GS do not cross-react with GS isolated from *A. cylindrica,* suggests that the cyanobacterial enzyme is considerably different from that of coliform bacteria.

Glutamate synthase. The ferredoxin-dependent GOGAT enzyme was originally isolated from pea chloroplasts (Lea and Miflin, 1974) and has now been demonstrated in chloroplasts from a wide range of plants (Wallsgrove et al., 1977), including both the bundle sheath and mesophyll cells of C-4 plants (Rathnam and Edwards, 1976; Harel et al., 1977). It was demonstrated in an alga known to contain glutamic dehydrogenase (GDH) with a low K_m for ammonia (McKenzie et al., 1979). Later experiments using chloroplasts isolated from pea leaf protoplasts suggest that all the GOGAT in the cell is located in the chloroplast (Wallsgrove et al., 1979a).

In the bacteria the enzyme is NAD(P)H-specific, and an NADH-dependent enzyme has been isolated from all the nonchlorophyllous tissues of higher plants tested (Stewart et al., 1980). In legume roots (Miflin and Lea, 1975; Fowler and Barker, 1979), root nodules (Awonaike, 1980), and tissue culture cells (Washitani and Sato, 1978; Emes and Fowler, 1978) the enzyme is localized within the plastid compartment of the cell.

Early attempts to detect the presence of a NAD(P)H-dependent GOGAT in cyanobacteria were either unsuccessful (Neilson and Doudoroff, 1973) or indicated very low levels of activity which could have been due to the action of two contaminating enzymes (Dharmawardene et al., 1972; Haystead et al., 1973). A ferredoxin-dependent enzyme was isolated from *Nostoc* and *Anabaena* species by Lea and Miflin (1975) and the presence of the enzyme was also demonstrated by Thomas et al. (1977) in *A. cylindrica.*

No data are available concerning the properties of GOGAT isolated from cyanobacteria, and, therefore, comparisons with the higher plant enzymes are impossible. Stewart et al. (1980) have reviewed the properties of all plant GOGAT enzymes so far examined. In general, the enzymes have a mol wt ranging from 145,000 to 235,000, and have an absolute specificity for glutamine and α-ketoglutarate. Neither asparagine ammonia, nor the keto-acids pyruvate, glyoxylate, and oxaloacetate act as substrates. Whether the enzyme from nonphotosynthetic tissues can use either NAD(P)H or reduced ferredoxin in vivo has not yet been determined.

Amino acid dehydrogenase. In crude extracts of higher plants only GDH is readily detectable, although the presence of AsDH and AlDH has been reported (Kretovich, 1965). The enzyme is predominantly NADH-dependent, and the majority of activity is located in the mitochondria which readily rupture on extraction and release the contents into the soluble fraction (Lea and Thurman, 1972). A number of workers have detected an NADP-dependent enzyme in the chloroplasts of *Vicia faba* (Leech and Kirk, 1968), lettuce (Lea and Thurman, 1972), *Eleusine coracana* (Rathnam and Das, 1974), spinach (Magalhaes et al., 1974), and in the green marine green algae *Caulerpa simpliciuscula* (Gayler and Morgan, 1976). Other workers have failed to detect any GDH activity in chloroplast preparations (Ritenour et al., 1967; Miflin, 1974a; Ehmke and Hartmann, 1976).

It is unlikely that higher plant chloroplastic GDH is involved in N assimilation, since the ammonia level required to saturate the enzyme would exert an uncoupling effect on the chloroplasts (see Miflin and Lea, 1976 for a discussion of this problem). In the alga *Caulerpa simpliciuscula* the K_m for ammonia for GDH is 0.4–0.7 mM, and therefore may be able to function at low levels of ammonia (Gayler and Morgan, 1976). The algal chloroplasts also contain ferredoxin-dependent GOGAT (McKenzie et al., 1979) and preliminary ^{15}N-feeding data suggest that the alga may assimilate ammonia by both pathways (Gayler, K. R., personal communication).

The cyanobacteria are one of the very few organisms that contain low levels of GDH, although both AsDH and AlDH have been detected (Hoare et al., 1967; Haystead et al., 1973; Batt and Brown, 1974). The AlDH has been further purified (Rowell and Stewart, 1976) and shown to have a high K_m for ammonia which is very pH dependent.

In conclusion, it would appear that there are strong similarities between cyanobacteria and chloroplasts in their complements of ammonia assimilatory enzymes. Both have high levels of GS with a low K_m for ammonia, and a ferredoxin-dependent GOGAT. Both also contain low levels of a dehydrogenase with a high K_m for ammonia.

Transfer of Nitrogen to Other Amino Acids

Although the amide nitrogen of glutamine cannot be transferred to an α-keto acid other than α-ketoglutarate, the nitrogen atom may be transferred directly to the non-amino groups of asparagine, histidine, arginine, and tryptophane. Of these only tryptophane synthesis has been detected in chloroplasts. The amino groups of all the amino acids must be formed by transamination from glutamate. Two major aminotransferases are present in chloroplasts: glutamate-oxoaloacetate and glutamate-pyruvate aminotransferase (Santarius and Stocking, 1969; Kirk and Leech, 1972; Givan, 1980). Kirk and Leech (1972) were able to show that, provided the α-keto acids are available, all the protein amino acids can be transaminated from aspartate and alanine in the chloroplast.

In the cyanobacteria high levels of glutamate-oxaloacetate and glutamate-pyruvate transminase have been detected (Hoare et al., 1967; Haystead et al., 1973; Batt and Brown, 1974).

Biosynthesis of Carbon Skeletons of Amino Acids

Although leaves and algae rapidly incorporate CO_2 into amino acids such as glycine, serine, alanine, aspartate, and glutamate, there is little evidence that chloroplasts are capable of catalyzing these reactions on their own. Glycine and serine are synthesized in a complex series of reactions involving chloroplasts, peroxisomes, and mitochondria in the process of photorespiration (Keys, 1980). From data obtained using reconstituted systems of chloroplasts and cytoplasm, Kirk and Leech (1972), Larsson (1979), and Larsson and Albertson (1974, 1979) concluded that recently fixed CO_2 was passed out of the chloroplasts as three-carbon sugar phosphates, and returned as the three α-keto acids pyruvate, oxaloacetate, and α-ketoglutarate.

It must be assumed that since cyanobacteria are capable of growing on CO_2 and an inorganic N source, they are capable of synthesizing the full range of amino acid skeletons. The cyanobacteria, however, do not contain a complete tricarboxylic acid cycle; the enzymes α-ketoglutarate dehydrogenase and succinyl CoA synthetase are missing (Smith, 1973). Thus, although α-ketoglutarate can be synthesized from citrate, it can not be converted to oxaloacetate. However, succinate (and presumably, therefore, oxaloacetate) may be synthesized via the splitting of isocitrate to glyoxylate, which also in the presence of acetyl CoA may be metabolized to malate (Smith, 1973).

Serine

Serine may be synthesized via 3-phosphoglycerate in chloroplasts (Larsson and Albertsson, 1979) as well as in the mitochondria during photorespiration. High levels of 3-phosphoglycerate dehydrogenase and phosphoserine phosphatase have been found in spinach chloroplasts but only low levels of the third enzyme phosphoserine aminotransferase have been detected. It is possible that an amino acid other than glutamate is used as the amino donor, or that the majority of phosphohydroxypyruvate is excreted from the organelle (Larsson and Albertsson, 1979).

Aromatic Amino Acids

Chloroplasts are able to convert CO_2 and shikimate into the aromatic amino acids phenylalanine, tyrosine, and tryptophane. Although the rates of aromatic acid synthesis were not as high as in the intact leaf, the addition of mitochondria, peroxisomes, and cytoplasm to the chloroplasts did not affect the rate of synthesis (Bickel et al., 1978; Buckholz et al., 1979). Such data suggest that chloroplasts are a major site of aromatic amino acid synthesis, which has been shown to be extensively regulated by feedback inhibition (Bickel and Schultz, 1979; Gilchrist and Kosuge, 1980). All the enzymes of tryptophane biosynthesis have been shown to be present in either etioplasts (Grosse, 1976) or chloroplasts (Feierabend and Brassel, 1977).

It has been suggested that the cyanobacteria, like higher plants, carry out the final steps of tyrosine synthesis via pretyrosine (Jensen and Pierson, 1975). However, plants are also able to synthesize tyrosine via 4-hydroxyphenylpyruvate in a manner similar to $E.$ $coli$ (Gilchrist and Kosuge, 1980). The precise localization of the pathways in plants has not yet been determined.

In higher plants three types of chorismate mutase exist, each of which shows differential activation by tryptophane, and inhibition by phenylalanine and tyrosine (Gilchrist and Kosuge, 1974); in the cyanobacteria the enzyme is apparently unaffected by any of the three end products. 3-Deoxy-D-arabino-heptulosorate-7-phosphate synthetase appears to undergo variable inhibition by either tyrosine or phenylalanine, depending on the species of cyanobacteria tested (Weber and Böck, 1969; Jensen et al., 1974). Tryptophane is able to inhibit anthranilate synthetase at very low concentrations ($1-10$ μM) in both cyanobacteria (Weber and Böck, 1969) and higher plants (Widholm, 1972), and this is probably the main site of regulation of tryptophane biosynthesis. However, tryptophane has been shown to repress all the enzymes required for its synthesis in the blue-green bacteria $Agmenellum$ (Ingram et al., 1972).

Glutamate Amino Acids

It has been shown in the previous sections of this chapter that α-ketoglutarate may be converted into glutamate and glutamine. There is little evidence that glutamate may be converted into proline, arginine, or ornithine in isolated chloroplasts, although in tobacco chloroplasts Δ'-pyroline-5-carboxylate may be reduced to proline by the action of light (Noguchi et al., 1968).

In cyanobacteria $^{14}CO_2$ is readily incorporated into the carbamyl group of citrulline (an intermediate in arginine synthesis) by the reaction of carbamyl phosphate and ornithine (Linko et al., 1957). Although arginine may have a slight feedback inhibitory action on N-acetylglutamate phosphokinase, there is no evidence of repression of the biosynthetic enzymes (Hood and Carr, 1971).

Aspartate Amino Acids

Probably the most significant recent advance in chloroplast amino acid metabolism has been the demonstration that aspartate may be converted to the nutritionally important amino acids lysine, threonine, methionine, and isoleucine (see Figure 2) in light-dependent reactions (Mills and Wilson, 1978a; Mills et al.,

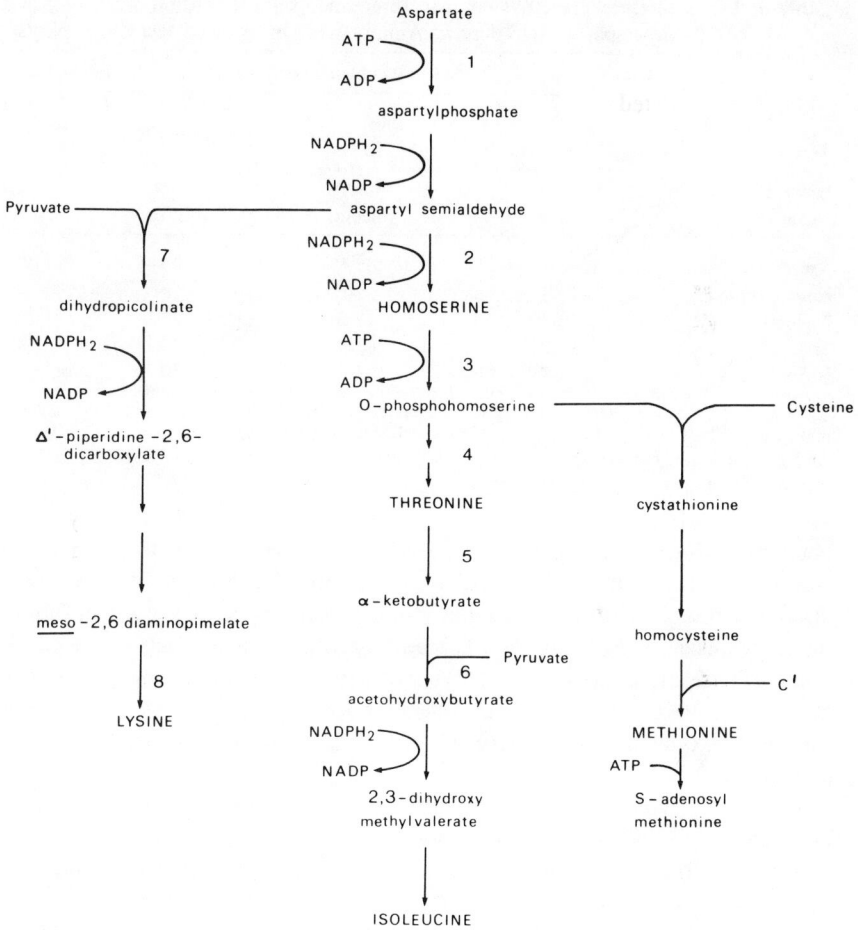

Figure 2. The pathway of biosynthesis of amino acids derived from aspartate. Enzymes: 1, aspartate kinase; 2, homoserine dehydrogenase; 3, homoserine kinase; 4, threonine synthase; 5, threonine deaminase; 6, acetolactate synthase; 7, dihydrodipicolinate synthase; 8, diaminopimelate decarboxylase.

1980). Few or no data are available on the pathways of aspartate metabolism in the cyanobacteria. It is generally assumed that the biosynthesis of the amino acids takes place by the same routes as determined for bacteria and plants (Fig. 2).

Table 3 shows that aspartate may be converted by isolated pea chloroplasts to soluble homoserine, lysine, threonine, and isoleucine and that all except homoserine are incorporated into protein (Mills et al., 1980). Lysine, threonine, and isoleucine all inhibit their own synthesis, as would be expected from data

Table 3. Effect of Isoleucine, Lysine, and Threonine on Incorporation of ^{14}C-Aspartic Acid into Aspartic Acid-derived Amino Acids by Isolated Pea Chloroplasts[a]

	Incubation conditions			
	No added amino acids Incorporation (cpm $\times$ 10^{-3})		+ Lysine	
Labeled compound	Soluble fraction	Protein fraction	Soluble fraction	Protein fraction
Aspartic acid[b]	1674.3	454.5	105	45
Homoserine	119.8	—	58	—
Lysine	10.4	12.2	27	0
Threonine	12.0	15.9	45	28
Isoleucine	7.4	1.8	50	0

[a]Aliquots (0.4 ml) containing 100 μg of chlorophyll and 2 μCi ^{14}C-aspartic acid were incubated in the light in KCl medium for 20 min. Isoleucine, lysine, and threonine were present at 2 mM each. Amino acids were extracted and analyzed by two-dimensional thin-layer chromatography. The data represent the means of two experiments.

obtained with whole plants (Bryan, 1980). Lysine also inhibits the synthesis of homoserine, threonine, and isoleucine, suggesting that it exerts its effect prior to homoserine formation. Threonine inhibits lysine synthesis less than homoserine synthesis, suggesting that its major site of action lies between aspartate semialdehyde and homoserine. The lack of inhibition of isoleucine synthesis by threonine is unexpected, and it must be assumed that aspartate (or oxaloacetate) was decarboxylated to yield pyruvate, which is a substrate for isoleucine synthesis.

There is no evidence that mitochondria are involved in the biosynthesis of the aspartate family of amino acids. A 66-fold increase in the contamination level of mitochondria caused a slight decrease in the synthesis of the amino acids.

Chloroplasts are also able to convert malate into aspartate-derived amino acids. Although the conversion of malate to aspartate took place in the dark, further metabolism was light-dependent. It was not possible to tell whether the malate to aspartate interconversion was taking place solely in the contaminating mitochondria or whether chloroplasts were also involved.

Threonine was converted to isoleucine in a light-stimulated reaction, which was inhibited by isoleucine and a combination of leucine and valine. The decarboxylation of diaminopimelate the last step in the synthesis of lysine (Bryan, 1980), was not light dependent, but the subsequent incorporation of lysine into protein was light dependent (Mills et al., 1980).

The initial report of Mills and Wilson (1978a), who utilized pea chloroplasts, suggested that there were higher incorporations of ^{14}C-aspartate into methionine than into homoserine. The differences in the two sets of data probably reflect differences in the variety or developmental age of the pea plants, or possibly in the cochromatography of homoserine (or a derivative) with methionine in the previous experiment.

Table 3. (continued)

		Incubation conditions			
+ Threonine		+ Lysine + threonine			
Incorporation as percent of that in absence of amino acids					+ Isoleucine
Soluble fraction	Protein fraction	Soluble fraction	Protein fraction	Soluble fraction	Protein fraction
105	58	105	59	94	96
26	—	3	—	88	—
89	110	18	0	92	76
10	0	6	0	84	74
83	65	69	52	28	0

[b]The label in aspartic acid derived from the protein fraction includes that present in the protein as asparagine.

Enzymes of the synthesis of aspartate-derived amino acids. Aspartate kinase has been isolated from a variety of higher plants and is regulated by the action of lysine, threonine, and methionine. The precise inhibitory effect of each end product apparently varies among species and is dependent upon the age of the tissue examined (Bryan, 1980).

Aspartate kinase from legume seedlings was originally described as being threonine sensitive (Aarnes and Rognes, 1974), however in young pea leaves the enzyme is sensitive to both lysine and threonine (Lea et al., 1979). Preliminary studies suggested that the enzyme was present in the chloroplasts, although the absolute distribution in the cell was not examined. The presence of aspartate kinase in spinach chloroplasts has also been reported (Wahnbaeck et al., 1979). The inhibition of aspartate kinase in the chloroplast by lysine readily explains the inhibition of lysine, threonine, homoserine, and isoleucine synthesis seen in Table 3. Aspartate kinase from *Anacystis nidulans* is very sensitive to threonine, 92% inhibition being detected at 8 mM. Lysine also inhibited the enzyme 32% at a concentration of 8 mM (Aarnes, 1974).

Dihydrodipicolinate synthase, the first enzyme unique to lysine synthesis, has been purified from wheat germ (Mazelis et al., 1977) and at least a proportion of the enzyme is in the chloroplast (Wallsgrove and Mazelis, 1980). Diaminopimelate decarboxylase has been shown to be localized in the chloroplast (Mazelis et al., 1976) and more recently 100% of the total cell activity was found to be plastid localized (Wallsgrove, R. M., and Mills, W. R., unpublished results), suggesting that all the lysine synthesis takes place in the chloroplast.

Homoserine dehydrogenase in maize has been suggested to be chloroplast located, although only 10% of the activity was shown to correlate with the organelle (Bryan et al., 1977). In pea leaves the enzyme is evenly distributed between the chloroplast and the cytoplasm. The chloroplast enzyme is sensitive to threonine, but the soluble enzyme is insensitive to threonine and sensitive to low levels of cysteine (R. M. Wallsgrove and J. K. Sainis, unpublished results). The inhibition by threonine of homoserine dehydrogenase explains the strong inhibition of homoserine synthesis demonstrated in chloroplasts in Table 3. The final two enzymes in the synthesis of threonine, homoserine kinase and threonine

Table 4. Comparison of the Nitrogen Metabolism of Chloroplasts and Cyanobacteria

Biochemical function	Chloroplasts	Blue-green cyanobacteria
Nitrate reductase	Probably not present, but may be loosely bound to outer membrane NADH-dependent	Firmly bound to membrane Ferredoxin-dependent
Nitrite reductase	All located in stroma Ferredoxin-dependent	Membrane bound Ferredoxin dependent
Glutamine synthetase	Mol wt 600,000; 12 subunits of 50,000 Not adenylated Inhibited by AMP + ADP Reversibly deactivated	Mol wt 330,000-390,000; 8 subunits of 45,000 Not adenylated Inhibited by AMP + ADP Reversibly deactivated
Glutamate synthase	Ferredoxin-dependent	Ferredoxin-dependent
Amino acid dehydrogenases	Low or zero glutamic dehydrogenase	Zero glutamic dehydrogenase, low levels of aspartic dehydrogenase and alanine dehydrogenase
Transaminases	Glutamate-oxaloacetate and glutamate pyruvate transaminases plus capability of forming the other protein amino acids	Glutamate-oxaloacetate and glutamate-pyruvate transaminases. No other enzymes investigated.
Synthesis of α-ketoacids	Cannot synthesize oxaloacetate or α-ketoglutarate; possibly able to synthesize some pyruvate via phosphenolpyruvate	Able to synthesize all α-ketoacids, but not via the Krebs cycle
Synthesis of carbon skeletons of amino acids	Able to synthesize serine, aromatic and aspartate-derived amino acids. No data available on glutamate-derived amino acids	Able to synthesize all amino acids
	Regulated by feedback inhibition, not repression	Regulated by feedback inhibition, not repression

synthase, have been shown to be localized in the chloroplasts of barley (Aarnes, 1979).

Chloroplasts have the enzymes required for O-phosphohomoserine synthesis and cysteine synthesis (see Chapter 10 of the present volume). The only other enzyme in methionine biosynthesis that has been studied is the final one responsible for the methylation of homocysteine. Homocysteine-dependent 5-methyltetrahydropteroyl glutamate transmethylase has been shown to be localized in the chloroplast (Shah and Cossins, 1970) and mitochondria (Clandinin and Cossins, 1974). The enzyme localization studies would tend to confirm the reports of Mills and Wilson (1978a, b) and Mills et al. (1980) that chloroplasts are capable of synthesizing methionine, although only at very low rates.

Threonine deaminase has been isolated from a number of higher plants (Kagan et al., 1969; Dougall, 1970; Sharma and Mazunder, 1970), green algae and cyanobacteria (Desai et al., 1972; Antia and Kripps, 1973). All the enzymes show feedback inhibition by isoleucine, although threonine deaminase from the cyanobacteria required 100 times more isoleucine to produce 50% inhibition compared with the enzyme from a green alga (Antia and Kripps, 1973). The inhibition of the *A. variabilis* enzyme is apparently relieved by valine (Hood and Carr, 1968); similarly, the higher plant enzyme is activated by valine (Kagan et al., 1969). No precise data are available concerning the localization of the enzyme within the plant cell, although it is known to be particulate (Kagan et al., 1969).

Acetolactate synthase is localized in root and leaf plastids (Miflin, 1974a), and is inhibited by leucine and valine (Miflin and Cave, 1972). The enzyme from *A. variabilis* is only inhibited 50% by valine, and it was suggested that two enzymes may exist (Hood and Carr, 1968). No evidence for feedback repression of any of the enzymes of the pathways of branched-chain amino acid synthesis has been detected in the cyanobacteria.

Conclusions

A summary of the nitrogen metabolism of chloroplasts and cyanobacteria can be seen in Table 4. The major difference appears to be in nitrate reductase, but it must be noted that there is still some discussion as to the localization of nitrate reductase within the leaf cell.

If the endosymbiotic hypothesis for the origin of the chloroplast is to be considered, it must be assumed that chloroplasts have lost the ability to carry out the two first steps in the assimilation of inorganic nitrogen: either the reduction of nitrogen gas or nitrate. The chloroplasts have also lost the property of converting the CO_2 that has been fixed in photosynthetic reactions, into α-keto acids which may be used for amino acid formation, and ultimately for protein synthesis.

We are indebted to Sue Wilson for her patience and considerable effort in the preparation of the manuscript.

168   P. J. Lea et al.

References

Aarnes, H. 1974. Aspartate kinase from some higher plants and algae. Physiol. Plant 32:400–402.

Aarnes, H., and Rognes, S. E. 1974. Threonine-sensitive aspartate kinase and homoserine dehydrogenase from *Pisum sativum*. Phytochemistry 13:2717–2724.

Aarnes, H. 1979. Comparative studies on regulatory mechanisms for the biosynthesis of the aspartate family of amino acids in seedlings of *Pisum sativum* L. and *Hordeum vulgare* L. Ph.D. thesis, University of Oslo.

Anderson, J. W. and Done, J. 1977a. Polarographic study of ammonia assimilation by isolated chloroplasts. Plant Physiol. 60:504–508.

Anderson, J. W., and Done, J. 1977b. A polarographic study of glutamate synthase activity in isolated chloroplasts. Plant Physiol. 60:354–359.

Anderson, J. W., and Done, J. 1978. Light-dependent assimilation of nitrite by isolated pea chloroplasts. Plant Physiol. 61:692–697.

Antia, N. J., and Kripps, R. S. 1973. Threonine dehydratase from blue-green algae. Arch. Microbiol. 94:29–46.

Aslam, M., Huffaker, R. C., Rains, W. D., and Rao, K. P. 1979. Influence of light and ambient carbon dioxide concentration on nitrate assimilation by intact barley seedlings. Plant Physiol. 63:1205–1209.

Atkinson, D. E. 1968. The energy charge of the adenylate pool as a regulatory parameter. Interaction with feedback modifiers. Biochemistry 7:4030–4034.

Awonaike, K. O. 1980. Studies on the development and properties of ammonia assimilating enzymes in *Phaseolus vulgaris* L. root nodules. Ph.D. thesis, University of London.

Batt, T., and Brown, D. H. 1974. The influence of inorganic nitrogen supply on amination and related reactions in the blue-green alga, *Anabaena cylindrica*. Planta 116:27–37.

Beevers, L., and Hageman, R. H. 1980. Nitrate and nitrite reduction. *In* The Biochemistry of Plants, Vol. 5, Amino Acids and Derivatives. B. J. Miflin (ed.). New York: Academic Press.

Bickel, H., Palme, L., and Schultz, C. 1978. Incorporation of shikimate and other precursors into aromatic amino acids and prenylquinones of isolated spinach chloroplasts. Phytochemistry 17:119–124.

Bickel, H., and Schultz, G. 1979. *In* Advances in the Biochemistry and Physiology of Plant Lipids. L.-A. Appelquist and C. Lilzenberg (eds). Amsterdam: Elsevier-North Holland, pp. 377–380.

Bryan, J. K., Lissik, E. A., and Matthews, B. F. 1977. Changes in enzyme regulation during growth of maize III. Intracellular localisation of homoserine dehydrogenase in chloroplast. Plant Physiol. 59:673–679.

Bryan, J. K. 1980. The synthesis of aspartate family and branched chain amino acids. *In* The Biochemistry of Plants, Vol. 5. Amino Acids and Derivatives. B. J. Miflin (ed.). New York: Academic Press.

Bucholz, B., Reupke, B., Bickel, H., and Schultz, G. 1979. Reconstitution of amino acid synthesis by combining spinach chloroplasts with other leaf organelles. Phytochemistry 18:1109–1113.

Butz, R. G., and Jackson, W. A. 1977. A mechanism for nitrate transport and reduction. Phytochemistry 16:409–417.

Campbell, W. H. 1976. Separation of soybean leaf nitrate reductases by affinity chromatography. Plant Sci. Lett. 7:239–247.

Candau, P., Manzano, C., and Losada, M. 1976. Bioconversion of light energy into chemical energy through reduction with water of nitrate to ammonia. Nature 262:715–717.

Canvin, D. T., and Atkins, C. A. 1974. Nitrate, nitrite and ammonia assimilation by leaves: effect of light, carbon dioxide and oxygen. Planta 116:207–224.

Canvin, D. T., and Woo, K. C. 1979. The regulation of nitrate reduction in spinach leaves. Can. J. Bot. 57:1155–1160.

Cheng, K. H., Miller, A. G., and Colman, B. 1972. An investigation of glycollate excretion in two species of blue-green algae. Planta 103:110–116.

Clandinin, M. T., and Cossins, E. A. 1974. Methionine biosynthesis in isolated *Pisum sativum* mitochondria. Phytochemistry 13:585–591.

Codd, G. A., and Stewart, W. D. P. 1973. Pathways of glycollate metabolism in the blue-green algae *Anabaena cylindrica*. Arch. Microbiol. 94:11–28.

Coupe, M., Champigny, M. L., and Moyse, A. 1967. On the intracellular localisation of nitrate reductase in the leaves and roots of barley. Physiol. Veg. 5:271–291.

Dalling, M. J., Tolbert, N. E., and Hageman, R. H. 1972. Intracellular location of nitrate reductase and nitrite reductase. I. Spinach and tobacco leaves. Biochim. Biophys. Acta 238:505–512.

Del Campo, F. A., Paneque, A., Ramirez, J. M., and Losada, M. 1963. Nitrate reduction in the light by isolated chloroplasts. Biochim. Biophys. Acta 66:450–452.

Desai, I. D., Laub, D., and Antia, N. J. 1972. Comparative characterisation of L-threonine dehydratase in seven species of unicellular marine algae. Phytochemistry 11:272–287.

Dharmawardene, M. W. N., Haystead, A., and Stewart, W. D. P. 1973. Glutamine synthase of the nitrogen-fixing alga *Anabaena cylindrica*. Arch. Microbiol. 90:281–295.

Dharmawardene, M. W. N., Stewart, W. D. P., and Stanley, S. 1972. NItrogenase activity, amino acid pool patterns and amination in blue-green algae. Planta 108:133–145.

Dougall, D. K. 1970. Threonine deaminase from Paul's scarlet rose tissue cultures. Phytochemistry 9:959–964.

Duel, T. F., and Stadtman, E. R. 1970. Some kinetic properties of *Bacillus subtilis* glutamine synthetase. J. Biol. Chem. 245:5206–5213.

Ehmke, A., and Hartmann, T. 1976. Properties of glutamate dehydrogenase from *Lemna minor*. Phytochemistry 15:1611–1617.

Emes, M. J., and Fowler, M. W. 1978. The intracellular localisation of nitrate assimilation in the apices of seedling pea roots. Planta 144:249–253.

Feierabend, J., and Brassel, M. 1977. Subcellular localisation of shikamate dehydrogenase in higher plants. Z. Pflanzenphysiol. 82:334–346.

Fowler, M. W., and Barker, D. J. 1979. Assimilation of ammonium in non-chlorophyllous tissue. *In* The Nitrogen Assimilation of Plants. E. J. Hewitt and C. V. Cutting (eds). London: Academic Press, pp. 489–500.

Gayler, K. R., and Morgan, W. R. 1976. An NADP-dependent glutamate dehydrogenase in chloroplasts from the marine green alga *Caulerpa simpliciuscula*. Plant Physiol. 58:283–287.

Gilchrist, D. G., and Kosuge, T. 1974. Regulation of aromatic amino acid biosynthesis in higher plants. Arch. Biochem. Biophys. 164:95–105.

Gilchrist, D. G., and Kosuge, T. 1980. Aromatic amino acid biosynthesis and its regulation. *In* The Biochemistry of Plants, Vol. 5. Amino Acids and Derivatives. B. J. Miflin (ed.). New York: Academic Press.

Givan, C. V. 1980. Aminotransferases in higher plants. *In* The Biochemistry of Plants, Vol. 5. Amino Acids and Derivatives. B. J. Miflin (ed). New York: Academic Press.

Givan, C. V., Givan, A., and Leech, R. M. 1970. Photoreduction of α-ketoglutarate to glutamate by *Vicia faba* chloroplasts. Plant Physiol. 45:624–630.

Givan, C. V. 1975. Light-dependent synthesis of glutamine in pea chloroplast preparations. Planta 122:281–291.

Grant, B. R., Atkins, C. A., and Canvin, D. T. 1970. Intracellular location of nitrate and nitrite reductase in spinach and sunflower leaves. Planta 94:60–72.

Grant, B. P., and Canvin, D. T. 1970. The effect of nitrate and nitrite on oxygen evolution and carbon dioxide assimilation and the reduction of nitrate and nitrite by intact chloroplasts. Planta 95:227–246.

Grodzinski, B. 1978. Glyoxylate decarboxylation during photorespiration. Planta 144:31–38.

Grosse, W. 1976. Enzymes of tryptophan biosynthesis in etioplasts of *Pisum sativum* L. Z. Pflanzenphysiol. 80:463–468.

Guerrero, M. G., Manzano, C., and Josada, M. 1974. Nitrite photoreduction by a cell free preparation of *Anacystis nidulans*. Plant Sci. Lett. 3:273–278.

Guis, C., Hirel, B., Shedlofsky, C., and Gadal, P. 1979. Occurrence and influence of light on the relative proportions of two glutamine synthetases in rice leaves. Plant Sci. Lett. 15:271–274.

Harel, E., Lea, P. J., and Miflin, B. J. 1977. The localisation of enzymes of nitrogen assimilation in maize leaves, and their activities during greening. Planta 134:195–200.

Hattori, A., and Uesugi, I. 1968. Ferredoxin-dependent photoreduction of nitrate and nitrite by subcellular preparations of *Anabaena cylindrica*. *In* Comparative Biochemistry and Biophysics of Photosynthesis. K. Shibata, A. Takamiya, A. T. Jagendorf, and R. C. Fuller (eds.). Tokyo: University of Tokyo Press, pp. 201–236.

Haystead, A. 1973. Glutamine synthetase in the chloroplasts of *Vicia faba*. Planta 111:271–274.

Haystead, A., Dharmawardene, M. W. N., and Stewart, W. D. P. 1973. Ammonia assimilation in a nitrogen-fixing blue-green algae. Plant Sci. Lett. 1:439–445.

Heber, U., and French, C. S. 1968. Effect of oxygen on the electron transport chain of photosynthesis. Planta 79:99–112.

Hewitt, E. J., Hucklesby, D. P., Mann, A. F., Notton, B., and Rucklidge, G. J. 1979. Regulation of nitrate assimilation in plants. *In* Nitrogen Assimilation in Plants. E. J. Hewitt and C. V. Cutting (eds.). London: Academic Press, pp. 255–288.

Hoare, D. S., Hoare, S. L., and Moore, R. B. 1967. The photoassimilation of organic compounds by autotrophic blue-green algae. J. Gen. Microbiol. 49:351–370.

Hood, W., and Carr, N. G. 1968. Threonine deaminase and acetolactate synthetase in *Anabaena variabilis*. Biochem. J., 109:4p.

Hood, W., and Carr, N. G. 1971. Apparent lack of control by repression of arginine metabolism in blue-green algae. J. Bact. 107:365–367.

Jackson, C., Drench, J. E., Morris, P., Lui, S. C., Hall, D. O., and Moore, A. L. 1979. Photorespiratory nitrogen cycling: Evidence for a mitochondrial glutamine synthetase. Biochem. Soc. Trans. 7:1122–1124.

Jensen, R. A., and Pierson, D. L. 1975. Evolutionary implications of different types of microbial enzymology for L-tyrosine biosynthesis. Nature 254:667–671.

Jensen, R. A., Stenmark-Cove, S., and Ingram, L. O. 1974. Mis-regulation of 3-deoxy-D-arabino-heptulosonate-7-phosphate synthetase does not account for growth inhibition by phenylalanine in *Agmenellum quadruplicatum*. J. Bact. 120:1124–1132.

Johansson, B. C., and Gest, H. 1977. Adenylation/deadenylation control of the glutamine synthetase of *Rhodopseudomonas capsulata*. Eur. J. Biochem. 81:365–371.

Jones, R. W., and Sheard, R. W. 1978. Accumulation and stability of nitrite in intact aerial leaves. Plant Sci. Lett. 11:285–291.

Kagan, Z. S., Sinel'nnikova, E. M., and Kretovich, W. L. 1969. α-Threonine dehydratases of flowering parasitic and saprophytic plants. Enzymologia 36:335–352.

Kanamori, T., and Matsumoto, H. 1972. Glutamine synthetase from rice plant roots. Arch. Biochem. Biophys. 152:404–412.

Keys, A. J. 1980. Synthesis and interconversion of glycine and serine. *In* The Biochemistry of Plants, Vol. 5, Amino Acid and Derivatives. B. J. Miflin (ed.). New York: Academic Press.

Keys, A. J., Sampaio, E. V. S. B., Cornelius, M. J., and Bird, I. F. 1977. Effect of temperature on photosynthesis and photorespiration of wheat leaves. J. Exp. Bot. 28:525–533.

Keys, A. J., Bird, I. F., Cornelius, M. J., Lea, P. J., Wallsgrove, R. M., and Miflin, B. J. 1978. Photorespiratory nitrogen cycle. Nature 275:741–743.

Kingdom, H. S. 1974. Feedback inhibition of glutamine synthetase from green pea seeds. Arch. Biochem. Biophys. 163:429–431.

Kirk, P. R., and Leech, R. M. 1972. Amino acid biosynthesis by isolated chloroplasts during photosynthesis. Plant Physiol. 50:228–234.

Kirkman, M. A., and Miflin, B. J. 1979. Nitrate content and amino composition of the xylem fluid of spring wheat throughout the growing season. J. Sci. Food Agric. 30:653–660.

Klepper, L. A., Flesher, D., Hageman, R. H. 1971. Generation of reduced nicotinamide adenine dinucleotide for nitrate reduction in green leaves. Plant Physiol. 48:580–590.

Kretovich, W. L. 1965. Some problems of amino acid and amide biosynthesis in plants. Annu. Rev. Plant Physiol. 16:141–154.

Ladha, J. K., Powell, P., and Stewart, W. D. P. 1978. Effects of 5'-hydroxylysine on acetylene reduction and NH4 assimilation in the cyanobacterium *Anabaena cylindrica*. Biochem. Biophys. Res. Commun. 83:688–696.

Larsson, C. 1979. $^{14}CO_2$ fixation and compartmentation of carbon metabolism in a recombined chloroplast-"cytoplasm" system. Physiol. Plant 46:221–226.

Larsson, C., and Albertsson, E. 1974. Photosynthetic $^{14}CO_2$ fixation by chloroplast populations isolated by a polymer two phase technique. Biochim. Biophys. Acta 357:412–419.

Larsson, C., and Albertsson, E. 1979. Enzymes related to serine synthesis in spinach chloroplasts. Physiol. Plant 45:7–10.

Lea, P. J., and Miflin, B. J. 1974. An alternative route for nitrogen assimilation in higher plants. Nature 251:614–616.

Lea, P. J., and Miflin, B. J. 1975. Glutamate synthase in blue-green algae. Biochem. Soc. Trans. 3:381–384.

Lea, P. J., and Miflin, B. J. 1979. Phytosynthetic ammonia assimilation. *In* The Encyclopaedia of Plant Physiology, Vol. 6. M. Gibbs and E. Latzko (eds.). Berlin, Heidelberg: Springer Verlag, pp. 445–456.

Lea, P. J., and Miflin, B. J. 1980. The transport and metabolism of asparagine and other nitrogen compounds within the plant. *In* The Biochemistry of Plants, Vol. 5, Amino Acids and Derivatives. B. J. Miflin (ed.). New York: Academic Press.

Lea, P. J., and Thurman, D. A. 1972. Intracellular location and properties of plant L-glutamate dehydrogenases. J. Exp. Bot. 23:440–449.

Lea, P. J., Mills, W. R., and Miflin, B. J. 1979. The isolation of a lysine sensitive aspartate kinase from pea leaves and its involvement in homoserine biosynthesis in isolated chloroplasts. FEBS Lett. 98:165–168.

Leech, R. M., and Kirk, P. R. 1968. NADP-dependent L-glutamate dehydrogenase from the chloroplasts of Vicia faba. Biochem. Biophys. Res. Commun. 32:685–690.

Lex, M., Silvester, W. B., and Stewart, W. D. P. (1972). Photorespiration and nitrogenase activity in the blue-green alga, Anabaena cylindrica. Proc. Roy. Soc. B 180:87–102.

Linko, P., Holm-Hansen, O., Bassham, J. A., and Calvin, M. 1957. Formation of radioactive citrulline during photosynthesis $C^{14}O_2$-fixation by blue-green algae. J. Exp. Bot. 8:147–154.

Lips, S. H., and Avissar, Y. 1972. Plant-leaf microbodies as the intracellular site of nitrate reductase and nitrite reductase. Eur. J. Biochem. 29:20–24.

Losada, M., and Guerrero, M. G. 1979. The photosynthetic reduction of nitrate and its regulation. In Photosynthesis in Relation to Model Systems J. Barker (ed.). New York: Elsevier-North-Holland, pp. 365–408.

McKenzie, G. H., Ch'ng, A. L., and Gayler, K. R. 1979. Glutamine synthetase/glutamine: α-ketoglutarate amino transferase in chloroplasts from the marine alga Caulerpa simpliciuscula. Plant Physiol. 3:578–582.

McParland, R. H., Guevara, J. G., Becker, R. R., and Evans, H. J. 1976. The purification and properties of glutamine synthetase from the cytosol of soya-bean root nodules. Biochem. J. 153:597–606.

Magalhaes, A. C., Neyra, C. A., and Hageman, R. H. 1974. Nitrite assimilation and amino nitrogen synthesis in isolated spinach chloroplasts. Plant Physiol. 53:411–415.

Magasanik, B. 1976. Classical and post-classical modes of regulation of the synthesis of degradative bacterial enzymes. Prog. Nucl. Acid. Res. Mol. Biol. 17:99–115.

Mann, A. F., Fentem, P. A., and Stewart, G. R. 1979. Identification of two forms of glutamine synthetase in barley (Hordeum vulgare) Biochem. Biophys. Res. Commun. 88:515–521.

Mann, A. F., Hucklesby, D. P., and Hewitt, E. J. 1978. Sources of reducing power for nitrate reduction in spinach leaves. Planta 140:261–263.

Mann, A. F., Hucklesby, D. P., and Hewitt, E. J. 1979. Effect of aerobic and anaerobic conditions on the in vivo nitrate reductase assay in spinach leaves. Planta 146:83–89.

Mazelis, M., Miflin, B. J., and Pratt, H. M. 1976. A chloroplast-localised diaminopimelate decarboxylase in higher plants. FEBS Lett. 64:197–200.

Mazelis, M., Whatley, F. R., and Whatley, J. 1977. The enzymology of lysine biosynthesis in higher plants. The occurrence, characteristics and some regulatory properties of dihydrodipicolinate synthase. FEBS Lett. 84:236–240.

Meeks, J. C., Wolk, C. P., Thomas, J., Lockau, W., Shaffer, P. W., Austin, S. M., Chien, W.-S. and Galonsky, A. 1977. The pathways of assimilation of $^{13}NH_4^+$ by the cyanobacterium Anabaena cylindrica. J. Biol. Chem. 252:7894–7900.

Miflin, B. J. 1974a. The location of nitrite reductase and other enzymes related to amino acid biosynthesis in the plastids of roots and beans. Plant Physiol. 54:550–555.

Miflin, B. J. 1974b. Nitrite reduction in leaves: studies on isolated chloroplasts. Planta 116:187–196.

Miflin, B. J. 1977. Modification controls in time and space. In The Regulation of Enzyme Synthesis and Activity in Higher Plants. H. Smith (ed.). London: Academic Press, pp. 23–40.

Miflin, B. J., and Cave, P. R. 1972. The control of leucine, isoleucine and valine biosynthesis in a range of higher plants. J. Exp. Bot. 23:511–516.

Miflin, B. J., and Lea, P. J. 1975. Glutamine and asparagine as nitrogen donors for reductant-dependent glutamate synthesis in pea roots. Biochem. J. 149:403–409.

Miflin, B. J., and Lea, P. J. 1976. The pathways of nitrogen assimilation in plants. Phytochemistry 15:873–885.

Miflin, B. J., and Lea, P. J. 1977. Amino acid metabolism. Annu. Rev. Plant Physiol. 28:299–329.

Miflin, B. J., and Lea, P. J. 1980. Ammonia assimilation. In The Biochemistry of Plants, Vol. 5, Amino Acids and Derivatives. B. J. Miflin (ed.). New York: Academic Press.

Mills, W. R., Lea, P. J., and Miflin, B. J. 1980. Photosynthetic formation of the aspartate family of amino acids in isolated chloroplast. Plant Physiol. 65:1166–1172.

Mills, W. R., and Wilson, K. G. 1978a. Amino acid biosynthesis in isolated pea chloroplast: metabolism of labelled aspartate and sulphate. FEBS Lett. 92:129–132.

Mills, W. R., and Wilson, K. G. 1978b. Effects of lysine, threonine and methionine on light-driven protein synthesis in isolated pea (Pisum sativum L.) chloroplasts. Planta 142:153–160.

Mitchell, C. A., and Stocking, C. R. 1975. Kinetics and energetics of light-driven chloroplast glutamine synthesis. Plant Physiol. 55:59–63.

Noguchi, M., Koiwai, A., Yokoyama, M., and Tamaki, E. 1968. Studies on nitrogen metabolism in tobacco plants. Plant Cell Physiol. 9:35–47.

Oaks, A. 1979. Nitrate reductase in roots and its regulation. In Nitrogen Assimilation in Plants. E. J. Hewitt and C. V. Cutting (eds.). London: Academic Press, pp. 217–226.

O'Neal, D., and Joy, K. W. 1973. Glutamine synthetase of pea leaves. I. Purification, stabilisation and pH optima. Arch. Biochem. Biophys. 159:113–122.

O'Neal, D., and Joy, K. W. 1973. Localisation of glutamine synthetase in chloroplasts. Nature New Biol. 246:61–62.

O'Neal, D., and Joy, K. W. 1975. Pea leaf glutamine synthetase: regulatory properties. Plant Physiol. 55:968–974.

Ortega, T., Castillo, F., and Cardenas, J. 1976. Photolysis of water coupled to nitrate reduction by Nostoc muscorum subcellular particles. Biochem. Biophys. Res. Commun. 71:885–891.

Ownby, J. D. 1977. Effects of amino acids on methionine sulphoximine-induced heterocyst formation in Anabaena. Planta 136:277–279.

Pate, J. S. 1973. Uptake, assimilation and transport of nitrogen compounds by plants. Soil Biol. Biochem. 5:109–123.

Plaut, Z., Lendzian, K., and Bassham, J. A. 1977. Nitrite reduction in reconstituted and whole spinach chloroplasts during carbon dioxide reduction. Plant Physiol. 59:184–188.

Plaut, Z., and Littan, A. 1975. Interaction between photosynthetic CO_2 fixation products and nitrate reduction in spinach and wheat leaves. In Proceedings of the Third International Congress on Photosynthesis. M. Avron (ed.). Amsterdam: Elsevier, pp. 1507–1516.

Rathnam, C. K. M. 1978. Malate and dihydroxyacetone phosphate-dependent nitrate reduction in spinach leaf protoplasts. Plant Physiol. 62:220–223.

Rathnam, C. K. M., and Das, V. S. R. 1974. Nitrate metabolism in relation to the aspartate-type C-4 pathway of photosynthesis in Eleusine coracana. Can. J. Bot. 52:2599–2605.

Rathnam, C. K. M., and Edwards, G. E. 1976. Distribution of nitrate-assimilating enzymes between mesophyll protoplasts and bundle sheath cells in leaves of three groups of C_4 plants. Plant Physiol. 57:881–885.

Rhodes, D., Sims, A. P., and Stewart, G. R. 1979. Glutamine synthetase and the control of nitrogen assimilation in *Lemna minor* L. *In* Nitrogen Assimilation of Plants. E. J. Hewitt and C. V. Cutting (eds.). London: Academic Press, pp. 501–520.

Ritenour, G. L., Joy, K. W., Bunning, J., and Hageman, R. H. 1967. Intracellular localisation of nitrate reductase and glutamic acid dehydrogenase in green leaf tissue. Plant Physiol. 42:233–237.

Rowell, P., Enticott, S., and Stewart, W. D. P. 1977. Glutamine synthetase and nitrogenase activity in the blue-green alga *Anabaena cylindrica*. New Phytol. 79:41–54.

Rowell, P., Sampaio, M. J. A. M., Ladha, J. K., and Stewart, W. D. P. 1979. Alteration of cyanobacterial glutamine synthetase activity *in vivo* response to light and NH_4^+. Arch. Microbiol. 120:195–200.

Sampaio, M. J. A. M., Rowell, P., and Stewart, W. D. P. 1979. Purification and some properties of glutamine synthetase from the nitrogen-fixing cyanobacteria *Anabaena cylindrica* and of *Nostoc* sp. J. Gen. Microbiol. 111:181–191.

Santarius, K. A., and Stocking, C. R. 1969. Intracellular localisation of enzymes in leaves and chloroplast membrane permeability to compounds involved in amino acid synthesis. Z. Naturforsch 24B:1170.

Sawhney, S. K., Naik, M. S., and Nicholas, D. J. D. 1978. Regulation of nitrate reduction by light, ATP and mitochondria respiration in wheat leaves. Nature 272:647–648.

Sawhney, S. K., and Nicholas, D. J. D. 1978. Some properties of glutamine synthetase from *Anabaena cylindrica*. Planta 139:289–299.

Shah, S. P. J., and Cossins, E. A. 1970. Pteroylglutamates and methionine biosynthesis in isolated chloroplasts. FEBS Lett. 7:267–270.

Sharma, R. K., and Mazunder, R. 1970. Purification, properties and feedback control of L-threonine dehydratase from spinach. J. Biol. Chem. 245:3008–3014.

Shen, T. C., Funkhouser, E. A., and Guerrero, M. G. 1976. NADH and NADH(P) nitrate reductase in rice seedlings. Plant Physiol. 58:292–294.

Sims, A. P., Toone, J., and Box, V. 1974a. The regulation of glutamine metabolism in *Candida utilis:* mechanisms of control of glutamine synthetase. J. Gen. Microbiol. 84:149–162.

Sims, A. P., Toone, J., and Box, V. 1974b. The regulation of glutamine synthesis in the food yeast *Candida utilis:* the purification and subunit structure of glutamine synthetase and aspects of enzyme deactivation. J. Gen. Microbiol. 80:485–489.

Smith, A. J. 1973. Synthesis of metabolic intermediates. *In* The Biology of the Blue-green algae. N. G. Carr and B. A. Whitton (eds.). Oxford: Blackwell, pp. 1–38.

Solomonson, L. P., Lorimer, G. H., Hall, R. L., Borchers, R., and Bailey, J. L. 1975. Reduced nicotinamide adenine dinucleotide-nitrate reductase of *Chlorella vulgaris*. J. Biol. Chem. 250:4120–4127.

Stacey, G., Tabita, F. R., and Van Baalen, C. 1977. Nitrogen and ammonia assimilation in the cyanobacteria: purification of glutamine synthetase from *Anabaena* sp. strain C.A. J. Bacteriol. 132:596–603.

Stewart, G. R., Mann, A. F., and Fentem, P. A. 1980. The enzymes of glutamate formation: Glutamate dehydrogenase glutamine synthetase and glutamate synthase. *In* The Biochemistry of Plants, Vol. 5, Amino Acids and Derivatives. B. J. Miflin (ed.). New York: Academic Press, pp. 271–327.

Stewart, G. R. and Rhodes, D. 1977. A comparison of the characteristics of glutamine synthetase and glutamate dehydrogenase from *Lemna minor* L.. New Phytol. 79:257–268.

Stewart, G. R., and Rhodes, D. 1977. Control of enzyme levels in the regulation of nitrogen assimilation. *In* The Regulation of Enzyme Synthesis and Activity. H. Smith (ed.). London: Academic Press, pp. 1–22.

Stewart, W. D. P., and Rowell, P. 1975. Effects of L-methionine-DL-sulphoximine on the assimilation of newly fixed NH_3, acetylene reduction and heterocyst production in *Anabaena cylindrica*. Biochem. Biophys. Res. Commun. 65:846–856.

Stewart, W. D. P., Haystead, A., and Dharmawardene, M. W. N. 1975. Nitrogen assimilation and metabolism in blue-green algae. *In* Nitrogen Fixation by Free-living Micro-organisms. W. D. P. Stewart (ed.). Cambridge, England: Cambridge University Press, I.B.P., Vol. 6, pp. 129–158.

Swader, J. A., and Stocking, C. R. 1971. Nitrate and nitrite reduction by *Wolfia arrhiza*. Plant Physiol. 47:189–191.

Tate, S. S., and Meister, A. 1973. Glutamine synthetases of mamalian liver and brain. *In* The Enzymes of Glutamine Metabolism. S. Prusiner and E. R. Stadtman (eds.). New York: Academic Press, pp. 77–127.

Thomas, J., Meeks, J. C., Wolk, C. P., Shaffer, P. W., Austin, S. M., and Chien, W.-S. 1977. Formation of glutamine from [^{13}N] ammonia, [^{13}N] dinitrogen, and [^{14}C] glutamate by heterocysts isolated from *Anabaena cylindrica*. J. Bact. 129:1545–1555.

Thomas, S. M., Hall, N. P., and Merrett, M. J. 1978. Ribulose 1,5-bisphosphate carboxylase/oxygenase activity and photorespiration during the ageing of the flag leaves of wheat. J. Exp. Bot. 29:1161–1125.

Tsukamoto, A. 1970. Reductive carboxylation and amination of keto acids by spinach chloroplasts. Plant Cell Physiol. 11:221–230.

Valentine, R. C., Shapiro, B. M., and Stadtman, E. R. 1968. Regulation of glutamine synthetase XII. Electron microscopy of the enzyme from *E. coli*. Biochemistry 7:2143–2152.

Vega, J. M., and Kamin, H. 1977. Spinach nitrite reductase. Purification and properties of a siroheme-containing iron-sulphur protein. J. Biol. Chem. 252:896–909.

Vennesland, B., and Guerrero, M. G. 1979. Reduction of nitrate and nitrite. *In* The Encyclopaedia of Plant Physiology, Vol. 6. M. Gibbs and E. Latzko (eds.). New York: Springer-Verlag, pp. 425–444.

Wallsgrove, R. M., and Mazelis, M. 1980. Complete localisation of the regulatory enzyme dihydrodipicolinate synthase in the chloroplasts of spinach leaves. FEBS Lett. 116:189–192.

Wallsgrove, R. M., Harel, E., Lea, P. J., and Miflin, B. J. 1977. Studies on glutamate synthase from the leaves of higher plants. J. Exp. Bot. 28:588–596.

Wallsgrove, R. M., Keys, A. J., Bird, I. F., Cornelius, M. J., Lea, P. J., and Miflin, B. J. 1980. The location of glutamine synthetase in leaf cells and its role in the reassimilation of ammonia. J. Exp. Bot. 31:1005–1017.

Wallsgrove, R. M., Lea, P. J., and Miflin, B. J. 1979a. The distribution of the enzymes of nitrogen assimilation within the pea cell. Plant Physiol. 63:232–236.

Wallsgrove, R. M., Lea, P. J., and Miflin, B. J. 1979b. The reduction of inorganic nitrogen and its assimilation into glutamine and glutamate in pea chloroplasts. *In* Nitrogen Assimilation in Plants. E. J. Hewitt and C. V. Cutting (eds.). London: Academic Press, pp. 431–433.

Washitani, I. and Sato, S. 1978. Studies on the function of proplastides in the metabolism of *in vitro* cultured tobacco-cells. II. Glutamine synthetase/glutamate synthase pathway. Plant Cell Physiol. 19:43–50.

Weber, H. L., and Böck, A. 1969. Regulation of the chorismic acid branch-point in aromatic acid synthesis in blue-green and green algae. Arch. Microbiol. 66:250–258.

Wedler, F. C., and Hoffmann, F. M. 1974. Glutamine synthetase of *Bacillus stearoth-ermophilus*. I. Purification and basic properties. Biochemistry 13:3207–3214.

Weissman, G. S. 1976. Glutamine synthetase regulation of energy charge in sunflower roots. Plant Physiol. 57:339–343.

Wolk, C. P., Thomas, J., Shaffer, P. W., Austin, S. M., and Galonsky, A. 1976. Pathway of nitrogen metabolism after fixation of ^{13}N-labelled nitrogen gas by the cyanobacterium, *Anabaena cylindrica*. J. Biol. Chem. 251:5027–5034.

Woolfolk, C. A., and Stadtman, E. R. 1967. Regulation of glutamine synthetase. III. Cumulative feedback inhibition of glutamine synthetase from *Escherichia coli*. Arch. Biochem. Biophys. 118:736–755.

Wray, J. L., Small, I. S., and Brown, J. S. 1979. A model for the subunit composition of higher-plant NADH-nitrate reductase. Biochem. Soc. Trans. 7:739–741.

Discussion of Presentation by Dr. Lea et al.

VIGIL: To what extent do enzyme activities in peroxisomes, e.g. glutamate amino transferase, contribute to or regulate ammonia metabolism in leaf cells exposed to light?

LEA: We have never looked at the role of peroxisomes in amino acid metabolism. There is no evidence that any aminotransferase enzymes are involved in regulation.

COHEN: Can you tell us what percentage of plant lysine is made in the chloroplast? Is S-adenosyl methionine made in the chloroplast?

LEA: Unpublished data of ours indicate that all the diaminopimelate decarboxylase in pea leaf cells is present in the chloroplasts. This would suggest that at least the final step of lysine biosynthesis is solely in the chloroplast. Mazelis and Wallsgrove (1980) have shown that the first enzyme of lysine synthesis, dihydrodipicolinate synthase, is located solely in the chloroplast. High methionine adenosyltransferase activity has been demonstrated in pea leaf chloroplasts (Selmer, J., Aarnes, H., and Rognes, S.E., unpublished work).

SCHIFF: In experiments measuring reduction of $^{35}SO_4^{2-}$ in *Chlorella pyrenoidosa*, although cysteine and gluthathione are formed, no methionine accumulates in the free pools; labeled methionine is, however, found in the proteins. S-Adenosyl methionine is found to be labeled (Schiff, 1959, 1964). In the same strain, methionine is toxic if supplied to the cells in white light; in red light the cells can grow on methionine as a sulfur source (Hodson et al., 1971). *Chlorella vulgaris*, I believe, accumulates free methionine from sulfate (Shrift, 1959).

SCHMIDT: Methionine was not found by me in chloroplasts using $^{35}SO_4^{2-}$.

LEA: Using chloroplasts from *Pisum sativum* var. Feltham First, we have demonstrated only very low incorporation of ^{14}C-aspartate and $^{35}SO_4^{2-}$ into methionine. I am not certain that we are recovering a significant amount of the methionine in the extracts. We intend to continue working on this problem.

BASSHAM: Your comment on the need for the chloroplast to obtain the keto acids, pyruvate, oxalacetate, and α-ketoglutarate from the rest of the cell bears on our earlier discussion about whether or not isolated chloroplasts could make fatty acids from CO_2. Clearly they could not do so if pyruvate is not formed in the chloroplast. Your comment that only the aromatic amino acids are labeled by isolated chloroplasts photosynthesizing with $^{14}CO_2$ suggests that the label could come via the erythrose moiety of the shikimic acid pathway. Would you agree?

LEA: We have never carried out any experiments ourselves on pyruvate synthesis in chloroplasts. We do know from the data of Larsson (1979) that there is very little incorporation of $^{14}CO_2$ into amino acids in purified chloroplasts.

The data of Bickel et al. (1978) suggest that the shikimic acid pathway is operating in the chloroplast.

STANIER: "Driven by light" is an ambiguous phrase. It could mean light-activated allosteric control mechanisms. On the other hand, it could be a simple question of energy charge, without any allosteric control.

LEA: We use the term to include reactions that take place in the chloroplast only when the light is turned on.

BASSHAM: The utilization of electrons from the light reactions for CO_2, sulfate, and nitrite reduction suggests some interesting questions about the regulation of these processes. Presumably CO_2 reduction to supply metabolites for the cell and plant has a high priority, reduction of CO_2 for starch storage in the chloroplasts, and reduction of NO_2^- and SO_4^{2-} an intermediate priority. Do you have any information on possible regulation or allocation of reducing power?

LEA: We have no information on this process. Presumably at high light intensities, there would be no problems with the supply of reducing power. At low light levels, careful regulation would be required to insure the correct balance of reduced metabolites.

SCHMIDT: There are some experimental indications that in fact this could happen; I refer to experiments on H_2S emission by leaves. I am, at the moment, in Dr. Filner's Laboratory in Michigan State University working on this system, and our data seems to indicate that CO_2 fixation is reduced when high H_2S emission occurs which could be interpreted as different allocations of electrons coming from ferredoxin.

SCHIFF: Is it really true that there is nitrate reductase in chloroplasts?

LEA: We have no evidence that the enzyme is localized in the chloroplast (Wallsgrove et al., 1979). There is evidence that nitrate reductase activity is correlated with the chlorophyll content of the leaf (Sawhney et al., 1972). The possibility that the enzyme is *very* loosely attached to the outer membrane of the chloroplast should be considered (Rathnam and Das, 1974).

VON WETTSTEIN: The location of the nitrate reductase concerned is not definitely established.

Dr. A. Kleinhofs of Washington State University has shown that one nitrate reductase in the barley leaf is coded by a nuclear gene (Warner and Kleinhofs, 1981). He has isolated mutants which form inactive enzyme as detected by reaction with antibodies raised against the wild type enzyme. The mutants are fully viable.

References to Discussion

Bickel, H., Palme, L., and Schultz, C. 1978. Incorporation of shikimate and other precursors into aromatic amino acids and prenylquinones of isolated spinach chloroplasts. Phytochemistry 17:119–124.

Hodson, R. C., Schiff, J. A., and Mather, J. P. 1971. Studies of sulfate utilization by algae. 10. Nutritional and enzymatic characterization of *Chlorella* mutants impaired for sulfate utilization. Plant Physiol. 47:306–311.

Larsson, C. 1979. $^{14}CO_2$ fixation and compartmentation of carbon metabolism in a recombined chloroplast–"cytoplasm" system. Physiol. Plant 46:221–226.

Rathnam, C. K. M., and Das, V. S. R. 1974. Nitrate metabolism in relation to the aspartate-type C-4 pathway of photosynthesis in *Eleusine coracana*. Can. J. Bot. 55:2599–2605.

Sawhney, S. K., Prakash, V., and Naik, M. S. 1972. Nitrate and nitrite reductase activities in induced chlorophyll mutants of barley. FEBS Lett. 22:200–202.

Schiff, J. A. 1959. Studies on sulfate utilization by *Chlorella pyrenoidosa* using sulfate-S^{35}; the occurrence of S-adenosyl methionine. Plant Physiol. 34:73–80.

Schiff, J. A. 1964. Studies of sulfate utilization by algae. II. Further identification of reduced compounds formed from sulfate by *Chlorella*. Plant Physiol. 39:176–179.

Selmer, J., Aarnes, H., and Roghes, S. E. Unpublished.

Shrift, A. 1959. Nitrogen and sulfur changes associated with growth uncoupled from cell division in *Chlorella vulgaris*. Plant Physiol. 34:505–512.

Wallsgrove, R. M., Lea, P. J., and Miflin, B. J. 1979. The distribution of the enzymes of nitrogen assimilation within the pea cell. Plant Physiol. 63:232–236.

Wallsgrove, R. M., and Mazelis, M. 1980. Complete localisation of the regulatory enzyme dihydrodificolinole syntase in the chloroplast of spinach leaves. FEBS Lett. 46:189–192.

Warner, R. L., and Kleinhofs, A. 1981. Nitrate utilization by nitrate reductase–deficient barley mutants. Plant Physiol. 67:740–743.

Assimilation of Sulfur

Ahlert Schmidt

The chloroplast of higher plants has the capacity to reduce sulfate in a light-dependent reaction for the formation of cysteine. Thus, the chloroplast provides not only reduced carbon and nitrogen compounds, but also provides the higher plant cell with reduced sulfur compounds for further metabolism. This was shown about 10 years ago during my thesis work with Dr. Trebst (Schmidt, 1968; Trebst and Schmidt, 1969; Schmidt and Trebst, 1969). In this chapter I will discuss the pathway of sulfate reduction and the further incorporation of reduced sulfur into amino acids on the enzymatic level and I will compare (where data are available) the characteristics of various enzymes from bacteria, algae, and higher plants in order to analyze relationships which might give information related to the origin of chloroplasts. In this chapter I will not discuss the problem of sulfolipid formation, since Dr. Mudd has provided a discussion of lipids in general, and this area is included in his chapter of the present volume.

Sulfate as the main sulfur source of plants and algae is at a valence state of $+6$ and must be reduced to the valence state of -2 before it can be incorporated into amino acids, proteins, and coenzymes. Therefore eight electrons are needed for reduction, and since sulfate reduction in the chloroplast is light-dependent and all necessary enzymes are present, sulfate reduction is clearly a light-driven process. This means that assimilatory sulfate reduction is coupled to the electron transport chain of chloroplasts either directly, by the flow of activated electrons with a sufficiently negative redox potential, or through ATP provided by photophosphorylation, which is needed as an energy source for sulfate reduction.

Address reprint requests to Dr. Ahlert Schmidt, Professor, Botanisches Institut, Universität München, Menzinger Strasse 67, D-8000 München 19, West Germany.

The process of sulfate reduction can be divided into four parts: (a) activation of sulfate to form either adenosine-5'-phosphosulfate (APS) or adenosine 3'-phosphate 5'-phosphosulfate (PAPS); (b) transfer of this activated sulfate to a suitable acceptor; (c) reduction of this bound sulfite to bound sulfide; (d) incorporation of the reduced sulfur into cysteine. The pathway in the chloroplast follows the following scheme:

$$SO_4^{2-} \rightarrow (P)APS \rightarrow R{-}SO_3^- \rightarrow R{-}S^- \rightarrow Cys{-}S^-$$

There is the possibility that in some bacteria (including members of the Chromatiaceae) a second pathway with free intermediates might be operative:

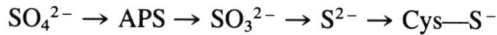

$$SO_4^{2-} \rightarrow APS \rightarrow SO_3^{2-} \rightarrow S^{2-} \rightarrow Cys{-}S^-$$

Note that the first pathway operates with bound intermediates, which implies that first a six-electron reduction step is involved and then a two-electron step, whereas in the second possibility (with free intermediates) the first reduction step needs two electrons for the reduction, followed by reduction to free sulfide, for which six electrons are needed. Later in this chapter I will discuss the enzymes involved in this scheme. I will compare enzymes from various organisms on which information is available related to the evolution of chloroplasts.

Prior to reduction, sulfate has to be activated to the sulfonucleotides APS and PAPS, since the redox potential of the sulfate-sulfite couple with -519 mV is too negative for reduction by ferredoxins, which have redox potentials of only -420 mV. After activation via the anhydride bond of either APS or PAPS this redox potential is shifted to -60 mV, which can be handled by living organisms (Siegel, 1975). This sulfate activation is achieved by coupling the two enzymes ATP-sulfurylase and APS-kinase together:

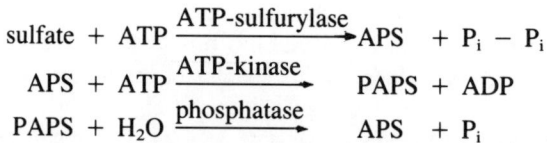

$$\text{sulfate} + \text{ATP} \xrightarrow{\text{ATP-sulfurylase}} \text{APS} + P_i - P_i$$
$$\text{APS} + \text{ATP} \xrightarrow{\text{ATP-kinase}} \text{PAPS} + \text{ADP}$$
$$\text{PAPS} + H_2O \xrightarrow{\text{phosphatase}} \text{APS} + P_i$$

The two enzymes ATP-sulfurylase and APS-kinase have been found in all organisms analyzed so far, including animals and humans (De Meio, 1975). However, APS-kinase is absent from dissimilatory sulfate reducers like *Thiobacillus* and *Desulfovibrio* (Peck, 1962). Phosphatases dephosphorylating PAPS to APS are mentioned, since such enzymes are of importance when sulfonucleotide specificities are discussed (Schmidt, 1979b). Two comments are relevant to the discussion of the sulfate activation sequence here. It was believed that PAPS formation was necessary to offset the unfavorable equilibrium constant for APS formation; the reaction can be pulled by reaction of APS with a second mole of ATP (De Meio, 1975) to form PAPS. However, other enzymes with about the same low K_m for APS and about the same change of free energy involved should be as effective as APS-kinase (Schmidt, 1976a). One example is the APS-sulfotransferase from photosynthetic organisms, and the second example is obtained from organisms like *Chromatium,* which have no APS-kinase but still grow on sulfate with APS as a metabolic intermediate (Schedel et al., 1979).

The first enzyme, ATP-sulfurylase, is found within the chloroplast based on observed specific activities in extracts from various chloroplast preparations (Schmidt, 1968; Balharry and Nicholas, 1970). A careful study of the distribution of this enzyme within plant organelles has not been done, but available data suggest that it can be found outside of the chloroplast as well. ATP-sulfurylase has been purified from leaf material and can also be found in roots (Cacco et al., 1977). Formation of PAPS has been demonstrated in bacteria, yeasts, molds, photosynthetic bacteria, cyanobacteria, and red and green algae, as well as in higher plants (Schmidt, 1979a,b). This demonstrates that the capacity for PAPS formation is more or less universal, with the exception of the dissimilatory sulfate reducers mentioned earlier. APS-kinase has been found in chloroplasts (Schmidt, 1968); this enzyme seems to be labile and at least partly attached to the chloroplast membrane (Mercer and Thomas, 1969; Schwenn et al., 1976). A purified APS-kinase has been obtained so far only from yeast. There are no data available that would permit a comparison of enzymes or provide information concerning the distribution of this enzyme within a plant. Thus it is not known whether this enzyme is found only in the chloroplast.

Reduction of sulfate by bacteria requires PAPS as the sulfate donor (Peck, 1962). The chloroplast of a higher plant, however, uses the sulfonucleotide APS specifically for further reduction (Schmidt, 1976a, 1979a). For this reason studies of the development of the APS system should give some information concerning the evolution of the chloroplast.

The enzyme involved in sulfate transfer from APS has been named *APS-sulfotransferase* (Schmidt, 1972a). It has been found in the green alga *Chlorella* by other workers (Tsang et al., 1971; Goldschmidt et al., 1975) and by myself using either spinach or *Chlorella* (Schmidt and Schwenn, 1971; Schmidt 1972a, b). This enzyme has the capacity to transfer the activated sulfate from APS to suitable acceptors forming a sulfonated compound, which has been identified as S-sulfoglutathione when glutathione was added as a thiol (Schmidt, 1972b; Tsang and Schiff, 1978b). This shows that the sulfo group of APS is transferred via a transferase to the thiol forming an organic thiosulfate:

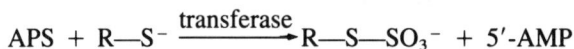

$$APS + R\!-\!S^- \xrightarrow{\text{transferase}} R\!-\!S\!-\!SO_3^- + 5'\text{-AMP}$$

This enzyme is nonspecific for the thiols used, however, the K_m values are different for each thiol (Schmidt, 1976a; Tsang and Schiff, 1976a). This enzyme has been detected in chloroplasts using isolated chloroplast preparations (Schmidt, 1972b; Schmidt, 1976a). Recent studies have shown that APS-sulfotransferase is found only in chloroplasts using spinach (Fankhauser and Brunold, 1978); it is not found in spinach mitochondria, microbodies, or the cytoplasm of the cell. There is some evidence that the mitochondrion might be involved in sulfate reduction in fungi and euglenoids (Bal et al., 1975; Brunold and Schiff, 1976); however, such data have not been obtained so far from green algae or higher plants. Studies in vivo have shown that the APS-sulfotransferase is accumulated during light-induced chloroplast development in etiolated plants (Schmidt, 1976b; Fankhauser, 1978). These data show that the APS-sulfotransferase is a typical chloroplast enzyme. Reduced sulfur compounds must be synthesized within the chloroplast, since one key enzyme for sulfate reduction is found exclusively within the chloroplast.

As I pointed out earlier in this chapter, bacterial sulfotransferase systems are specific for the sulfonucleotide PAPS. The finding that the sulfotransferase of higher plants and green algae is specific for APS led to the speculation that all oxygenic photosynthetic organisms might be of the plant type in their specificity towards APS (Tsang and Schiff, 1975) However, data obtained in my laboratory demonstrate that this generalization cannot be made. My data are in agreement with the observation of Tsang and Schiff (1975) that APS systems can be found in eukaryotic organisms. However, within the prokaryotic photosynthetic organisms—in the cyanobacteria and the photosynthetic bacteria—different sulfotransferases can be found.

Let me first consider the situation in bacteria. It has been shown that sulfate is activated to PAPS, and that PAPS is converted to a bound intermediate in yeast (Torii and Bandurski, 1967). This intermediate has also been found in *E. coli* (Tsang and Schiff, 1976b; 1978a). This reaction requires thioredoxin, but the exact role of this coenzyme has not been determined. At the time of writing it is believed that all nonphotosynthetic organisms with assimilatory sulfate reduction use PAPS as the sulfate donor for further reduction:

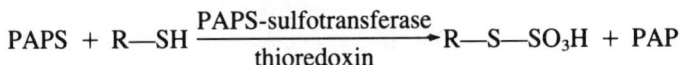

$$\text{PAPS} + \text{R—SH} \xrightarrow[\text{thioredoxin}]{\text{PAPS-sulfotransferase}} \text{R—S—SO}_3\text{H} + \text{PAP}$$

According to the work of Wilson and Bierer (1976) thioredoxin is the electron donor for the reduction of a small peptide which will accept the sulfo group from PAPS. The amino acid composition of this small "carrier" is not known.

Within the photosynthetic bacteria sulfotransferase systems with specificities for either APS or PAPS have been found (Schmidt, 1977; Schmidt and Trüper, 1977). However, these enzymes have not been well characterized, so we cannot comment on their thioredoxin dependence.

We have more information about various systems in cyanobacteria (Table 1). Sulfotransferase reactions can be found in these organisms which are similar to those obtained from bacterial systems as far as their specificity for PAPS and their requirement for thioredoxin are concerned (Schmidt, 1977; Schmidt and Christen, 1978; Wagner et al., 1978). Such a PAPS system has been found in *Synechococcus* 6301 and *Synechocystis* 6714. The *Synechococcus* PAPS-sulfotransferase is nonspecific for thioredoxins of various origins; it will accept, for example, thioredoxins from *E. coli,* other cyanobacteria, green algae, and higher plants (Wagner et al., 1978). During our screening of cyanobacteria with strains

Table 1. Thioredoxin Requirements of Various Sulfotransferases

A PAPS-sulfotransferase specific for thioredoxin, found in *Synechococcus* 6301	Relative activity (percent)[a]
PAPS + thioredoxin	1700
PAPS − thioredoxin	100
An APS-sulfotransferase specific for thioredoxin, found in *Chroococcidiopsis* 7203	
APS + thioredoxin	2200
APS − thioredoxin	100
An APS-sulfotransferase not requiring thioredoxin, found in *Plectonema* 73110	
APS + thioredoxin	165
APS − thioredoxin	100

[a]"Relative activity" is defined as acid-volatile radioactivity formed from ^{35}S-labeled APS or PAPS using partially purified enzyme systems. The activity obtained in the presence of dithioerythritol is set as 100%; the stimulation obtained by the addition of thioredoxin is shown.

obtained from Dr. Stanier, we have come across APS-sulfotransferases which differ in their requirement for thioredoxin, as is shown in Table 1. *Chroococcidiopsis* 7203 is given as an example, but thioredoxin-dependent APS sulfotransferases were also found in *Synechococcus* 6312 and in *Pseudanabena* 6901 (Schmidt and Christen, 1979). Besides these thioredoxin-dependent APS-sulfotransferases, the plant type APS-sulfotransferases are found in cyanobacteria as well (Tsang and Schiff, 1975), for instance in *Plectonema* 73110, which is active also in the absence of thioredoxin. We have demonstrated other APS-sulfotransferases in the LPP-group (*Lyngbya, Phormidium,* and *Plectonema*) as well; however, these sulfotransferases have not yet been characterized.

The APS-sulfotransferase from *Plectonema* does not need thioredoxin for activity, although it is stimulated to some extent by thioredoxin. These data clearly demonstrate that, within the cyanobacteria, different types of sulfotransferases can be found with characteristics of the bacterial type of sulfate reduction, or with characteristics of the chloroplast sulfotransferase of higher plants. A summary of sulfotransferases from different photosynthetic organisms is given in Table 2.

A few remarks about thioredoxins would be appropriate, since they are cofactors for some of the enzymes studied. We have isolated and characterized thioredoxins from *Synechococcus* and spinach using the *Synechococcus* PAPS-dependent sulfotransferase assay. Studies with these thioredoxins have shown that the fructose bisphosphatase from *Synechococcus* 6301 is activated by thioredoxin (Schmidt, 1979c), and the *Synechococcus* thioredoxin is also active in cyanobacterial ribonucleotide reductase. The activation of fructose bisphosphatase indicates that thioredoxin seems to have regulatory functions in cyanobacteria. Therefore it might have a role like the thioredoxins isolated from spinach chloroplasts (Buchanan et al., 1979). Spinach chloroplasts contain two different

Table 2. Properties of Sulfotranstenases of Sulfate Reduction from Various Organisms

Organism	Spinacia oleracea	Chlorella pyrenoidosa	Plectonema 73110	Rhodospirillum rubrum
Sulfonucleotide needed	APS	APS	APS	APS
Without thiol	1	1	4	0.3
+ Dithiothreitol	100	100	100	100
+ BAL (2,3 dithiopropane-1-ol)	64	1	81	48
+ Reduced Glutathione	98	99	22	136
+ Cysteine	22	2	14	4
+ Mercaptoethanol	13	2	31	3
Thioredoxin dependence	No	No	No	No
Inhibitory nucleotide	5'-AMP	5'-AMP	5'-AMP	5'-AMP

thioredoxins; we have recently found a second thioredoxin in *Synechococcus* (Schmidt, 1979); spinach and *Synechococcus* thioredoxins differ in their molecular weights.

The next enzyme to be mentioned is an APS-reductase, which catalyzes the reduction of APS with two electrons to free sulfite:

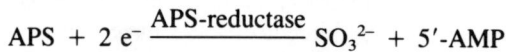

$$\text{APS} + 2\ e^- \xrightarrow{\text{APS-reductase}} SO_3^{2-} + 5'\text{-AMP}$$

This enzyme will reduce APS with appropriate electron donors (certain cytochromes) to free sulfite and 5'-AMP (Trüper, 1975). This reaction is reversible, which leads, under conditions of sulfite oxidation, to the formation of APS from sulfite and 5'-AMP. The energy stored in APS can be used for ATP formation (Peck, 1962); this might have been the earliest form of substrate level phosphorylation. This APS-reductase is found in *Thiobacillus, Desulfovibrio,* and in the Chromatiaceae—that is, *Chromatium, Thiocapsa,* and *Thiocystis*—which indicates a close relationship among these groups of organisms.

Let us come back to our main scheme of assimilatory sulfate reduction in chloroplasts. The next step in this sequence is the reduction of bound sulfite to bound sulfide. We have used the name "bound intermediate pathway," which was based on our finding that chloroplasts reduce sulfate without free intermediates; however, bound intermediates could be detected (Schmidt and Schwenn, 1971). The nature of the actual acceptor and of the chemical binding site still needs further investigation. Different low-molecular-weight proteins have been isolated or suggested to fill such a role in this bound intermediate pathway; rigorous proof for any of these postulated intermediates based on enzymological investigations is lacking (Schmidt, 1979a). The difficulty with these systems is

Table 2. (continued)

Porphyridium cruentum	Cyanophora paradoxa	Chroococcidiopsis 7203	Synechococcus 6312	Synechococcus 6301	Synechocystis 6714
APS	PAPS	APS	APS	PAPS	PAPS
3	4	1.4	0	0	0.7
100	100	100	100	100	100
34	142	50	31	5	15
80	22	4	2	1	6
15	33	4	5	1	4
18	4	5	2	1	7
No	No	Yes	Yes	Yes	Yes
?	5'-AMP 5'-AMP APS	5'-AMP	5'-AMP	3'-5'-ADP	?

the reactivity of organic thiosulfates and perhaps of other groups which we might have failed to isolate up to the time of writing. Some of my own data, based on isotopic exchange characteristics, suggest that the linkage may not be to a thiol. However, large-scale preparations of such fractions has not been possible so far and, as I have said before, these isolated fractions have to cooperate with the enzymes needed.

We suggest from data obtained from spinach and *Chlorella* that the bound sulfite is reduced to bound sulfide with an enzyme named *thiosulfonate reductase* (or *organic thiosulfate reductase*) (Schmidt, 1973; Schmidt et al., 1974), which catalyzes the following reaction:

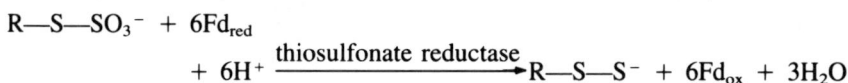

$$R—S—SO_3^- + 6Fd_{red} + 6H^+ \xrightarrow{\text{thiosulfonate reductase}} R—S—S^- + 6Fd_{ox} + 3H_2O$$

Model reactions suggest that the sulfur is bound to glutathione (Tsang and Schiff, 1978b); however, one should keep in mind that other linkage groups have not been ruled out, so this linkage to a thiol is only a suggestion. Progress in this area is small, and data indicate only that there is a reduction of bound sulfite to bound sulfide with the nature of the carrier unknown. All data obtained so far for chloroplasts indicate a small molecule with a mol wt between 1000 and 10,000. Ferredoxin is used in this reduction of bound sulfite to bound sulfide, which clearly shows that it is connected with the electron transport chain of the chloroplast (Schmidt and Schwenn, 1971; Schmidt, 1973; Schmidt et al., 1974). A bound intermediate pathway has been found in spinach and *Chlorella* as well as in *E. coli* (Tsang and Schiff, 1976) and yeast (Torii and Bandurski, 1967; Wilson and Bierer, 1976); in the two last organisms NADPH is used as electron donor for the 6-electron step. Evidence from mutants of *E. coli* suggests a reduction via sulfite reductase with free sulfite and sulfide as intermediates (Tsang and Schiff, 1978a).

The existence of this bound intermediate pathway leads to the question of siroheme sulfite reductases, which have been found in plants, algae, and phototrophic prokaryotes (Siegel, 1975). There is some evidence that sulfite reductases are involved in sulfide oxidation. This can be shown in *Chromatium vinosum* (Schedel et al., 1979). If this organism is grown on CO_2 and H_2S, high levels of sulfite reductase are found. Two percent of the total protein of the cells is a siroheme-sulfite reductase under these conditions. However, when *Chromatium vinosum* is grown on malate and sulfate, no siroheme-sulfite reductase is found, showing that this enzyme has dissimilatory functions in sulfur oxidation, and that this enzyme is not involved in assimilatory sulfate reduction. Unfortunately, such data are not available for cyanobacteria, algae, or higher plants. However, one has to be aware of these possibilities.

The last step in this scheme of assimilatory sulfate reduction is the incorporation of sulfur at the level of H_2S into cysteine. The enzyme involved is called o-acetyl-L-serine-sulfhydrylase, or cysteine synthase (Schmidt, 1979), and catalyzes the following reaction:

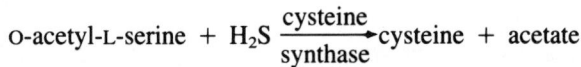

$$\text{o-acetyl-L-serine} + H_2S \xrightarrow[\text{synthase}]{\text{cysteine}} \text{cysteine} + \text{acetate}$$

Cysteine synthases have been isolated from a variety of higher plants and algae, and it has been shown that the molecular weights from different sources are between 60,000 and 70,000. In higher plants two different cysteine synthases have been detected; one has been shown to be present within the chloroplast, and the other is localized in the cytoplasm (Ng and Anderson, 1978; Fankhauser and Brunold, 1979). Isoenzymes of cysteine synthases have also been found in cyanobacteria by myself, and have been found in phototrophic bacteria which do not have chloroplasts (Hensel and Trüper, 1976). This raises the question of whether other important reactions are catalyzed by one of these enzymes. It seems possible that one cysteine synthase might be involved in thioether formation. This seems to be of no value at first glance, but I would like to remind the reader that a variety of cytochromes and the biliprotein chromophore of the cyanobacteria are linked to the corresponding proteins by thioether bridges. The loss of this ability to form thioethers could perhaps give the cell another mechanism to control the chloroplast. This is speculative at the time of writing, but I hope that it might stimulate some new experiments.

In Figure 1 I have incorporated the data obtained so far on the comparative biochemistry of sulfotransferases involved in assimilatory sulfate reduction. These data suggest relationships among various organisms, and give some hints for the phylogenetic origin of chloroplasts. I am aware that this is the most speculative area of my presentation and that I have neglected all other evidence

BACTERIA   PHOTOSYNTHETIC   CYANOBACTERIA   GREEN ALGAE
                        BACTERIA   AND HIGHER PLANTS

PAPS-   PAPS-   PAPS-
Sulfotransferase   Sulfotransferase   Sulfotransferase
 (TR-dependent)———————▶ (TR-dependent?)———————▶ (TR-dependent)
                          1. Rhodopseudomonas   1. Synechococcus
                          2. Rhodospirillum   2. Synechocystis

                                                APS-
                                                Sulfotransferase
                                                 (TR-dependent)

                                                  1. Synechococcus
                                                  2. Pseudanabaena

APS-   APS-   APS
Sulfotransferase   Sulfotransferase   Sulfotransferase
 (TR-independent)——————▶ (TR-independent)——————▶ (TR-independent)
   1. Rhodospirillum   1. Plectonema
      rubrum

APS-Reductase——————————▶APS-Reductase
  1. Thiobacillus   1. Chromatiacease
  2. Desulfovibrio
  3. Desulfomaculatum

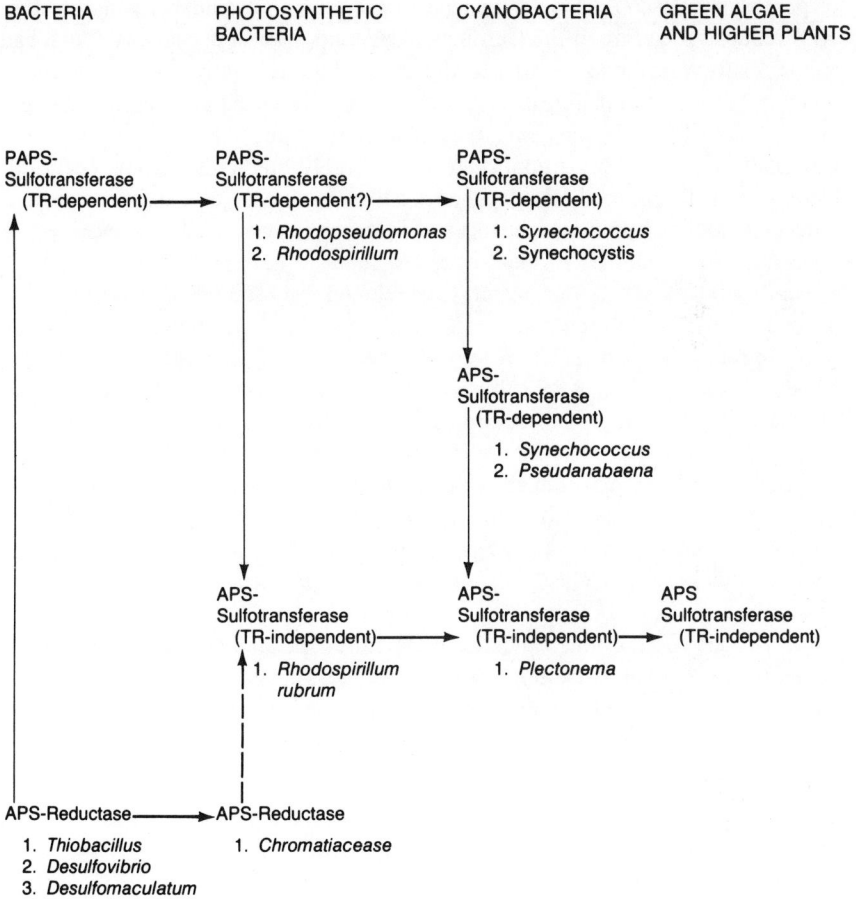

Figure 1. Possible relationships among the sulfotransferases for sulfate reduction from various prokaryotes.

obtained so far concerning other pathways. I have constructed a phylogenetic tree which is based only on the properties of different sulfonucleotide-dependent reactions. It is conceivable that the first enzyme which used sulfonucleotides was an APS-reductase. This might have been the first mechanism for a substrate-level phosphorylation, when the oxidation of reduced sulfur was used as an energy source. When sulfur supplies in the reduced form were exhausted, organisms had to adapt to other energy sources, and these organisms then needed mechanisms to reduce sulfur as needed. Thus the bacteria invented the enzymes known to us as *PAPS-sulfotransferases*. One would suggest that the invention of photosystem I could have happened twice: once in organisms which had an APS-reductase leading to the Chromatiaceae, and once in organisms which had a PAPS-sulfotransferase leading to the other phototrophic bacteria. It is likely that the PAPS-sulfotransferase changed into an APS-sulfotransferase, or it could be that the APS-sulfotransferases developed from the APS-reductases in the Chromatiaceae. The invention of photosystem II might then have occurred, leading from the thioredoxin-dependent sulfotransferases of the Rhodospirilla-ceae or Rhodopseudomonadaceae to such cyanobacteria as *Synechococcus* or *Synechocystis*. It can be further suggested that the development of the thiore-doxin-dependent APS-sulfotransferase developed from sources with thioredoxin-dependent PAPS-sulfotransferases. It might be more likely, however, that there was a direct development from the APS-dependent sulfotransferases in some Rhodospirillaceae to the APS-sulfotransferases in cyanobacteria like *Plectonema*, in which thioredoxin is not required for this activity. Since this type of APS-sulfotransferase is found in the chloroplasts of higher plants and in green algae, one would consider an organism with such a sulfotransferase within the cyano-bacteria to have a close relationship to the ancestors of higher-plant chloroplasts. Thus we should search in various groups of cyanobacteria for pathways or other features which have a close relationship to the properties of the higher-plant chloroplast.

I have tried to give you a picture of the sulfur metabolism in chloroplasts of higher plants. I certainly cannot prove that my ideas are correct. I hope, however, that further research will show that some points are correct and that others are not too far away from reality; in any case, through such work we will learn more about the origin of the chloroplast.

References

Bal, J., Maleszka, R., Stepien, P., and Cybis, J. 1975. Subcellular mislocation of cysteine synthase in a cysteine auxotroph of *Aspergillus nidulans*. FEBS Lett. 58:164–166.

Balharry, G. J. E., and Nicholas, D. J. D. 1970. ATP-sulphurylase in spinach leaves. Biochim. Biophys. Acta 220:513–524.

Brunold, C., and Schiff, J. A. 1976. Studies of sulfate utilization by algae. 15. Enzymes of assimilatory sulfate reduction in *Euglena* and their cellular localization. Plant Physiol. 57:430–436.

Buchanan, B. B., Wolosiuk, R. A., and Schürmann, P. 1979. Thioredoxin and enzyme regulation. TIBS 4:93–96.

Cacco, G., Saccomani, M., and Ferrari, G. 1977. Development of sulfate uptake capacity and ATP-sulfurylase activity during root elongation in maize. Plant Physiol. 60:582–584.

De Meio, R. H. 1975. Sulfate activation and transfer. *In* The Metabolism of Sulfur Compounds. M. Greenberg (ed.). New York, San Francisco, London: Academic Press.

Fankhauser, H. 1978. Untersuchungen zur Lokalisierung der assimilatorischen Sulfatre-duktion bei Spinat (*Spinacia oleracea* L.) Thesis, University of Bern.

Fankhauser, H., and Brunold, C. 1978. Localization of adenosine-5'-phosphosulfate sulfotransferase in spinach leaves. Planta 143:285–289.

Fankhauser, H., and Brunold, C. 1979. Localization of O-acetyl-L-serine sulfhydrylase in *Spinacia oleracea* L. Plant Sci. Lett. 14:185–192.

Goldschmidt, E. E., Tsang, M. L-S., and Schiff, J. A. 1975. Studies of sulfate utilization by algae. 13. Adenosine-5'-phosphosulfate (APS) as an intermediate in the conversion of adenosine-3'-phosphate-5'-phosphosulfate (PAPS) to acid-volatile radioactivity. Plant Sci. Lett. 4:293.

Hensel, G., and Trüper, H. G. 1976. Cysteine and S-sulfocysteine biosynthesis in phototrophic bacteria. Arch. Microbiol. 109:101–103.

Mercer, E. I., and Thomas, G. 1969. The occurrence of ATP-adenylsulphate 3'-phosphotransferase in the chloroplasts of higher plants. Phytochemistry 8:2281–2285.

Ng, B. H., and Anderson, J. W. 1978. Chloroplast cysteine synthases of *Trifolium repens* and *Pisum sativum*. Phytochemistry 17:879–885.

Peck, H. D., Jr. 1962. Comparative metabolism of inorganic sulfur compounds in microorganisms. Bacteriol. Rev. 26:67–94.

Schedel, M., Vanselow, M., and Trüper, H. G. 1979. Siroheme sulfite reductase isolated from *Chromatium vinosum*. Purification and investigation of some of its molecular and catalytic properties. Arch. Microbiol. 121:29–36.

Schmidt, A. 1968. Untersuchungen zum Mechanismus der photosynthetischen Sulfatreduktion isolierter Chloroplasten. Thesis, University of Göttingen.

Schmidt, A. 1972a. On the mechanism of photosynthetic sulfate reduction. An APS-sulfotransferase from *Chlorella*. Arch. Mikrobiol. 84:77–86.

Schmidt, A. 1972b. Über Teilreaktionen der photosynthetischen Sulfatreduktion in zellfreien Systemen aus Spinatchloroplasten und *Chlorella*. Z. Naturforsch. 27b:183–192.

Schmidt, A. 1973. Sulfate reduction in a cell-free system of *Chlorella*. The ferredoxin-dependent reduction of a protein-bound intermediate by a thiosulfonate reductase. Arch. Mikrobiol. 93:29–52.

Schmidt, A. 1976a. The adenosine-5'-phosphosulfate sulfotransferase from spinach (*Spinacia oleracea* L.). Stabilization, partial purification, and properties. Planta 130:257–263.

Schmidt, A. 1976b. Development of the adenosine-5'-phosphosulfate sulfotransferase in sunflower (*Helianthus annuus* L.). Z. Pflanzenphysiol. 78:164–168.

Schmidt, A. 1977. Assimilatory sulfate reduction via 3'-phosphoadenosine-5'-phosphosulfate (PAPS) and adenosine-5'-phosphosulfate (APS) in blue-green algae. FEMS Microbiol. Lett. 1:137–140.

Schmidt, A. 1979a. Different types of sulfotransferases in cyanobacteria. Third International Symposium on Photosynthetic Prokaryotes (abs.). Oxford, B-36.

Schmidt, A. 1979b. Photosynthetic assimilation of sulfur compounds. *In* The Encyclopedia of Plant Physiology (n.s.), Vol. 6. M. Gibbs and E. Latzko (eds.) Berlin, Heidelberg, New York: Springer-Verlag, pp. 481–496.

Schmidt, A. 1979c. A thioredoxin-activated fructose bisphosphatase in the cyanobacterium *Synechococcus* 6301. Plant Physiol. 63:S-11.

Schmidt, A., and Christen, U. 1978. A factor-dependent sulfotransferase specific for 3'-phosphoadenosine-5'-phosphosulfate (PAPS) in the cyanobacterium *Synechococcus* 6301. Planta 140:239–244.

Schmidt, A., and Christen, U. 1979. A PAPS-dependent sulfotransferase in *Cyanophora paradoxa* inhibited by 5'-AMP, 5'-ADP, and APS. Z. Naturforsch. 34c:222–228.

Schmidt, A., and Schwenn, J.-D. 1971. On the mechanism of photosynthetic sulfate reduction. Second International Congress on Photosynthesis, Stresa. Pp. 507–513.

Schmidt, A., and Trebst, A. 1969. The mechanism of photosynthetic sulfate reduction by isolated chloroplasts. Biochim. Biophys. Acta 180:529–535.

Schmidt, A., and Trüper, H. G. 1977. Reduction of adenylylsulfate and 3'-phosphoadenylylsulfate in phototrophic bacteria. Experientia 33:1008–1009.

Schmidt, A., Abrams, W. R., and Schiff, J. A. 1974. Studies of sulfate utilization by algae. Reduction of adenosine-5'-phosphosulfate (APS) to cysteine in extracts from Chlorella and mutants blocked for sulfate reduction. Eur. J. Biochem. 47:423–434.

Schwenn, J. D., Depka, B., and Henies, H. H. 1976. Assimilatory sulfate reduction in chloroplasts: Evidence for the participation of both stromal and membrane-bound enzymes. Plant and Cell Physiol. 17:165–176.

Siegel, L. M. 1975. Biochemistry of the sulfur cycle. In The Metabolism of Sulfur Compounds. D. M. Greenberg (ed.). New York, San Francisco, London: Academic Press.

Torii, K., and Bandurski, R. S. 1967. An intermediate in the reduction of 3'-phosphoryl-5'-adenosinephosphosulfate to sulfite. Biochim. Biophys. Acta 136:286–295.

Trebst, A., and Schmidt, A. 1969. Photosynthetic sulfate and sulfite reduction by chloroplasts. Prog. Photosynthesis Res.: 1510–1516.

Trüper, H. G. 1975. The enzymology of sulfur metabolism in phototrophic bacteria—a review. Plant Soil 43:29–39.

Tsang, M. L-S., and Schiff, J. A. 1975. Studies of sulfate utilization by algae. 14. Distribution of adenosine-3'-phosphate-5'-phosphosulfate (PAPS) and adenosine-5'-phosphosulfate (APS) sulfotransferases in assimilatory sulfate reducers. Plant Sci. Lett. 4:301–307.

Tsang, M. L-S., and Schiff, J. A. 1976a. Studies of sulfate utilization by algae. 17. Reactions of the adenosine-5'-phosphosulfate (APS) sulfotransferase from Chlorella and studies of model reactions which explain the diversity of side products with thiols. Plant and Cell Physiol. 17:1209–1220.

Tsang, M. L-S., and Schiff, J. A. 1976b. Sulfate-reducing pathway in Escherichia coli involving bound intermediates. J. Bacteriol. 125:923–933.

Tsang, M. L-S., and Schiff, J. A. 1978a. Assimilatory sulfate reduction in an Escherichia coli mutant lacking thioredoxin activity. J. Bacteriol. 134:131–138.

Tsang, M. L-S., and Schiff, J. A. 1978b. Studies of sulfate utilization by algae. 18. Identification of glutathione as a physiological carrier in assimilatory sulfate reduction by Chlorella. Plant Sci. Lett. 11:177–183.

Tsang, M. L-S., Goldschmidt, E. E., and Schiff, J. A. 1971. Adenosine-5'-phosphosulfate as an intermediate in the conversion of 3'-phosphate-adenosine-5'-phosphosulfate to acid-volatile radioactivity. Plant Physiol. 47:S-20.

Wagner, W., Follmann, H. and Schmidt, A. 1978. Multiple functions of thioredoxins. Z. Naturforsch. 33c:517–520.

Wilson, L. G., and Bierer, D. 1976. The formation of exchangeable sulfite from adenosine-3'-phosphate-5'-sulphatophosphate in yeast. Biochem. J. 158:255–270.

Discussion of Presentation by Dr. Schmidt

STANIER: What are the respective roles of the thioredoxins?

SCHMIDT: As far as is known, the functions of thioredoxins in cyanobacteria are: (a) activation of PAPS-sulfotransferase activity in Synechococcus 6301; (b) activation of fructose bisphosphatase in S. 6301; (c) acting as electron donor for ribonucleotide reduction in *Anabaena* 7119; and (d) activation of APS-sulfotransferases in *Chroococcidiopsis* 7203. One might speculate that they activate glutamine synthetase and inactivate glucose-6-phosphate dehydrogenase.

COHEN: Will the synthesis of adenosine 3'-phosphate-5'-phosphosulfate be affected by 5'-phosphates of nucleosides modified in the 3' position; for example, xylosyladenosine or 3'-deoxyadenosine (cordycepin)?

SCHMIDT: We have not analyzed this possibility, but I welcome the suggestion and will try to explore some of these possibilities.

Additional Comments:
Evolution of Pathways for Assimilatory Sulfate Reduction
Jerome A. Schiff

I would like to add several comments to Dr. Schmidt's interesting presentation in Chapter 10 of the present volume in order to suggest how the various contemporary pathways of sulfate reduction may have evolved and how one of these may have become localized in the chloroplasts. I will be brief since many of these ideas have been developed more extensively elsewhere (Hodson and Schiff, 1973; Schiff, 1980, 1981; Schiff and Fankhauser, 1981).

Two major pathways of assimilatory sulfate reduction appear to be present in contemporary organisms. One pathway begins with adenosine 5'-phosphosulfate (APS) as the activated sulfo donor (Figure 1). This sulfo group is thought to be transferred to a thiol carrier (which may be glutathione in *Chlorella*, or a somewhat larger molecule in spinach) to form an organic thiosulfate (R—S—SO_3^-). The sulfo group is thought to be further reduced on this carrier to form the persulfide (R—S—S^-). The thiol group of the persulfide is then transferred reductively to *O*-acetylserine via *O*-acetylserine sulfhydrase to form cysteine. This system is found in spinach chloroplasts where the reductant ferredoxin is readily available. Since the *Chlorella* system also uses ferredoxin as the reductant, and since the green algae are closely related to the higher plants, this may also be a plastid-localized system. In *Euglena*, where this sulfate reducing system is localized in the mitochondria and microbodies, the reductant is NADPH. This may explain why *Euglena* can lose its chloroplasts so readily. Unlike green algae and higher plants, in which the reduction of carbon dioxide, nitrite, and sulfate as well as polysaccharide synthesis are localized in the plastids, in *Euglena* only carbon dioxide reduction is found in the plastids.

The other major pathway of assimilatory sulfate reduction is found in non-photosynthetic organisms such as *Escherichia coli* and yeast. This pathway begins with adenosine 3'-phosphate 5'-phosphosulfate (PAPS) as the activated sulfo donor (Figure 2). Although a bound intermediate is thought to exist, free sulfite appears to be the major product of the transfer reaction and thioredoxin or another similar compound is required; whether thioredoxin is required as a

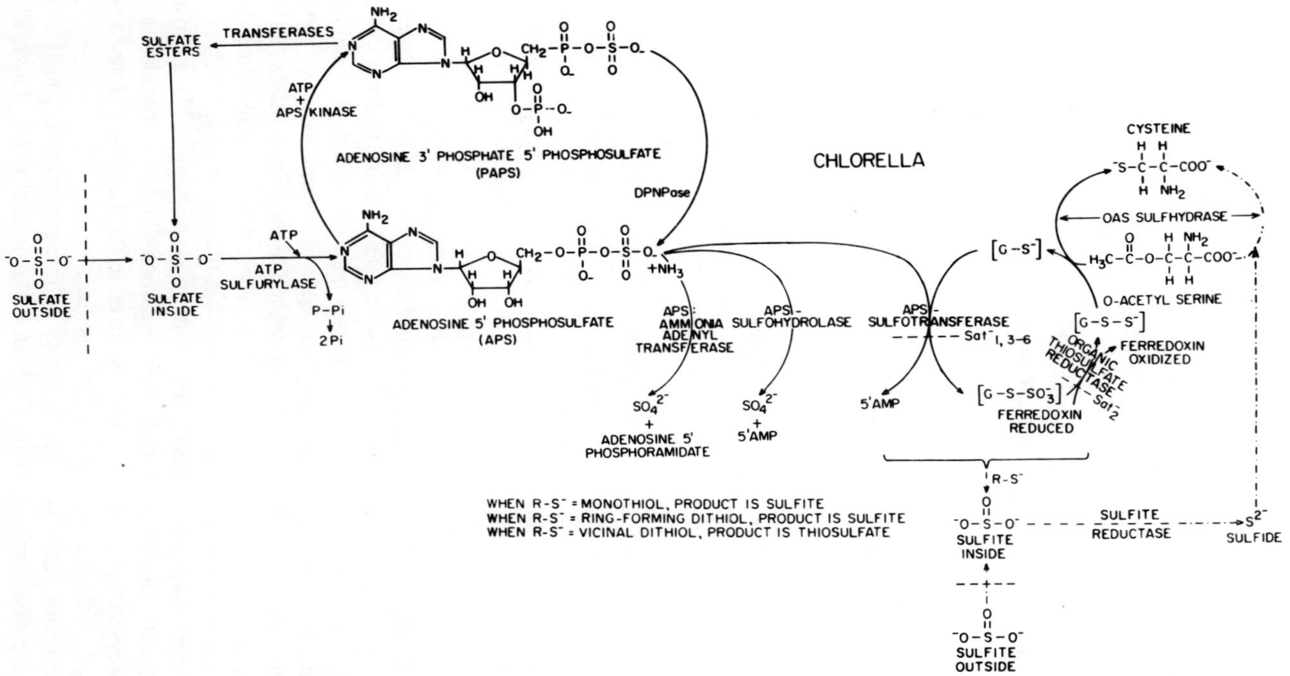

192

Figure 1. APS pathway of sulfate reduction in *Chlorella*. Reactions thought to be on the main path of sulfate reduction in vivo are shown as solid lines. Reactions blocked in various mutants that cannot grow on sulfate (*Sat⁻* mutants) are shown by dashes through the reactions deleted. Side-reactions are shown by alternate dashes and dots. Sulfite (or thiosulfate) is only produced through chemical reactions requiring thiols, and probably only occurs in vivo when sulfite (or thiosulfate) is available to the cells from outside. *G-S⁻* designates reduced glutathione.

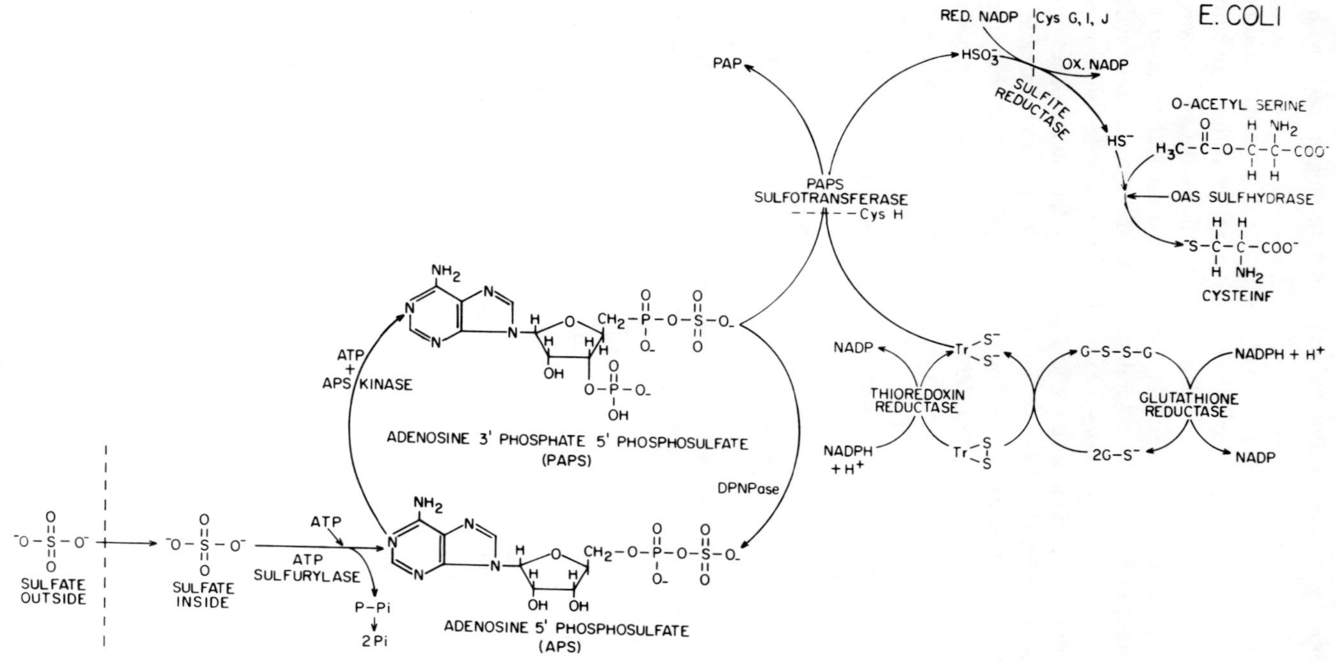

Figure 2. PAPS pathway of sulfate reduction in *E. coli*. Thioredoxin is required in the PAPS sulphotransferase reaction and formation of sulfite, but the reactions it undergoes have not been elucidated fully. Mutants (*Cys*) blocked in various enzyme reactions are shown by dashes through the reactions deleted. *G-S*⁻ designates reduced glutathione. $Tr \lessgtr^{S^-}_{S^-}$ designates reduced thioredoxin.

reductant, a carrier, or both has not been clearly established. Sulfite is then reduced to sulfide via NADPH-sulfite reductase and the sulfide reacts with *O*-acetyl serine to form cysteine.

How did these systems evolve in the various groups of organisms and how did they acquire their various properties? The most primitive pathway of sulfate reduction in contemporary organisms seems to be the dissimilatory pathway in *Desulfovibrio*. These organisms are strict anaerobes and use sulfate rather than oxygen as an oxidant for cellular respiration. As a consequence, they reduce sulfate to sulfide (Figure 3). In this system, APS is the activated sulfo donor, ferredoxin is a reductant, and free sulfite and sulfide are the intermediates in the pathway of sulfate reduction. Free sulfite and sulfide are autooxidizable by molecular oxygen, but this would present no problems to a strict anaerobe. I would suggest that the dissimilatory sulfate-reducing system evolved very early while the environment was still free of oxygen. Perhaps these organisms served to complete a primitive anaerobic sulfur cycle in which they reduced the sulfate formed from the photosynthetic oxidation of sulfide by anoxygenic photosynthetic bacteria. In any case, dissimilatory sulfate reduction serves two functions, first, as an electron system for respiration and oxidative phosphorylation, and, second, as a source of reduced sulfur to form the thiol-containing amino acids and coenzymes.

When oxygen came into the atmosphere through the activities of the early oxygenic prokaryotes, such as cyanobacteria and prochlorophytes, we suppose that a division of labor occurred. Respiration, because oxygen was available, no longer required sulfate, but sulfate reduction was still required to form the reduced sulfur required for the synthesis of amino acids and coenzymes. At this point assimilatory sulfate reduction evolved as a pathway distinct from respiration. In its primitive state it retained APS as the sulfo donor and ferredoxin as a reductant. With the evolution of oxygenic photosynthesis, however, autooxidizable intermediates such as sulfite and sulfide would be susceptible to oxidation by molecular oxygen. The evolution of a sulfate-reducing system having bound, nonautooxidizable intermediates such as organic thiosulfates and persulfides would have been highly adaptive. Such a system is found in some cyanobacteria and, so far, is the only system found in eukaryotic organisms carrying out oxygenic photosynthesis, such as the algae and higher plants.

Dr. Schmidt's discussion has served to remind us that the prokaryotes were a testing ground for many variations on the basic theme of assimilatory sulfate reduction. Since the APS-ferredoxin system employing bound intermediates is found in some cyanobacteria and in eukaryotic algae and higher plant chloroplasts, this can be added to the accumulating evidence supporting the prokaryotic nature of chloroplasts, and would be consistent with an endosymbiotic origin for the plastid. But, as Dr. Schmidt reminds us, some contemporary cyanobacteria have the PAPS-thioredoxin system for sulfate reduction. This system is also found in bacteria such as *Escherichia coli*. It is this system which is also found

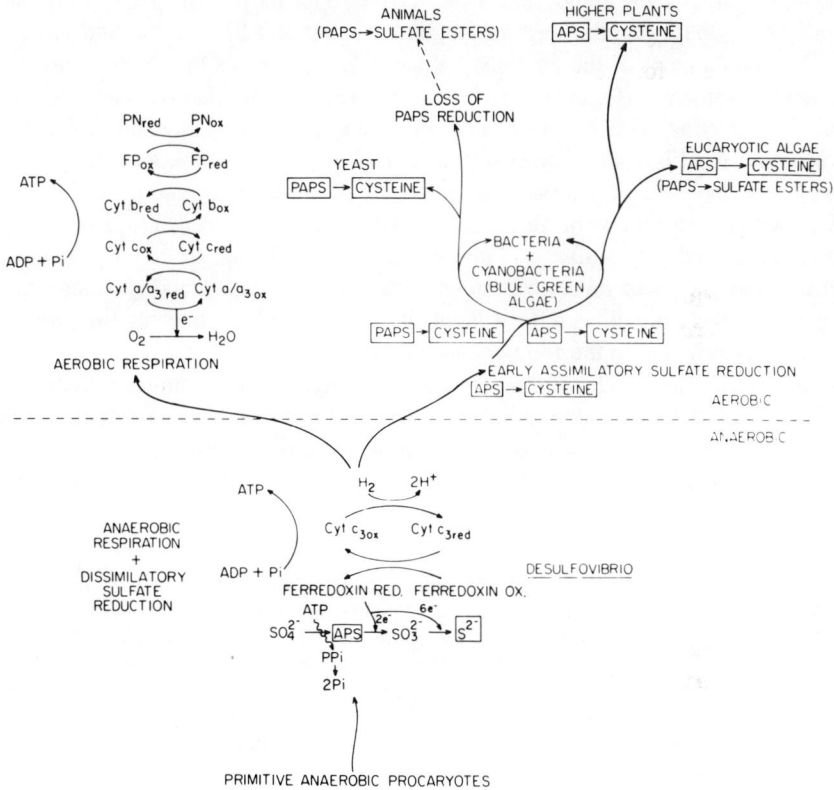

Figure 3. Suggested evolutionary relationships of sulfate-reducing pathways in various groups. Dissimilatory sulfate reduction, as part of anaerobic respiration, probably arose in the anaerobic phase of the origin of life, and has persisted in the modern prokaryotic anaerobes *Desulfovibrio* and *Desulfotomaculum,* which use sulfate as an oxidant in respiration, forming ATP and sulfide. This pathway begins with APS. With the release of oxygen into the atmosphere, oxygen became the electron acceptor in respiration and a separation of respiration and sulfate reduction occurred, leading to the establishment of assimilatory sulfate reduction, still using APS as the activated sulfate for reduction. For as yet unknown reasons the PAPS pathway for reduction evolved from the APS pathway and modern prokaryotes (bacteria and blue-green algae or cyanobacteria) have one or the other. The APS pathway has persisted in the evolution of the eukaryotic photosynthetic oxygen evolvers, including the eukaryotic algae and higher plants. The PAPS pathway leads to yeast, among the fungi. Animals have lost the ability to reduce PAPS to the thiol level of the amino acids, but retain PAPS as a donor of the sulfo group in esterification reactions via specific sulfotransferases. Eukaryotic algae also use PAPS as the donor of the sulfo group in esterification reactions, although they use APS as the substrate for reduction.

in yeast, among the fungi. Since this system appears to form free sulfite and sulfide, it would function best in facultative anaerobes like *E. coli* and yeast.

In the evolution of the animals, sulfate reduction seems to have disappeared almost completely (Figures 3 and 4) along with carbon dioxide reduction and nitrate reduction. This may be have been due to the loss of photosynthesis. These reduction reactions are very costly in terms of energy requirements (Figure 4) and with the loss of photosynthesis would be difficult to maintain. Instead, predation became a way of life for the animals, energy was needed for movement, and reduced carbon, sulfur, and nitrogen could be obtained by eating organisms that could carry out these reductions, such as plants and plantlike microorganisms, or the animals which eat them. In this way, plants became the primary producers originating the food chains we know today.

Although sulfate reduction was lost in the animal line of evolution, PAPS was retained as the sulfate donor for esterification reactions important for the formation of sulfated polysaccharides, phenol sulfates, steroid sulfates, etc. Although the algae use APS as the sulfo donor for sulfate reduction, they use PAPS

Figure 4. Loss of energetically expensive reduction reactions in various contemporary groups of organisms. Dissimilatory sulfate reduction is restricted to two small genera of anaerobic bacteria (*Desulfovibrio* and *Desulfotomaculum*). The clear areas indicate organisms that can reduce sulfate, nitrate, and carbon dioxide (a small group of aerobic chemosynthetic bacteria [not shown] are capable of reducing carbon dioxide, given oxidizable inorganic substrates as energy sources). The various patterns indicate the loss of these reduction reactions in various groups. All three reduction pathways have been lost in the evolution of the protozoa and higher animals.

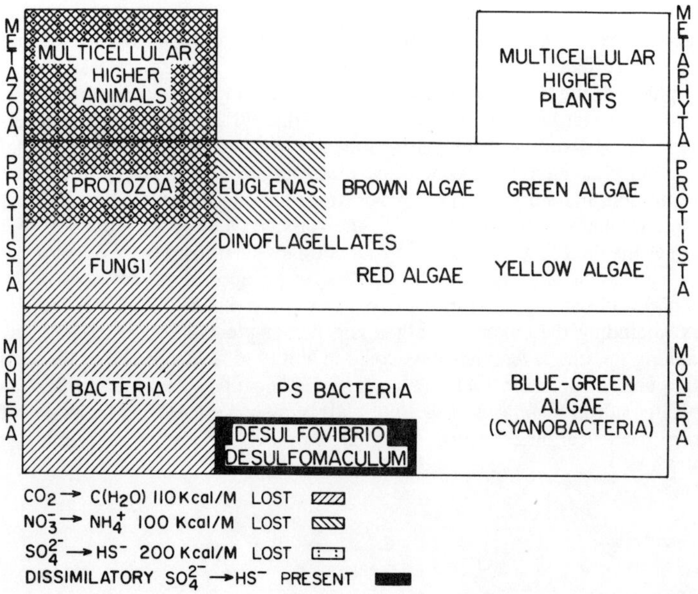

as the sulfate donor for esterification reactions. Although many different combinations of sulfate-reducing reactions were present among the primitive prokaryotes, the APS-ferredoxin-bound intermediate system prevailed in the evolution of chloroplasts, while the PAPS-thioredoxin system was adopted (for as yet unknown reasons) in the evolution of yeast, among the fungi.

References to Comments

Hodson, R. C., and Schiff, J. A. 1973. The metabolism of sulfate. Annu. Rev. Plant Physiol. 24:381–414.

Schiff, J. A. 1980. Pathways of assimilatory sulfate reduction in plants and microorganisms. *In* Sulphur in Biology, Ciba Foundation Symposium No. 72 (n.s.) Amsterdam: Excerpta-Medica, Elsevier-North Holland, pp. 49–69.

Schiff, J. A. 1981. Reduction and other metabolic reactions of sulfate. In *Encyclopedia of Plant Physiology* (n.s.), Vol. 12. A. Läuchli and R. L. Bieleski (eds.). Springer-Verlag (in press).

Schiff, J. A., and Fankhauser, H. 1981. Assimilatory sulfate reduction. *In* Symposium on the Biochemistry of Nitrogen and Sulphur Metabolism. A. Trebst (ed.). Springer-Verlag (in press).

Biosynthesis of Chlorophyll *a*

Paul A. Castelfranco and B. M. Chereskin

Figure 1 shows the pathway for the synthesis of chlorophyll *a*. While there is a lot of fascinating biochemistry going on at each of these steps, in this chapter we wish to focus on only three of them: (a) the formation of δ-aminolevulinic acid; (b) the insertion of Mg; and (c) the reduction of the 7-8 double bond. These are probably the steps that are most intimately concerned with the physiological controls of the biosynthesis of chlorophyll *a*. At the first of these steps (ALA synthesis), tetrapyrrole metabolism is separated from the general metabolism. At the second one (Mg insertion) chlorophyll biosynthesis is separated from cytochrome biosynthesis. The third one (ring D reduction) is probably the point at which light regulates chlorophyll biosynthesis in angiosperms; in addition, this step shows interesting taxonomic differences and illustrates some differences between wild-type organisms and mutants.

Finally we shall conclude by discussing some observations on the control of the pathway which may have some biogenetic implications.

ALA Biosynthesis

The coarse control of the whole pathway appears to be at the ALA biosynthesis level. The evidence comes from a variety of experimental observations and is on the whole very compelling, but we do not have space to examine it here (Harel, 1978a; Castelfranco and Beale, 1981).

Experimental work in the authors' laboratory supported by grant nos. PCM-75 10957 and PCM-78 13250 from the National Science Foundation.

Address reprint requests to Dr. Paul A. Castelfranco, Professor of Botany, Department of Botany, University of California, Davis, California 95616.

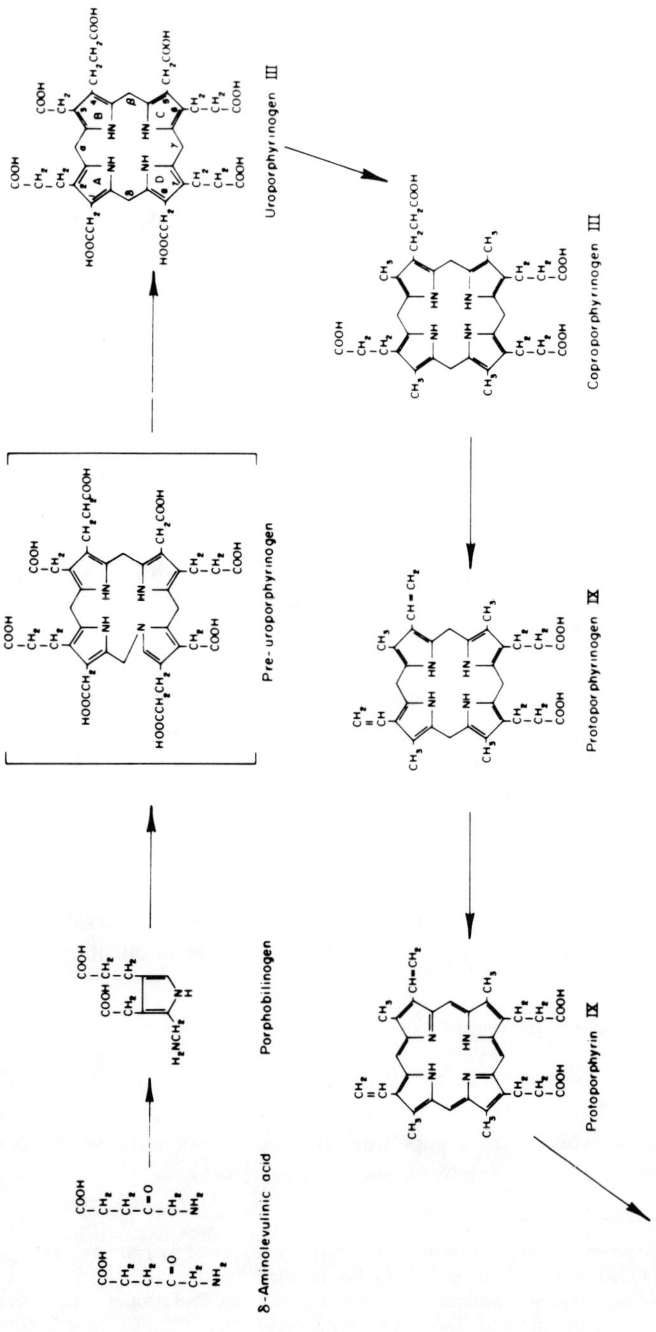
δ-Aminolevulinic acid
Porphobilinogen
Pre-uroporphyrinogen
Uroporphyrinogen III
Coproporphyrinogen III
Protoporphyrinogen IX
Protoporphyrin IX

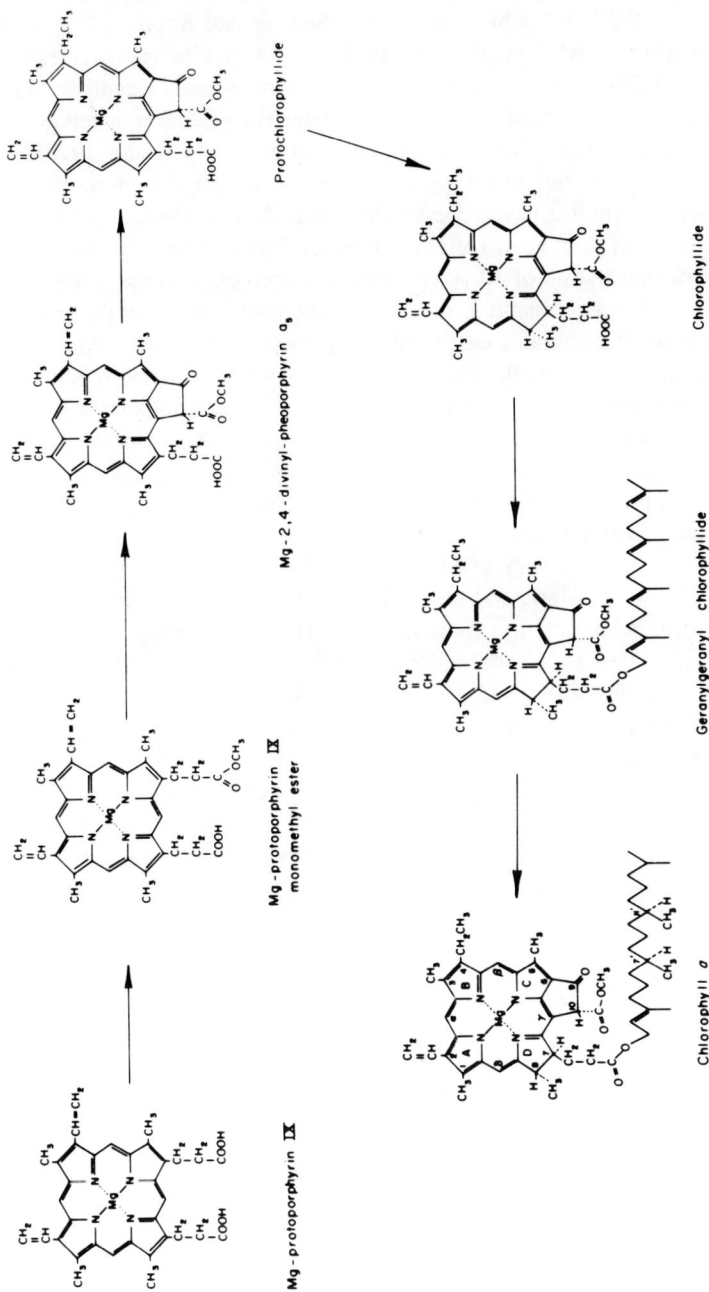

Figure 1. Pathway from δ-aminolevulinic acid (ALA) to chlorophyll *a*. *Source:* From Castelfranco and Beale (1981).

An enzyme which catalyzes the synthesis of ALA has been known since 1958 (Gibson et al., 1958; Kikuchi et al., 1958; Sawyer and Smith, 1958) and is commonly known as ALA synthetase (succinyl-CoA-glycine succinyl transferase, E.C.2.3. 1.37) (Figure 2). It has been shown to increase rapidly in warm-blooded animals, yeast, and photosynthetic bacteria whenever a tetrapyrrole needs to be synthesized in massive amounts (Granick and Beale, 1978). The systems producing hemoglobin, P_{450}, and bacteriochlorophyll have received the greatest attention. On this basis, one would expect ALA synthetase to be highly active in extracts of dark-grown angiosperm seedlings that are exposed to light. In fact, the demonstration of ALA synthetase in such extracts has failed.

Levulinic acid is a competitive inhibitor of the next enzyme in the pathway: ALA dehydrase. It penetrates easily into plant tissues and allows ALA to accumulate (Beale, 1970). Beale and Castelfranco (1973) attempted to show that the isotopic labeling of ALA accumulated in vivo by greening plant material in the presence of levulinic acid was consistent with the ALA synthetase reaction. These experiments produced unexpected results. One would predict that glycine -1-^{14}C is a poor precursor of ALA in the ALA synthetase reaction; glycine 2-^{14}C is an excellent precursor; glutamate-1-^{14}C is a poor precursor; and glutamate-3,4-^{14}C (or -U-^{14}C) is a very good precursor. Instead, ^{14}C incorporation into ALA from glycine was low, and the position of the label in the substrate made little difference, while ^{14}C-incorporation into ALA from glutamate was high, and again the position of the label made little difference (Beale and Castelfranco, 1973, 1974). On the other hand, when a tissue known to contain the classical ALA-synthetase was tested, namely whole avian blood, the predicted pattern of 14-C-incorporation was obtained (Beale and Castelfranco, 1974).

Subsequently Beale et al. (1975) showed that glutamate-1-^{14}C gave rise to ALA labeled in carbon 5, while glutamate-3,4-^{14}C gave rise to ALA labeled in carbons 1–4. These studies established the likelihood that the glutamate carbon skeleton is incorporated intact into ALA. Several possible mechanisms were

Figure 2. Reaction catalyzed by ALA synthetase (E.C.2.3. 1.37: succinyl COA: glycine succinyltransferase [decarboxylating]).

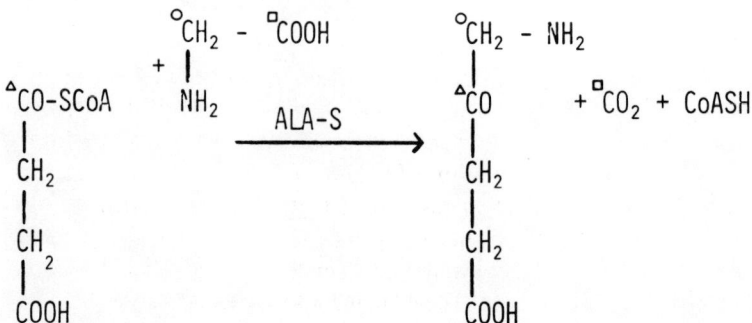

COOH
|
CH_2
|
CH_2 reductase
| (d)
$CHNH_2$
|
COOH

Glutamate

(g) glutamate dehydrogenase

COOH
|
CH_2
|
CH_2 reductase
| (b)
C=O
|
COOH

α-Keto-glutarate

(f) α-ketoglutarate dehydrogenase

CO_2

Succinyl CoA

COOH
|
CH_2
|
CH_2
|
$CHNH_2$
|
CHO

Glutamate-1-semialdehyde

intramolecular transaminase (e)

COOH
|
CH_2
|
CH_2
|
C=O
|
CHO

γ,δ-Dioxo-valeric acid

transaminase (c)

COOH
|
CH_2
|
CH_2
|
C=O
|
$CH_2 NH_2$

ALA

ALA synthetase (a)

CO_2

COOH
|
CH_2
|
CH_2
|
COSCoA

Glycine
$CH_2 NH_2$
|
COOH

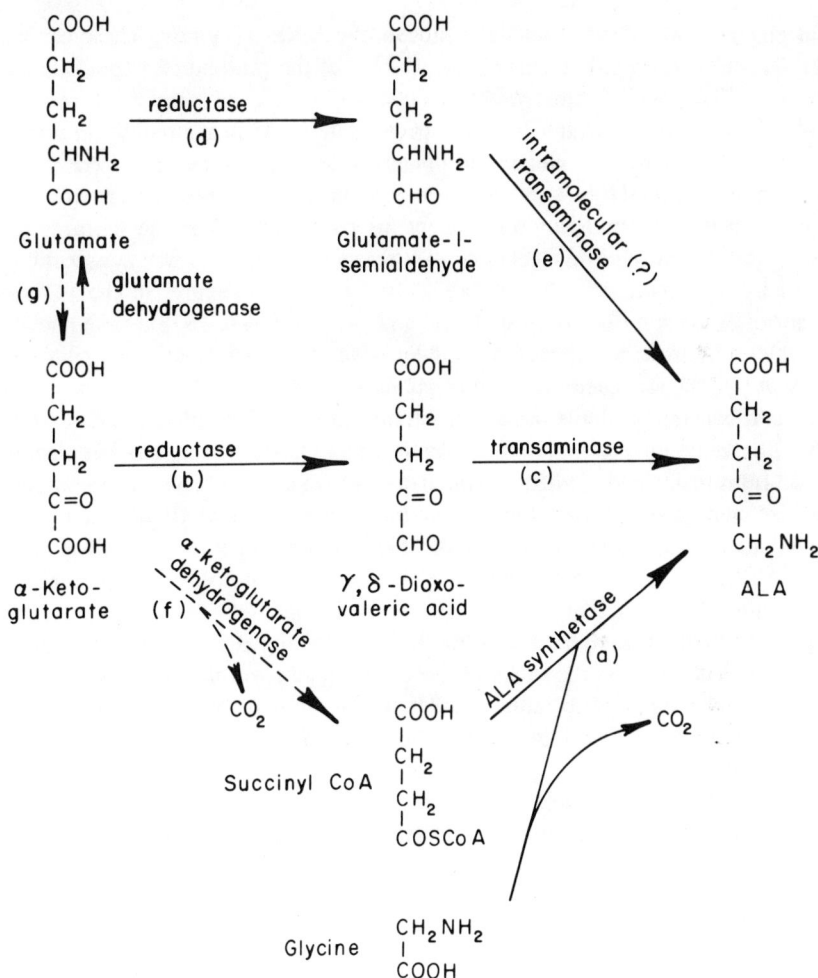

Figure 3. Proposed schemes for ALA formation in plants. *Source*: From Castelfranco and Beale (1981).

proposed which subsequent research has whittled down in number. The most likely mechanisms are shown in Figure 3. The two sequences d, e and g, b, c would explain the labeling of ALA in vivo from glutamate, while the sequence of g, f, a, involving α-ketoglutaric dehydrogenase and ALA synthetase, is excluded (at least in greening angiosperm seedlings) by the ^{14}C-incorporation data.

Castelfranco and Jones (1975) demonstrated that : (a) ^{14}C from glutamate went into the chlorophyllide moiety of newly synthesized chlorophyll by greening barley seedlings, (b) the same label went into a small heme (protoheme) fraction extractable with 2% HCl in acetone. The label incorporated in these two fractions from glycine -1- or -2-^{14}C, under these conditions was very low. These experiments raised the possibility that at least some of the protoheme is made from the same ALA pool as chlorophyll.

We would now like to discuss the current state of ALA synthetase (succinyl CoA glycine succinyl transferase) in plants. It definitely does not appear to be involved in chlorophyll biosynthesis in higher plants. However, the enzyme has been found in a nongreening soybean callus culture by Wider de Xifra et al. (1971, 1978), and also in peels of cold-stored potatoes (another nongreening tissue) by Ramaswamy and Nair (1973, 1974, 1976). Finally, in the yellow mutant of *Scenedesmus obliquus*, Klein and Senger (1977, 1978), have shown that chlorophyll can be labeled either from glycine and succinate, according to the ALA synthetase reaction, or from glutamate.

Several cell-free systems have been reported for the formation of ALA from either glutamate or α-ketoglutarate. The α-ketoglutarate system has been solubilized from maize and studied by Harel et al. (1978). The glutamate system has been isolated in a cell-free form from *Euglena* (Salvador, 1978), cucumber cotyledons (Weinstein and Castelfranco, 1978) and barley (Gough and Kannangara, 1977). The requirements of the glutamate-utilizing preparations from the three sources are remarkably similar (Harel, 1978b); on the other hand, the glutamate-requiring system differs markedly from the α-ketoglutarate system (Table 1). Both systems require Mg^{2+} and a reduced pyridine nucleotide. However, the glutamate system requires ATP, while the α-ketoglutarate system requires pyridoxal phosphate plus an amino donor. Furthermore, the pH optima differ widely.

The α-ketoglutarate system is not inhibited by the addition of unlabeled glutamate (Harel et al., 1978), nor is the glutamate system inhibited by addition of unlabeled α-ketoglutarate (Kannangara and Gough, 1977). From these experiments we may conclude that both substrates are converted to ALA by two separate routes without being interconverted to one another.

We would like to postulate a mechanism for the conversion of α-ketoglutarate to ALA, which takes into account the observed cofactor requirement (Figure 4). This scheme consists of three enzymes: (a) an enzyme similar to lactic dehydrogenase; (b) an enzyme similar to glyoxylase; and (c) a transaminase. Jerzykowski et al. (1973) have shown that γ,δ-dioxovaleric acid can act as the substrate for yeast and ox liver glyoxylase. However, they failed to demonstrate the reverse reaction starting with α-hydroxyglutaric acid.

The transaminase activity (reaction 3) has been demonstrated in a variety of

Table 1. Requirements of Cell-free Reaction Mixtures Forming ALA

Substrate	Source	Cofactors	Optimum pH
α-Ketoglutarate	Maize	Mg, NADH, pyridoxal phosphate, alanine	6.2
L-glutamate	Barley	Mg, NADPH, ATP	7.9

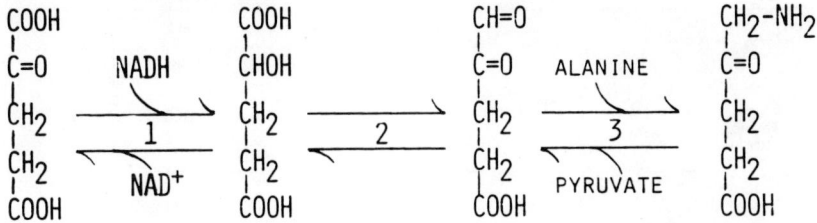

COOH   COOH   CH=O CH_2-NH_2
C=O   NADH   CHOH   C=O   ALANINE   C=O
CH_2 → CH_2 → CH_2 → CH_2
     1   2   3
CH_2 NAD^+ CH_2 CH_2 PYRUVATE CH_2
COOH   COOH   COOH   COOH

1. REACTION ANALOGOUS TO LACTIC DEHYDROGENASE.

2. REACTION ANALOGOUS TO GLYOXYLASE.

3. TRANSAMINASE.

Figure 4. Postulated scheme for ^{14}C-incorporation into ALA from ^{14}C-α-ketoglutarate by soluble plant extracts.

plant, animal, and bacterial tissues, including many which never form chlorophyll. The substrate for reaction 3, γ,δ-dioxovaleric acid, has not to our knowledge been found in any biological material. Finally, the whole scheme could be reversible: the presence of NADH and amino donors would favor ALA synthesis, while the presence of NAD and amino acceptors would favor ALA degradation. A very similar sequence of ractions was postulated by Shemin et al. (1955) as part of their succinate-glycine cycle. The only difference is that Shemin's group did not consider that the oxidation of γ,δ-dioxovaleric acid to α-ketoglutaric acid could be reversed.

A proposed scheme for ALA synthesis from glutamate (Kannangara and Gough, 1978) follows (Figure 5): Steps 1 and 2 are analogous to the γ-glutamyl kinase and the γ-glutamyl semialdehyde dehydrogenase which are involved in proline and ornithine synthesis (Miflin and Lea, 1977), except that in the proposed scheme the α-carboxyl is attacked instead of the γ-carboxyl. Step 3 must be catalyzed by an internal transaminase since it is unaffected by the addition of amino donors (Kannangara and Gough, 1978). The α-glutamyl semialdehyde intermediate has not been isolated, but the synthetic semialdehyde is converted to ALA by the enzyme. Finally, the α-ketoglutarate system is 2–3 times more active in dark-grown tissue than in tissue which has been greened for several hours (Harel et al., 1978), while the glutamate system has very low activity in barley and cucumber etioplasts, is very active in greening chloroplasts and much less active in chloroplasts isolated from fully green leaves (Kannangara and Gough, 1979; Weinstein, 1979).

In conclusion, it seems probable that the glutamate system is the normal system for the synthesis of ALA in higher plants, while the α-ketoglutarate sequence normally runs in the opposite direction (i.e., it degrades ALA to α-ketoglutarate), but under special experimental conditions, can be reversed to make ALA.

$$
\begin{array}{ccccccc}
\text{COOH} & & \text{COOP} & & \text{CHO} & & \text{CH}_2\text{NH}_2 \\
| & & | & & | & & | \\
\text{CHNH}_2 & & \text{CHNH}_2 & & \text{CHNH}_2 & & \text{C} \cdot \text{O} \\
| & \text{ATP,Mg}^{2+} & | & \text{NADPH} & | & & | \\
\text{CH}_2 & \xrightarrow{\quad} & \text{CH}_2 & \xrightarrow{\quad} & \text{CH}_2 & \xrightarrow{\quad} & \text{CH}_2 \\
| & \text{I} & | & \text{II} & | & \text{III} & | \\
\text{CH}_2 & & \text{CH}_2 & & \text{CH}_2 & & \text{CH}_2 \\
| & & | & & | & & | \\
\text{COOH} & & \text{COOH} & & \text{COOH} & & \text{COOH}
\end{array}
$$

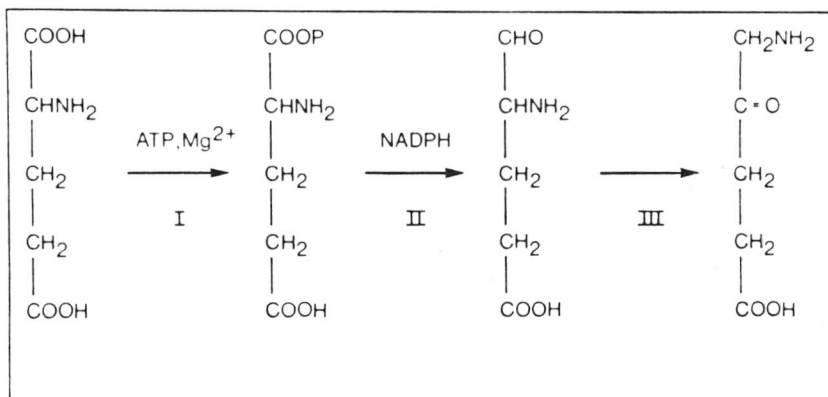

Figure 5. Postulated scheme for ^{14}C-incorporation of L-glutamate into ALA by isolated chloroplasts and chloroplast extracts. *Source:* From Kannangara and Gough (1978).

The Mg Insertion Step

The Mg insertion has eluded biochemists for many years. We have now what looks like a promising cell-free system (Castelfranco et al., 1979a, b; Pardo et al., 1980).

1. A developing chloroplast pellet is quickly isolated from greening cucumber cotyledons. This preparation is free from mitochondrial contamination, but is still contaminated with glyoxysomes and starch grains.

2. The chelation of Mg requires protoporphyrin and ATP. ADP and other nucleoside triphosphates cannot substitute for ATP.

3. The active particle preparation has a vigorous ATPase activity; therefore we must use high initial ATP concentrations to observe the formation of Mg-protoporphyrin.

4. ADP is very slightly inhibitory but AMP is strongly inhibitory (Figure 6).

5. The addition of an ATP regenerating system (like PEP and pyruvate kinase) enables us to drop the initial ATP concentration down to a reasonable level (Table 2) (Castelfranco et al., 1979a; Pardo et al., 1980).

6. There does not appear to be any problem with the penetration of the two rather bulky substrates protoporphyrin and ATP. This bit of negative evidence suggests to us that Mg chelatase may be located in the chloroplast envelope. This, of course, is mere speculation.

7. Mg chelatase activity has a very pronounced requirement for organelle integrity. It is lost very quickly if the particles are treated with Triton, sonicated very gently, shocked osmotically, or freeze-thawed.

8. Developmentally, Mg chelatase is low but not absent in etiolated tissue, goes up during greening, and reaches a plateau after 12–20 hr of illumination. In light-grown seedlings, it is still present after 6 days of germination, but is undetectable after 14 days.

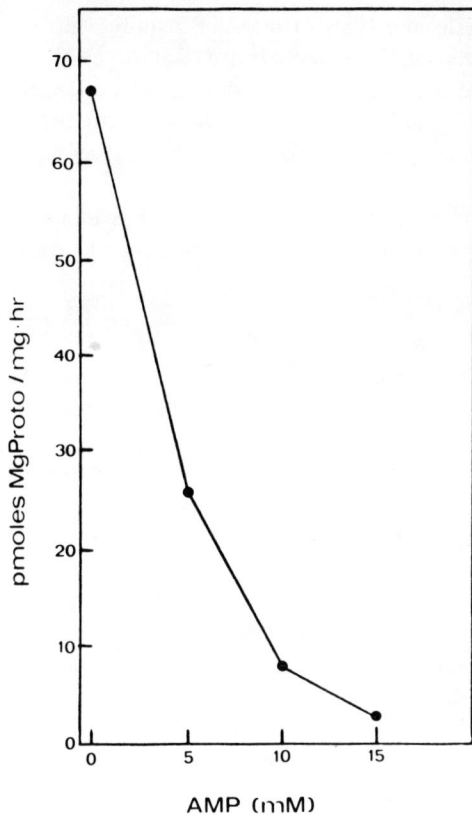

Figure 6. Inhibition of Mg chelatase by AMP. The reaction mixtures contained 10 mM ATP.

Table 2. Effect of an ATP Regenerating System on Mg Chelatase Activity in Purified Plastids

Additions to the standard system[a]	Mg-protoporphyrin IX formed (pmol)
Experiment 1	
1.5 mM ATP	3.5
1.5 mM ATP plus PEP + PK	820
10 mM ATP	625
10 mM ATP + PEP + PK	922
Experiment 2	
ATP + PEP + PK	566
PEP + PK	6.4
ATP + PK	1.5
ATP + PEP	65.6

[a]For the composition of the standard assay system, see Pardo et al. (1980). Experiment 1 contained 8.8 mg of purified plastid protein per 1.0 ml of incubation mixture; and experiment 2 contained 6.8 mg. ATP concentration in experiment 2 was 1.5 mM. Other concentrations: phosphoenol pyruvate (PEP), 20 mM; pyruvate kinase (PK), 20 µg/ml. Incubation was for 2 hr at 30°C.

9. While we do not know the molecular basis of the ATP requirement, we do know that there is no such requirement for ferrochelatase (Dailey, 1977). A branching point in which one of the branches—the introduction of Fe leading to cytochromes—does not require ATP, while the other branch—the introduction of Mg leading to chlorophyll—does require ATP, could have interesting regulatory possibilities.

10. Mg chelatase has a sulfhydryl requirement of some sort (it is inhibited by *p*-chloromercuribenzoate,) and is easily destroyed by strong light and air.

Reduction of Ring D

The reduction of the 7–8 double bond in ring D can be brought about either by a photochemical reaction or by a dark reaction. In angiosperm seedlings the photochemical reaction seems to be universal.

Gymnosperm seedlings and most algae normally carry out the conversion of protochlorophyll(ide) to chlorophyll(ide) equally well in the dark and in the light. It is possible to manipulate these tissues physiologically or genetically so that they are no longer able to carry out the reduction in the dark, but can still do so in the light. Thus, mutants of *Chlorella, Chlamydomonas,* and *Scenedesmus* are known that require light to green. Likewise, it has been shown by Wolwertz (1978) that Jeffrey pine seedlings germinated at 10°C do not form chlorophyll in the dark, but if they are germinated at 23°C they do. However, they form chlorophyll at either temperature if germinated in the light.

Griffiths (1978) has presented good evidence that phototransformable protochlorophyllide is a ternary complex of protochlorophyllide, NADPH, and a reusable reductase; when the complex sees light, hydrogen is transferred from NADPH to ring D of the tetrapyrrole forming $NADP^+$ and chlorophyllide, and the complex falls apart. The reductase can then be reused when supplied with new protochlorophyllide and NADPH molecules. Griffiths and Mapleston (1978) found that plants which form chlorophyll in the dark, like spruce seedlings and wild-type *Chlamydomonas reinhardtii*, contain the photochemical protochlorophyllide reductase activity which can be tested for in vitro. However, photosynthetic bacteria (which likewise do not require light for bacteriochlorophyll synthesis) did not contain the protochlorophyllide reductase.

In summary, it appears that those gymnosperms and algae which can form chlorophyll in the dark must contain two mechanisms for the reduction of ring D: (a) the photoreductase which is also present in angiosperm seedlings; and (b) another mechanism which does not require light and must therefore be a regular, thermo-activated enzymatic reaction.

Regulation of Chlorophyll *a* Biosynthesis

As soon as ring D is reduced, the resynthesis of protochlorophyll(ide) starts up again, beginning with ALA. How is the message transmitted from the region of protochlorophyll(ide) reduction to the region of ALA synthesis? Conceptually, the simplest mechanism would be feedback inhibition of ALA synthesis by protochlorophyllide itself (Schiff and Epstein, 1966).

We should like briefly to summarize the situation in the photosynthetic bacterium *Rhodopseudomonas spheroides*. This organism can grow either as a heterotrophic aerobe or as a photoheterotrophic anaerobe; it synthesizes bacteriochlorophyll in light or dark provided the O_2 tension is low (Lascelles, 1978).

In this organism ALA is made from glycine and succinyl CoA, according to the ALA synthetase reaction.

The site of O_2 inhibition is not completely understood. O_2 inhibits Mg-protophyrin biosynthesis in whole cells; in addition, ALA synthetase activity is dependent on endogenous trisulfides; O_2 can cause the destruction of trisulfides and therefore the decrease in ALA synthetase activity in homogenates. ALA synthetase in vitro is strongly inhibited by heme (Lascelles and Hatch, 1969) and Mg-protoporphyrin (Yubisui and Yoneyama, 1972). In whole cells iron deficiency imposed by the addition of o-phenanthroline, α,α'-dipyridyl, or similar aromatic chelating agents to the medium causes an overaccumulation of free porphyrins and Mg-porphyrins.

As we turn our attention to plants we find many similarities with the bacterial system.

1. Treatment of plant tissues with aromatic chelating agents such as o-phenanthroline and α,α'-dipyridyl causes an abnormal production of porphyrins and Mg-porphyrins. (Granick, 1960; Duggan and Gassman, 1974; Gassman and Duggan, 1974).

2. While there is little net heme synthesis during the first few hours of greening there is a fairly rapid heme turnover in greening bean (Duggan and Gassman, 1974) and barley (Castelfranco and Jones, 1975).

3. α,α'-Dipyridyl in vitro causes a two- to threefold stimulation in the synthesis of ALA from glutamate by a cell-free system from greening cucumber cotyledons, but α,α'-dipyridyl does not have any effect on the reactions beyond ALA, that is, on the conversion of ALA to protoporphyrin and on Mg chelation (Castelfranco and Chereskin, unpublished data).

4. Heme, Mg-protoporphyrin, and protoporphyrin inhibit the cell-free synthesis of ALA from glutamate (Table 3; Figure 7). The concentration required for 50% inhibition is of the order of 2.0 μM for heme, and 2.5 μM for Mg-protoporphyrin. Free protoporphyrin is less inhibitory than the two metalloporphyrins since the inhibition levels off at around 45% (Figure 7). The inhibition by heme is not overcome by addition of α,α'-dipyridyl, the inhibition by protoporphyrin is totally overcome, and the inhibition by Mg-protoporphyrin is partially overcome (Table 3). These findings may indicate that Fe chelation is necessary before protoporphyrin can become inhibitory. The partial recovery of the Mg-protoporphyrin inhibition in the presence of α,α'-dipyridyl could mean that this inhibition also involves an interaction with Fe or some other transition metal cation.

Castelfranco and Jones (1975) published a scheme for the regulation of chlorophyll biosynthesis in higher plants. This scheme was based on three facts:

Table 3. Effect of α, α'-Dipyridyl, Heme, Mg-protoporphyrin-IX, and Protoporphyrin-IX on the Conversion of L-Glutamate to ALA

Additions to the standard system[a]	ALA (pmol/hr)
Experiment 1	
Control	1324
α, α'-Dipyridyl	2292
Protoporphyrin IX	771
Protoporphyrin IX + α, α'-dipyridyl	1653
Heme	258
Heme + α, α'-dipyridyl	297
Experiment 2	
Control	1142
1.4 μM Mg-protoporphyrin	628
1.4 μM Mg-protoporphyrin + α, α'-dipyridyl	873
2.8 μM Mg-protoporphyrin	430
2.8 μM Mg-protoporphyrin + α, α'-dipyridyl	847
4.6 μM Mg-protoporphyrin	442
4.6 μM Mg-protoporphyrin + α, α'-dipyridyl	812

[a]For the composition of the standard assay system, see Weinstein and Castelfranco (1978). Experiment 1 contained 8.5 mg of plastid protein per 1.0 ml of incubation mixture; and experiment 2, 9.9 mg. Protoporphyrin-IX and heme in experiment 1 were 10 μM. α, α'-Dipyridyl was 0.2 mM. Incubation was for 1 hr at 28°C.

(a) the observed heme turnover in greening barley and bean tissues; (b) Dugganand Gassman's (1974) report that treatment of bean leaves with Fe chelators stimulates the synthesis of Mg porphyrins; and (c) Lascelles' published work on *Rhodopseudomonas spheroides* (Lascelles and Hatch, 1969). Our current hypothesis for ALA regulation in higher plants, which takes into consideration the recent experimental findings, is summarized in Figure 8. As in photosynthetic bacteria, ALA synthesis is regulated by heme and Mg-protoporphyrin feedback inhibition.

The plant system differs from the bacterial system in two important respects: (a) the nature of the reactions leading to ALA formation; and (b) the nature of the mechanism causing the accumulation of inhibitory metalloporphyrin concentrations. In photosynthetic bacteria, it is the blocking of Mg-chelatase by O_2 which results in the high concentration of protoprophyrin that gives rise to heme by Fe chelation. In plants, it is the accumulation of protochlorophyllide in the dark that causes the accumulation of protoporphyrin, either by the progressive filling up of the small, intermediate, membrane-bound pools between protoporphyrin and protochlorophyllide, or by direct feedback inhibition of Mg-chelatase by protochlorophyllide. However, to date we have failed to obtain direct evidence for feedback inhibition of Mg-chelatase by protochlorophyllide. The excess protoporphyrin is then coverted to heme, which causes the inhibition of ALA synthesis.

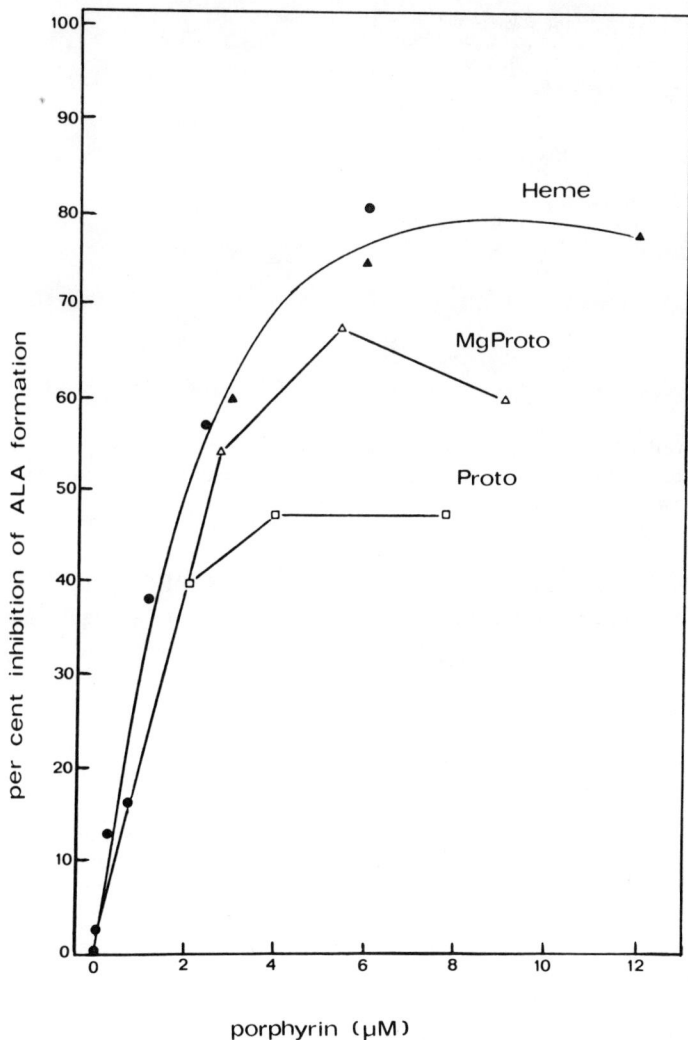

Figure 7. Inhibition of ALA synthesis from L-glutamate by heme, Mg-protoporphyrin IX ("Mg Proto"), and protoporphyrin IX ("Proto").

Our last comment pertains to the breakdown of chlorophyll intermediates that has been observed in vivo and in vitro by several authors; for example, Rosalía Frydman (Frydman et al., 1972, 1973) and Merrill Gassman (Gassman et al., 1978). These systems have porphobilinogen oxygenases. Vlcek and Gassman (1979) have observed the breakdown of Mg-protoporphyrin (Me) in vivo. As you have seen, we interpret the incorporation of label from α-ketoglutarate into ALA as evidence for an ALA degradation pathway. We should like to propose that these systems for the degradation of chlorophyll biosynthetic intermediates

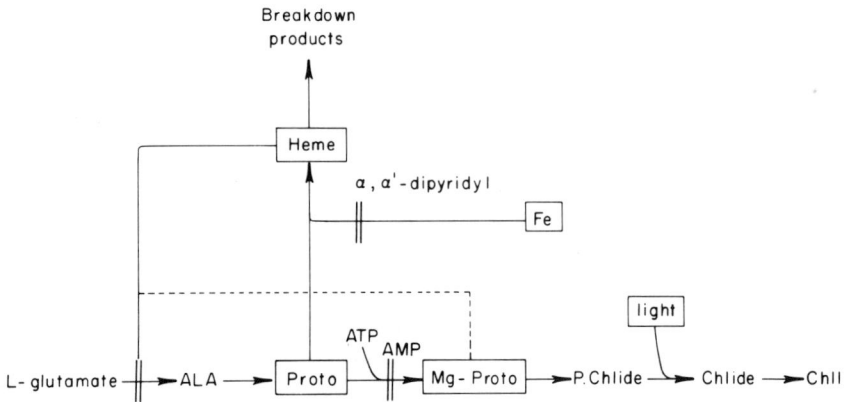

Figure 8. Postulated scheme for the regulation of chlorophyll *a* biosynthesis in higher plants. Abbreviations: *Proto,* protoporphyrin IX; *P. chlide,* protochlorophyllide; *Chlide,* chlorophyllide; *Heme,* protoheme.

are part of the overall regulatory strategy. Otherwise, because of the photodynamic properties of many of these compounds, a slight leakiness in the regulation of ALA synthesis could cause serious damage to the cell.

References

Beale, S. I., 1970. The biosynthesis of δ-aminolevulinic acid in *Chlorella.* Plant Physiol. 45:504–506.

Beale, S. I., and Castelfranco, P. A. 1973. [14]C incorporation from exogenous compounds into δ-aminolevulinic acid by greening cucumber cotyledons. Biochem. Biophys. Res. Commun. 52:143–149.

Beale, S. I., and Castelfranco, P. A. 1974. The biosynthesis of δ-aminolevulinic acid in higher plants. II. Formation of ^{14}C-δ-aminolevulinic acid from labeled precursors in greening plant tissues. Plant Physiol. 53:297–303.

Beale, S. I., Gough, S. P., and Granick, S. 1975. Biosynthesis of δ-aminolevulinic acid from the intact carbon skeleton of glutamic acid in greening barley. Proc. Nat. Acad. Sci. USA 72:2719–2723.

Castelfranco, P. A., and Beale, S. I. 1980. Chlorophyll biosynthesis. *In* The Biochemistry of Plants: A Comprehensive Treatise, Vol. VIII. P. K. Stumpf and E. E. Conn, (eds.). New York: Academic Press, pp. 375–421.

Castelfranco, P. A., and Jones, O. T. G. 1975. Protoheme turnover and chlorophyll synthesis in greening barley tissue. Plant Physiol. 55:485–490.

Castelfranco, P. A., Pardo, A. D., Chereskin, B. M., and Weinstein, J. D. 1979a. Mg-chelatase in developing chloroplasts. Plant Physiol. 63:S-98.

Castelfranco, P. A., Weinstein, J. D., Schwarcz, S., Pardo, A. D., and Wezelman, B. E. 1979b. The Mg insertion step in chlorophyll biosynthesis. Arch Biochem Biophys 192:592–598.

Dailey, H. A. 1977. Purification and characterization of the membrane-bound ferrochelatase from *Spirillum itersonii.* J. Bacteriol. 132:302–307.

Duggan, J., and Gassman, M. 1974. Induction of porphyrin synthesis in etiolated bean leaves by chelators of iron. Plant Physiol. 53:206–215.

Frydman, R. B., Tomaro, M. L., Wanschelbaum, A., and Frydman, B. 1972. The enzymatic oxidation of porphobilinogen. FEBS Lett. 26:203–206.

Frydman, R. B., Tomaro, M. L., Wanschelbaum, A., Anderson, E. M., Awruch, J., and Frydman, B. 1973. Porphobilinogen oxygenase from wheat germ: isolation, properties, and products formed. Biochemistry 12:5253–5262.

Gassman, M. L., and Duggan, J. X. 1974. Chemical induction of porphyrin synthesis in higher plants. *In* Proceedings of the Third International Congress on Photosynthesis. M. Avron (ed.). Amsterdam: Elsevier, pp. 2105–2113.

Gassman, M. L., Duggan, J. X., Stillman, L. C., Vlcek, L. M., Castelfranco, P. A., and Wezelman B. 1978. Oxidation of chlorophyll precursors and its relation to the control of greening. *In* Chloroplast Development. G. Akoyunoglou and J. H. Argyroudi-Akoyunoglou (eds.). Amsterdam: Elsevier/North Holland, pp. 167–181.

Gibson, K. D., Laver, W. G., and Neuberger, A. 1958. Initial stages in the biosynthesis of porphyrins. 2. The formation of δ-aminolaevulinic acid from glycine and succinyl-coenzyme A by particles from chicken erythrocytes. Biochem. J. 70:71–81.

Gough, S. P., and Kannangara, C. G. 1977. Synthesis of Δ-aminolevulinate by a chloroplast stroma preparation from greening barley leaves. Carlsberg Res. Commun. 42:459–464.

Granick, S. 1960. Magnesium porphyrins in chlorophyll biosynthesis. Fed. Proc. 19I:330.

Granick, S., and Beale, S. I. 1978. Hemes, chlorophylls and related compounds: biosynthesis and metabolic regulation. Adv. Enzymol. 46:33–203.

Griffiths, W. T. 1978. Reconstitution of chlorophyllide formation by isolated etioplast membranes. Biochem. J. 174:681–692.

Griffiths, W. T., and Mapleston, R. E. 1978. NADPH-Protochlorophyllide oxidoreductase. *In* Chloroplast Development. G. Akoyunoglou and J. H. Argyroudi-Akoyunoglou (eds.). Amsterdam: Elsevier/North Holland, pp. 99–104.

Harel, E. 1978a. Chlorophyll biosynthesis and its control. *In* Progress in Phytochemistry, Vol. 5. L. Reinhold, J. B. Harbone, and T. Swain, (eds.). Oxford: Pergamon Press, pp. 127–180.

Harel, E. 1978b. Initial steps in chlorophyll synthesis—problems and open questions. *In* Chloroplast Development. G. Akoyunoglou and J. H. Argyroudi-Akoyunoglou (eds.). Amsterdam: Elsevier/North Holland, pp. 33–44.

Harel, E., Meller, E., and Rosenberg, M. 1978. Synthesis of 5-aminolevulinic acid-[^{14}C] by cell-free preparations from greening maize leaves. Phytochemistry 17:1277–1280.

Jerzykowski, T., Winter, R., and Matuszewski, W. 1973. γδ-dioxovalerate as a substrate for the glyoxalase enzyme system. Biochem J 135:713–719.

Kannangara, C. G., and Gough, S. P. 1977. Synthesis of Δ-aminolevulinic acid and chlorophyll by isolated chloroplasts. Carlsberg Res. Commun. 42:441–457.

Kannangara, C. G., and Gough, S. P. 1978. Biosynthesis of Δ-aminolevulinate in greening barley leaves: glutamate 1-semialdehyde aminotransferase. Carlsberg Res. Commun. 43:185–194.

Kannangara, C. G., and Gough, S. P. 1979. Biosynthesis of Δ-aminolevulinate in greening barley leaves. II. Induction of enzyme synthesis by light. Carlsberg Res. Commun. 44:11–20.

Kikuchi, G., Kumar, A., Talmage, P., and Shemin, D. 1958. The enzymatic synthesis of δ-aminolevulinic acid. J. Biol. Chem. 233:1214–1219.

Klein, O., and Senger, H. 1977. Biosynthetic pathways of δ-aminolevulinic acid induced by blue light in the pigment mutant C-2A of *Scenedemus obliquus*. Photochem. Photobiol. 27:203–208.

Klein, O., and Senger, H. 1978. Two biosynthetic pathways to δ-aminolevulinic acid in a pigment mutant of the green alga, *Scenedemus obliquus*. Plant Physiol. 62:10–13.

Lascelles, J. A. 1978. Regulation of pyrrole synthesis. *In* The Photosynthetic Bacteria. R. K. Clayton and W. R. Sistrom (eds.). New York: Plenum Press, pp. 795–808.

Lascelles, J. A., and Hatch, T. P. 1969. Bacteriochlorophyll and heme synthesis in *Rhodopseudomonas spheroides*: possible role of heme in regulation of the branched biosynthetic pathway. J. Bacteriol. 98:712–720.

Miflin, B. J., and Lea, P. J. 1977. Amino acid metabolism. Annu. Rev. Plant Physiol. 28:299–329.

Pardo, A. D., Chereskin, B. M., Castelfranco, P. A., Francheschi, V. R., and Wezelman, B. E. 1981. ATP requirement for Mg chelatase in developing chloroplasts. Plant Physiol. 65:956–960.

Ramaswamy, N. K., and Nair, P. M. 1973. δ-aminolaevulinate synthetase in extracts of cultured soybean cells. Biochim. Biophys. Acta 293:269–277.

Ramaswamy, N. K., and Nair, P. M. 1974. Temperature and light dependency of chlorophyll synthesis in potatoes. Plant Sci. Lett. 2:249–256.

Ramaswamy, N. K., and Nair, P. M. 1977. Pathway for the biosynthesis of Δ-aminolevulinic acid in greening potatoes. Indian J. Biochem. Biophys. 13:394–397.

Salvador, G. F. 1978. δ-aminolevulinic acid synthesis from γ-δ-dioxovaleric acid by acellular preparations of *Euglena gracilis*. Plant Sci. Lett. 13:351–355.

Sawyer, E., and Smith, R. A. 1958. δ-aminolevulinate synthesis in *Rhodopseudomonas spheroides*. Bacteriol. Proc., p. 111.

Schiff, J. A., and Epstein, H. T. 1966. The replicative aspect of chloroplast continuity in *Euglena*. *In* The Biochemistry of Chloroplasts, Vol. 1. T. W. Goodwin (ed.). New York: Academic Press, pp. 341–353.

Shemin D., Russel, C. S., and Abramsky, T. 1955. The succinate-glycine cycle. I. The mechanism of pyrrole synthesis. J. Biol. Chem. 215:613–626.

Vlcek, L. M., and Gassmann, M. L. 1979. Reversal of α-α'-dipyridyl-induced porphyrin synthesis in etiolated and greening red kidney bean leaves. Plant Physiol. 64:393–397.

Weinstein, J. D. 1979. Ph.D. dissertation, University of California, Davis.

Weinstein, J. D., and Castelfranco, P. A. 1978. Mg-protoporphyrin-1X and δ-aminolevulinic acid synthesis from glutamate in isolated greening chloroplasts. δ-Aminolevulinic acid synthesis. Arch. Biochem. Biophys. 186:376–382.

Wider de Xifra, E. A., Batlle, A. M. del C., and Tigier, H. 1971. δ-Aminolaevulinate synthetase in extracts of cultured soybean cells. Biochim. Biophys. Acta 235:511–517.

Wider de Xifra, E. A., Stella, A. M., and Batlle, A. M. del C. 1978. Porphyrin biosynthesis—immobilized enzymes and ligands. IX. Studies on δ-aminolaevulinate synthetase from cultured soybean cells. Plant Sci. Lett. 11:93–98.

Wolwertz, M.-R. 1978. Two alternative pathways of chlorophyll biosynthesis in *Pinus jeffreyi*. *In* Chloroplast Development. G. Akoyunoglou and J. H. Argyroudi-Akoyunoglou (eds.). Amsterdam: Elsevier/North Holland, pp. 111–118.

Yubisui, T., and Yoneyama, Y. 1972. δ-Aminolevulinic acid synthetase of *Rhodopseudomonas spheroides*: purification and properties of the enzyme. Arch. Biochem. Biophys. 150:77–85.

Discussion of Presentation by Drs. Castelfranco and Chereskin

TRENCH: Much of the discussion has been centered around chlorophyll synthesis in the developing plastid. Are the pathways the same in the mature plastid? How much chlorophyll turnover is there in a mature chloroplast?

CASTELFRANCO: Chlorophyll in mature chloroplasts turns over slowly, if at all. Developing plastids, which are undergoing a net synthesis of chlorophyll, are used in order to get reasonable rates of enzyme activity that can be measured.

KAMEN: What is the present status of knowledge on the biosyntheses of chlorophylls and hemes in strictly anaerobic systems? It would seem important from the standpoint of evolution to update this area of investigation.

CASTELFRANCO: I am not aware of any recent work on such organisms.

VON WETTSTEIN: Drs. Kanangara and Harel have established that the two enzyme systems, one converting dioxovaleric acid into δ-aminolevulinate and the other converting glutamate into δ-aminolevulinate, occur in maize as well as in barley. They agree that the former enzyme system probably is involved primarily in degradation, whereas the latter is involved in chlorophyll synthesis during greening of chloroplasts in the two species.

CASTELFRANCO: Fine, no further comment.

VON WETTSTEIN: Does anyone know about an enzyme which carries out an internal transamination like the one apparently accomplished by the glutamate-1-semialdehyde aminotransferase?

CASTELFRANCO: I don't know of any example. Someone pointed out that this so-called "internal transamination" could also be a reaction between two glutamic α-semialdehyde molecules.

KLEIN: As you mentioned, in 8- to 10-day-old greening barley and maize leaves two C-5 pathways for δ-aminolevulinate synthesis exist, one of which is light-dependent while the other is not. In young cells protochlorophyllide accumulates in the dark; in some cases (for instance, as in very young bean leaves) to a relatively large amount. Do both systems for δ-aminolevulinate synthesis (from α-ketoglutarate and from glutamate) occur already in the young meristematic cells, or is it possible that in those cells only one pathway exists, while the second one may develop only during later stages of etioplast formation?

VON WETTSTEIN: The enzyme converting glutamate into δ-aminolevulinate via glutamate-1-seminaldehyde is present in etioplasts, and its amount increases about three- to fivefold upon a brief illumination.

SCHIFF: In *Euglena* (and other organisms) several control points appear to exist in the chlorophyll pathway. There is certainly control by protochlorophyllide at the level of δ-aminolevulinate and since Mg-protoporphyrin IX accumulates in the dark (Frey et al., 1979) there may be a control point just beyond this molecule; regulation by heme is well known. In a branched pathway like this one would expect regulation by the two end products (heme and protochlorophyllide) on the first step (δ-aminolevulinate formation); one would also expect some regulation near the branch point.

When I first suggested the internal transamination pathway to Sam Beale many years ago in Woods Hole he didn't care for it because of the lack of analogous enzymatic reactions; however, he must have thought better of it since he included it as one possibility in his review. I'm glad to see that there is now experimental evidence for this pathway.

Although Granick suggested that heme and chlorophyll are made by a common pathway, we really don't know if this is so. It is quite possible that heme is made by one pathway in the mitochondrion and that heme and chlorophyll are made by another in the plastid. Maybe this is why some organisms seem to have two different δ-aminolevulinate-synthesizing systems in the same cell.

CASTELFRANCO: Concerning the last comment, as a partial answer I will cite the work of Castelfranco and Jones (1975) and the work of Troxler and Offner (1979). The first of these studies indicates that heme and chlorophyll are made from a common ALA pool (which is synthesized by the C-5 route). The second indicates that ALA is made by a C-5 pathway even in an organism which makes no chlorophyll and (probably) no plastids.

COHEN: It has been suggested that dioxovalerate, α-β-dicarbonyl, is an intermediate in the C-5 glutamate pathway. This would be a very reactive compound and a substance such as o-phenylenediamine, which would form a quinoxaline with α,β-dicarbonyls (Lanning and Cohen, 1951), ought to trap dioxovalerate. Has such a trapping effect been sought in the C-5 pathway?

CASTELFRANCO: O-Phenenediamine was used by several workers to trap and to detect dioxovalarate (Kissel and Heilmeyer, 1969; Jerzykowski et al., 1973).

KLEIN: Dr. Castelfranco mentioned the existence of the "classical" pathway for δ-aminolevulinate biosynthesis (from succinyl CoA and glycine) in photosynthetic bacteria, and of the "new" C-5 pathway in higher plants and in Scenedemus. There is, however, information about the occurrence of these pathways for a number of additional photosynthetic organisms, as shown in Table 1.

In the representatives of the cyanobacteria, algae, and bryophytes which have been investigated so far, both pathways seem to exist. In greening leaves with massive chlorophyll formation, the C-5 pathway is dominant, while in higher plant tissues without prominent chlorophyll synthesis, the classical pathway is found. It has been suggested that this latter system may be related to mitochondrial porphyrin synthesis. It is pertinent to the subject of this book

Table 1. The Occurrence in Various Photosynthetic Organisms of "Classical" and "C-5" Pathway(s) for δ-Aminolevulinate Synthesis

Organism	Group	"Classical"[a]	"C-5"[b]	Reference
Rhodopseudomonas	Photosynthetic bacteria	+		Nandi and Shemin, 1976
Fremyella diplosiphon	Cyanobacteria	+	+	Meller and Harel, 1978
Cyanidium caldarium	Red algae	(+)	+	Jurgensen et al., 1975, 1976
Euglena gracilis	Euglenoids	+	+	Harel, 1978
Chlorella vulgaris	Green algae	+	+	Meller and Harel, 1978
Scenedemus obliquus	Green algae	+	+	Klein and Senger, 1978
Camptothecium lutescens	Mosses	+	+	Harel, 1978
Spinach, bean, barley, and maize (greening leaves)	Flowering plants		+	Gough and Kannangara, 1977; Ramaswamy and Nair, 1973; Beale et al., 1975; Meller et al., 1975
Potato tubers	Flowering plants	+		Ramaswamy and Nair, 1973
Soybean tissue culture	Flowering plants	+		Wider de Xifra et al., 1978

[a]The Shemin pathway from succinyl CoA and glycine.
[b]The pathway from glutamate and α-ketoglutarate.

that the new C-5 pathway, which appears to be involved in chlorophyll synthesis, already exists in the cyanobacteria.

References to Discussion

Beale, S. I., Gough, S. P., and Granick, S. 1975. Biosynthesis of δ-aminolevulinic acid from the intact carbon skeleton of glutamic acid in greening barley. Proc. Nat. Acad. Sci. USA 72:2719–2723.

Castelfranco, P. A., and Jones, O. T. G. 1975. Protoheme turnover and chlorophyll synthesis in greening barley tissue. Plant Physiol. 55:485–490.

Frey, M., Alberte, R. S., and Schiff, J. A. 1979. Studies by fluorescence of protochlorophyll(ide) and its phototransformation in dark-grown *Euglena gracilis* var. *bacillaris*. Biol. Bull. 157:368–369.

Gough, S. P., and Kannangara, C. G. 1977. Synthesis of Δ-aminolevulinate by a chloroplast stroma preparation from greening barley leaves. Carlsberg Res. Commun. 42:459–464.

Harel, E. 1978. Initial steps in chlorophyll synthesis—problems and open questions. *In* Chloroplast Development. G. Akoyunoglou and J. H. Argyroudi-Akoyunoglou (eds.). New York: Elsevier-North Holland Biomedical Press, pp. 33–44.

Jerzikowski, T., Winter, R., and Matuszewiski, W. 1973. γ,δ-Dioxovalerate as a substrate for the glyoxylase enzyme system. Biochem. J. 135:713–719.

Jurgensen, J. E., Beale, S. I., Troxler, R. R., Bartolf, M. M., Fitzgerald, M. P., Ramus, J., and Schiff, J. A. 1975. Biosynthesis of δ-aminolevulinic acid in *Cyanidium caldarium* and other algae. Biol. Bull. 149:432.

Jurgensen, J. E., Beale, S. I., and Troxler, R. F. 1976. Biosynthesis of δ-aminolevulinic acid in the unicellular rhodophyte *Cyanidium caldarium*. Biochem. Biophys. Res. Commun. 69:149–157.

Kissel, H. J., and Heilmeyer, L., Jr. 1969. Nachweis und Bestimmung von γ,δ-Diox-ovaleriansäure reversibele Umwandlung von γ,δ-Dioxovaleriansäure und δ-Aminolävulinsäure in Ratten. Biochim. Biophys. Acta 177:78–87.

Klein, O., and Senger, H. 1978. Biosynthetic pathways to δ-aminolevulinic acid induced by blue light in the pigment mutant C-2A of *Scenedemus obliquus*. Photochem. Photobiol. 27:203–208.

Lanning, M. C., and Cohen, S. S. 1951. The detection and estimation of 2-ketohexonic acids. J. Biol. Chem. 189:109–114.

Meller, E., and Harel, E. 1978. The pathway of δ-aminolevulinic acid synthesis in *Chlorella vulgaris* and in *Fremyella diplosiphon*. *In* Chloroplast Development. G. Akoyunoglou and J. H. Argyroudi-Akoyunoglou, (eds.). New York: Elsevier-North Holland Biomedical Press, pp. 51–57.

Meller, E., Belkin, S., and Harel, E. 1975. The biosynthesis of δ-aminolevulinic acid in greening maize leaves. Phytochemistry 14:2399–2402.

Nandi, D. L., and Shemin, D. 1976. Quartenary structure and mechanism of action of δ-aminolevulinic acid synthetase. Fed. Proc. 35:1522.

Ramasawamy, N. K., and Nair, P. M. 1973. δ-Aminolevulinic acid synthetase from cold-stored potatoes. Biochim. Biophys. Acta 293:269–277.

Troxler, R. F., and Offner, G. D. 1979. δ-Aminolevulinic acid synthesis in a *Cyanidium caldarium* mutant unable to make chlorophyll *a* and phycobiliproteins. Arch. Biochem. Biophys. 195:53–55.

Wider de Xifra, E. A., Battle, A. M. C., and Tigier, H. A. 1971. δ-Amino-levulinate synthetase in extracts of cultured soybean cells. Biochim. Biophys. Acta 235:511–517.

Rapporteur's Summary:
Origin and Evolution
of Chloroplast Metabolism

R. Wayne Hendren

The metabolic capacity of modern-day chloroplasts reflects the evolution over several billion years of pathways which originated in prokaryotic cells prior to the development of organelles. These metabolic pathways consist of sequential enzyme-catalyzed reactions in which the products of one reaction serve as the reactants for the next; the pathways are frequently branched or cyclical. Possible mechanisms for the adaptation of primordial metabolic pathways to selective environmental pressures include addition or atrophy of branches and completion or disruption of cycles. Environmental pressures have been applied directly to the evolutionary development of chloroplast metabolism in ancestral prokaryotes, but these pressures act indirectly since the enclosure of chloroplasts within eukaryotic cells.

One method of developing theories concerning the origin and evolution of chloroplasts is the inductive approach based on geological and paleontological evidence. The earliest known bacterium-like fossil structures are more than 3.1 billion years old (Barghoorn and Schopf, 1966). These early organisms are thought to have obtained energy by fermentation of reduced organic compounds present in the primitive hydrosphere (Broda, 1975; Chai, 1976). Geological evidence indicates that the primitive atmosphere was reducing in nature and contained no free oxygen until approximately 2 billion years ago (Cloud, 1968). The accumulation of oxygen in the atmosphere is presumed to have resulted from photosynthesis and correlates well with the first appearance 3.0 billion

Address reprint requests to Dr. R. Wayne Hendren, Senior Biochemist, Chemistry and Life Sciences Division, Research Triangle Institute, P. O. Box 12194, Research Triangle Park, North Carolina 27709.

years ago of fossil structures bearing a strong morphological resemblance to modern day oxygen-evolving cyanobacteria (Schopf, 1974). Microfossil evidence suggests that eukaryotic cells originated approximately 1.4 billion years ago (Schopf and Oehler, 1976). The temporal sequence thus established, in which oxygen-evolving photosynthesis appeared in prokaryotes prior to the development of eukaryotes, is a key element in current theories of chloroplast evolution.

A second approach to understanding the origin and evolution of chloroplasts is the deductive method based on comparative analyses of present-day organisms and metabolic pathways, particularly in photosynthetic prokaryotes and eukaryotic chloroplasts. Comparisons of chloroplast metabolic pathways may involve the structure and sequence of intermediates, the structure of enzymes catalyzing the component reactions, and the mechanisms by which these enzymes are regulated. Additional insights into the evolution of photosynthetic eukaryotes may be derived from the degree to which chloroplasts are metabolically dependent on the cytoplasm and vice versa.

Comparisons of extant organisms have resulted in the establishment of numerous similarities among prokaryotes, mitochondria, and chloroplasts, and in particular between cyanobacteria and chloroplasts (Cohen, 1973; Bogorad, 1975; Grun, 1976). These observations have prompted the development of two opposing theories to explain the origin of eukaryotic chloroplasts and mitochondria. One theory holds that organelles arose by the internal partition and subsequent compartmentalization of DNA in an evolving protoeukaryotic cell (Raff and Mahler, 1972; Uzzell and Spolsky, 1974); one statement of this principle is the cluster-clone hypothesis (Bogorad, 1975). According to the second theory, the ancestors of chloroplasts were free-living photosynthetic prokaryotes which became endosymbionts within primitive eukaryotes and subsequently lost their genetic autonomy (Margulis, 1970; Stanier, 1970, 1974).

The discussions of the origin and evolution of chloroplast metabolism presented in Part II have been developed along both inductive and deductive lines. Each chapter has focused on one of five major areas of chloroplast metabolism; carbon fixation, lipid metabolism, sulfur assimilation, nitrogen assimilation and amino acid synthesis, and pigment biosynthesis.

J. A. Bassham has compared the metabolism of carbon fixation in cyanobacteria, eukaryotic algae, and higher plants, and has proposed a scenario for the evolution of the reductive pentose phosphate pathway. Two important components of this proposal are the evolution of the reductive pentose phosphate pathway from the oxidative pentose phosphate pathway and the appearance of a ribulose 1,5-bisphosphate carboxylase prior to the development of oxygenic photosynthesis. The oxidative pentose phosphate pathway is considered to have developed very early in response to needs for pentoses in the synthesis of nucleic acids and for reduced nicotinamide adenine dinucleotide phosphate in membrane lipid synthesis. Once the concentrations of reduced organic compounds in the primitive hydrosphere were nearly depleted by fermentation, cyclic photophosphorylation arose to harness sunlight as an energy source. As the need to use atmospheric CO_2 as a source of carbon became more critical, net incorporation of CO_2 was achieved by the development of phosphoribulokinase, ribulose 1,5-bisphosphate carboxylase, and sedoheptulose diphosphatase, the only enzymes of the reductive pentose phosphate pathway not present in the preexisting oxidative pentose phosphate and glycolytic pathways. The presence of the reductive pentose phosphate cycle in modern-day chemoautotrophs further suggests the antiquity of this pathway, and raises the possibility that it may have as easily evolved in a primitive chemoautotroph as in a photosynthetic bacterium.

The evolution of photosystem II allowed water to be used as the reductant in noncyclic photophosphorylation and thereby introduced molecular oxygen into the primitive atmosphere. Ribulose 1,5-bisphosphate carboxylase also catalyzes the oxidation of ribulose 1,5-bisphosphate with molecular oxygen, and thus accounts for at least a portion of photorespiration (Bowes, et al., 1971). The oxidation reaction catalyzed by ribulose 1,5-bisphosphate carboxylase may be gratuitous since no definite function has yet been assigned to photorespiration and since the carboxylase is postulated to have evolved during a period in which no oxygen was present in the atmosphere. Although carbon dioxide was present in the primordial atmosphere, its concentration may have never been much larger than the current level of 0.03%; otherwise, the oceans would have been acidic and more erosion of rocks would have been expected than has actually occurred (Broda, 1975). The dramatic change of the ratio of oxygen and carbon dioxide concentrations in the atmosphere which followed the evolution of oxygenic photosynthesis may well have marked the appearance of photorespiration. Pathways such as the C-4 carboxylation pathway and Crassulacean acid metabolism (Hatch and Slack, 1970) which minimize photorespiration are considered to have evolved much more recently. Similarities have been observed in the mechanisms by which the carbon fixation pathways are regulated in the cyanobacterium *Aphanocapsa* and in higher-plant chloroplasts. In addition, thioredoxin activates the diphosphatases in the cyanobacterium *Synechococcus* as in the chloroplasts of higher plants.

J. B. Mudd has surveyed and compared lipid metabolism in cyanobacteria and higher plants. Lipids, excluding pigments, comprise approximately 25% of the dry weight of chloroplasts (Galliard, 1973) and are vital components of the thylakoid membranes in which the photophosphorylation and electron transport components are embedded. The principal lipids of chloroplast membranes in higher plants are the monogalactosyl and digalactosyl diglycerides (MGDG and DGDG respectively), phosphatidyl glycerol, and the sulfolipid sulfoquinovosyl diglyceride. These four lipids are also the only membrane lipids found in cyanobacteria (Nichols, 1973). The major chloroplast lipids are characteristically esterified with polyunsaturated fatty acids, principally α-linolenic acid (Nichols, 1970). In animals, γ-linolenic acid is formed by desaturation of dietary linoleic acid, but neither linoleic nor α-linolenic acid can be synthesized de novo. In general, prokaryotes contain lipids esterified only with saturated and monounsaturated fatty acids (Fulco, 1974). Cyanobacteria exhibit exceptional diversity with respect to the synthesis of unsaturated fatty acids; some cyanobacteria contain only saturated or monounsaturated fatty acids, some contain linoleic acid but not linolenic acid, and others resemble chloroplasts in that they contain high concentrations of α- or γ-linolenic acids (Stanier and Cohen-Bazire, 1977).

The chloroplast is the major, if not the sole, site of de novo synthesis of saturated fatty acids in higher plants. The fatty acid synthetases present in chloroplasts of higher plants and eukaryotic algae require an acyl carrier protein (ACP) for activity and thus resemble the fatty acid synthetase of the bacterium *Escherichia coli* (Volpe and Vagelos, 1973). The fatty-acyl-ACP products of the chloroplast fatty acid synthetase serve as substrates for synthesis of the galactosyl diglycerides (Renkonen and Bloch, 1969). The cyanobacterium *Phormidium luridum* has also been found to contain an ACP-dependent fatty acid synthetase (H. K. Lin and K. Bloch, unpublished observations.

Dependence on exogenous ACP distinguishes the chloroplast fatty acid synthetases from the ACP-independent fatty acid synthetase multienzyme complexes found in animals, yeast, and certain bacteria such as *Mycobacterium smegmatis* (Volpe and Vagelos, 1976). Furthermore, the component activities of the fatty acid synthetase multienzyme complexes are covalently linked in large polyfunctional polypeptides (Schweizer et al., 1973; Stoops et al., 1975); in contrast, the ACP-dependent fatty acid synthetases from *Escherichia coli* and the eukaryotic alga *Euglena gracilis* consist of six separable and distinct component activities (Volpe and Vagelos, 1976; Hendren and Bloch, 1980). Levels of ACP and all six of the component activities of the *Euglena* ACP-dependent fatty acid synthetase are very low in dark-grown cells but increase dramatically as functional chloroplasts develop when etiolated cells are illuminated (R. W. Hendren and K. Bloch, unpublished observations).

Euglena gracilis also contains an ACP-independent fatty acid synthetase multienzyme complex which is located in the cytoplasm and is present in cells grown in either the dark or light (Delo et al., 1971; Goldberg and Bloch, 1972). The eukaryotic alga *Chlamydomonas reinhardi* contains only an ACP-dependent fatty acid synthetase, thereby resembling higher plants (Sirevag and Levine, 1972).

Light-induced expression of the *Euglena* chloroplast fatty acid synthetase component activities is regulated by protein synthesis on both chloroplast and cytoplasm ribosomes. Nevertheless, certain *Euglena* mutants which lack chloroplast DNA, and presumably therefore lack functional chloroplast ribosomes, contain significant levels of ACP-dependent fatty acid synthetase (Ernst-Fonberg et al., 1974). In *Euglena*, therefore, the cytoplasm has maintained some autonomy with respect to fatty acid biosynthesis while exercising partial control over the expression of fatty acid synthesis in the chloroplast. In *Chlamydomonas* and higher plants, however, the cytoplasm appears to have lost its independent role in fatty acid synthesis.

The metabolic pathway leading to galactosyl diglycerides has been dispersed between the cytoplasm, which supplies uridine diphosphate galactose, and the chloroplast, where fatty-acyl ACP and diacylglycerol are synthesized. Studies of lipid metabolism have thus provided examples of situations in which either segments of metabolic pathways or regulation of pathways may be dispersed between cytoplasm and chloroplast; ribulose 1,5-bisphosphate carboxylase illustrates a third dispersal mechanism in which synthesis of the subunits of a multimeric enzyme is divided between cytoplasm and chloroplast (see the chapter by Wildman in the present volume).

In discussing sulfur assimilation in chloroplasts, Ahlert Schmidt in Chapter 10 of this volume has described three general types of sulfotransferase: one type requires adenosine-3'-phosphate-5'-phosphosulfate (PAPS) and thioredoxin; a second type requires adenosine-5'-phosphosulfate (APS) and thioredoxin; and a third type requires only APS. Some species of cyanobacteria contain the first type of sulfotransferase, some species contain the second type, and others contain the third type, but only the third type is found in chloroplasts of higher plants. The phylogenetic distribution of sulfotransferase types is analogous to the previously described existence among different cyanobacteria of three polyunsaturated fatty acid patterns, only one of which is found in higher-plant chloroplasts.

Since the discovery of chloroplast-specific DNA, RNA, and ribosomes, much attention has been focused on determining the extent to which chloroplast metabolism is autonomous. P. J. Lea has addressed this question in his discussion of nitrogen assimilation and amino acid synthesis in chloroplasts. Cyanobacteria can grow on either nitrate or ammonia as the sole nitrogen source and can synthesize all amino acids de novo. Although nitrite reductase is present in the chloroplasts of higher plants, nitrate reductase is located in the cytoplasm. Chloroplasts of higher plants retain the ability to synthesize most, if not all, of the amino acids. Carbon dioxide fixed in the chloroplast is incorporated directly into aromatic amino acids, but the cytoplasm supplies the pyruvate and oxaloacetate required for the synthesis of the remaining amino acids. The relatively inefficient conversion of carbon dioxide into fatty acids by isolated chloroplasts suggests that chloroplast fatty acid synthesis may be similarly dependent on a cytoplasmic substrate such as pyruvate.

In his discussion of chlorophyll biosynthesis, Paul Castelfranco described two distinct pathways for δ-aminolevulinate biosynthesis. The first pathway, which proceeds from glutamate to δ-aminolevulinate without involving glycine as an intermediate, is normally present in greening barley seedlings and in higher plants in general. The second pathway involves glycine and succinyl coenzyme A as intermediates and is found in bacteria, including the photosynthetic bacteria. Both pathways are present in algae, cyanobacteria, and some mosses.

The discussions of the origin and evolution of chloroplast metabolism which are summarized here are based primarily on comparisons of extant metabolic pathways in photosynthetic prokaryotes and eukaryotic chloroplasts and, to some extent, on rationalizations of these comparisons with the available geological and paleontological evidence. The data presented emphasize the similarities

between modern-day cyanobacteria and chloroplasts and are thus in congruence
with the endosymbiotic theory of chloroplast evolution. However, the results can
also be explained in terms of opposing theories such as the cluster-clone hypothesis; in general, objective criteria for distinguishing an external versus internal prokaryotic origin for chloroplasts do not appear to be currently available.
Perhaps the most important contributions of these studies have been toward
understanding the complex genetic, metabolic, and regulatory interactions among
organelles within eukaryotic cells.

References

Barghoorn, E. S., and Schopf, J. W. 1966. Microorganisms three billion years old from
the Precambrian of South Africa. Science 152:758–763.

Bogorad, L. 1975. Evolution of organelles and eukaryotic genomes. Science
188:891–898.

Bowes, G., Ogren, W. L., and Hageman, R. H. 1971. Phosphoglycolate production
catalyzed by ribulose diphosphate carboxylase. Biochem. Biophys. Res. Commun.
45:716–722.

Broda, E. 1975. The Evolution of the Bioenergetic Processes. New York: Pergamon
Press.

Chai, C. K. 1976. Genetic Evolution. Chicago: University of Chicago Press.

Cloud, P. E. 1968. Atmospheric and hydrospheric evolution on the primitive earth.
Science 160:729–736.

Cohen, S. S. 1973. Mitochondria and chloroplasts revisited. Am. Sci. 61:437–445.

Delo, J., Ernst-Fonberg, M. L., and Bloch, K. 1971. Fatty acid synthetases from *Euglena
gracilis*. Arch. Biochem. Biophys. 143:384–391.

Ernst-Fonberg, M. L., Dubinskas, F., and Jonak, Z. L. 1974. Comparison of two fatty
acid synthetases from *Euglena gracilis* variety *bacillaris*. Arch. Biochem. Biophys.
165:646–655.

Fulco, A. J. 1974. Metabolic alterations of fatty acids. Annu. Rev. Biochem. 43:215–241.

Galliard, T. 1973. Phospholipid metabolism in photosynthetic plants. *In* Biochimica et
Biophysica Acta Library, Form and Function of Phospholipids. G. B. Ansell, J. N.
Hawthorne, and R. M. C. Dawson, (eds.). New York: Elsevier Scientific Publishing
Co, vol. 3, pp. 253–288.

Goldberg, I., and Bloch, K. 1972. Fatty acid synthetases in *Euglena gracilis*. J. Biol.
Chem. 247:7349–7357.

Grun, P. 1976. Cytoplasmic Genetics and Evolution. New York: Columbia University
Press.

Hatch, M. D. and Slack, C. R. 1970. Photosynthetic CO_2—fixation pathways. Annu.
Rev. Plant Physiol. 21:141–162.

Hendren, R. W., and Bloch, K. 1980. Fatty acid synthetases from *Euglena gracilis*.
Separation of component activities of the ACP-dependent fatty acid synthetase and
partial purification of the β-ketoacyl ACP synthetase. J. Biol. Chem. 255:1504–1508.

Margulis, L. 1970. Origin of Eukaryotic Cells. New Haven: Yale University Press.

Nichols, B. W. 1970. Comparative lipid biochemistry of photosynthetic organisms. *In*
Phytochemical Phylogeny. J. B. Harborne (ed.) New York: Academic Press, pp.
105–118.

Nichols, B. W. 1973. Lipid composition and metabolism. *In* The Biology of the Blue-Green Algae. N. G. Carr and B. A. Whitton (eds.). Berkeley: University of California Press, pp. 144–161.

Raff, R. A. and Mahler, H. R. 1972. The nonsymbiotic origin of mitochondria. Science 177:575–582.

Renkonen, O., and Bloch, K. 1969. Biosynthesis of monogalactosyl diglycerides in photoauxotrophic *Euglena gracilis*. J. Biol. Chem. 244:4899–4903.

Schopf, J. W. 1974. Paleobiology of the Precambian: The age of blue-green algae. Evol. Biol. 7:1–43.

Schopf, J. W., and Oehler, D. Z. 1976. How old are the eukaryotes? Science 193:47–49.

Schweizer, E., Kniep, B., Castorph, H., and Holnzer, U. 1973. Pantetheine-free mutants of the yeast fatty acid synthetase complex. Eur. J. Biochem. 39:353–362.

Sirevag, R., and Levine, R. P. 1972. Fatty acid synthetase from *Chlamydomonas reinhardi*. Sites of transcription and translation. J. Biol. Chem. 247:2586–2591.

Stanier, R. Y. 1970. Some aspects of the biology of cells and their possible evolutionary significance. *In* Organization and Control in Prokaryotic and Eukaryotic Cells. Twentieth Symposium of the Society for General Microbiology. H. P. Charles and B. C. J. G. Knight (eds.). Cambridge, England: Cambridge University Press, pp. 1–38.

Stanier, R. Y. 1974. The origins of photosynthesis in eurkaryotes. *In* Evolution in the Microbial World. Twenty-fourth Symposium of the Society for General Microbiology. M. J. Carlile and J. J. Skehel (eds.). Cambridge, England: Cambridge University Press, pp. 219–240.

Stanier, R. Y., and Cohen-Bazire, G. 1977. Phototrophic prokaryotes: The cyanobacteria. Annu. Rev. Microbiol. 31:225–274.

Stoops, J. K., Arslanian, M. J., Oh, Y. H., Aune, K. C., Vanaman, T. C., and Wakil, S. J. 1975. Presence of two polypeptide chains comprising fatty acid synthetase. Proc. Nat. Acad. Sci. USA 72:1940–1944.

Uzzell, T., and Spolsky, C. 1974. Mitochondria and plastids as endosymbionts: A revival of special creation? Am. Sci. 62:334–343.

Volpe, J. J., and Vagelos, P. R. 1976. Mechanisms and regulation of biosynthesis of saturated fatty acids. Physiol. Rev. 56:339–417.

PART III
ORIGIN AND EVOLUTION
OF PLASTID PROTEINS

Except during the nine months before he draws his first
breath, no man manages his affairs as well as a tree does.

—George Bernard Shaw

Further Aspects
of Fraction-1 Protein Evolution

S. G. Wildman

In a collection of chapters seeking answers to the question of how eukaryotes acquired chloroplasts, analysis of fraction-1 (F-1) protein evolution is an appropriate topic, because this protein composes half of the total soluble proteins of chloroplasts in some eukaryotes. I have commented on aspects of F-1 protein evolution (Wildman, 1979) so my intention here is to concentrate on aspects previously omitted or treated superficially.

F-1 protein is found in all organisms that contain chlorophyll a, and is the enzyme ribulose 1-5 bisphosphate carboxylase-oxygenase (RuBPcase) which catalyzes the combination of atmospheric carbon dioxide with ribulose 1-5 bisphosphate (RuBP) in plant cells to form phosphoglyceric acid during photosynthesis. The enzyme also catalyzes the combination of oxygen with RuBP.

As a physical entity, F-1 protein is a ponderable particle of symmetrical shape with an average diameter of about 120 Å as seen with the electron microscope. The F-1 protein macromolecule has a molecular weight of about 550,000 daltons, and is composed of two kinds of subunits: eight large subunits, each having a molecular weight of about 56,000 daltons, combined with eight small subunits. A plausible model of the F-1 protein of tobacco leaves is shown in Figure 1, in which each of the small subunits has a molecular weight of about 12,500 daltons. In other plants, the quaternary structure appears to be the same, but the small subunits may have slightly less or greater molecular weights. In the crystalline state, F-1 protein is composed entirely of amino acids. The large and small subunits unite to form the quaternary structure without benefit of covalent

Address reprint requests to Dr. S. G. Wildman, Emeritus Professor of Biology, Department of Biology, University of California, Los Angeles, Los Angeles, California 90024.

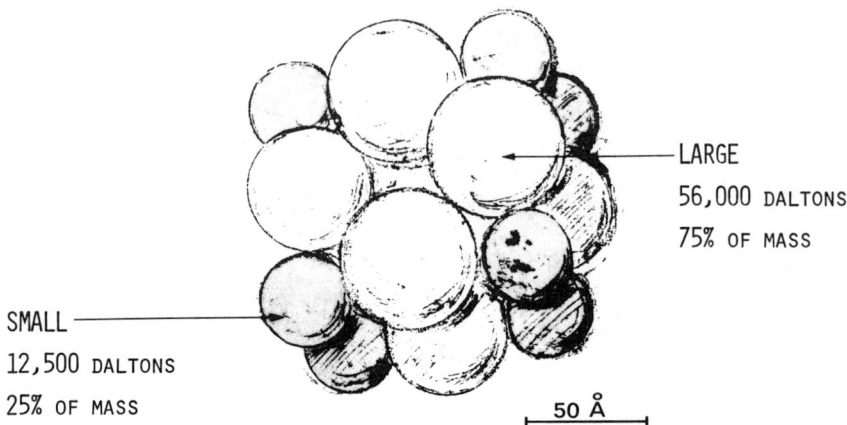

LARGE
56,000 DALTONS
75% OF MASS

SMALL
12,500 DALTONS
25% OF MASS

50 Å

Figure 1. A plausible model of fraction-1 protein. *Source:* Adapted from Baker et al. (1977).

bonds—not even disulfide linkages. Consequently, the large subunits are readily dissociated and easily separated from the small subunits by agents such as sodium dodecyl sulfate (SDS), urea, and high pH, which disrupt hydrogen bonds.

Numerous examples based on studies of inheritance in angiosperms have shown extranuclear DNA to contain the coding information for the large subunit of F-1 protein. Fragments of chloroplast DNA have been isolated and have been shown to have genes coding for the large subunit. Coding information for the small subunits is contained in nuclear DNA. A similar division of coding information has been found for the unicellular eukaryotes *Euglena* and *Chlamydomonas*.

From this brief overview of the structure and function of F-1 protein in eukaryotes, we can proceed to pose some questions as to what might have happened to F-1 protein as prokaryotic organisms evolved into eukaryotes, followed by the subdivision of the latter into the phyla of the plant kingdom.

Evolution of the Quaternary Structure of Fraction 1 Protein

The basic quaternary structure of F-1 protein now observed in angiosperms can be traced back at least 3.5 billion years. Fossils of blue-green algae are this old, and it is hard to distinguish the morphology of the fossils from species living at this moment. It seems a safe assumption that the organisms leaving these fossilized remains depended upon an RuBPcase for their photoautotrophic existence, and that its quaternary structure was the same as that presently found in living species.

Two laboratory groups, that of Akazawa et al. (1978), and that of McFadden and Purohit (1978), have extensively studied the RuBPcases of blue-green algae and photosynthetic bacteria. They are in agreement that the quaternary structure of RuBPcase isolated from blue-green algae consists of eight large and eight

small subunits combined to produce a macromolecule like that found in angiosperms. The angiosperm type of RuBPcase is probably of much earlier origin
than 3.5 billion years. The same two laboratories have found an "eight-large-
+ -eight-small-subunit" type of structure in *Chromatium,* a photolithotropic bacterium. *Hydrogenomonas,* a chemolithotropic bacterium, also contains RuBPcase
of similar structure (McFadden and Purohit, 1978). These bacteria are considered
to be of more ancient origin than blue-green algae.

The RuBPcase of *Rhodospirillum rubrum,* a photosynthetic bacterium, is of
special interest, since the Akazawa and McFadden groups agree that it is a dimer
of two large subunits unaccompanied by small subunits. The bacterial enzyme
shows no immunological relatedness to angiosperm or *Chromatium* RuBPcase,
whereas prokaryotic and eukaryotic RuBPcases with the eight large + eight
small subunits structure display immunological similarities. The two groups of
investigators have arrived at the attractive notion that the earliest form of
RuBPcase was similar to the "two-large-subunit" structure now extant in *R.
rubrum.* Because of immunological unrelatedness however, the Akazawa group
postulate that the *R. rubrum* RuBPcase has undergone extensive alteration in
amino acid composition since its derivation from an ancestral form composed
of two large subunits. As evolution of prokaryotes advanced toward the time
when microorganisms could use water as the predominant substrate for photosynthesis, and oxygen became part of the earth's atmosphere, evolution of the
ancestral two-large-subunit form of RuBPcase may have progressed from two
to six and then to eight large subunits. The McFadden group has identified
microorganisms which have RuBPcase composed of six or eight large subunits.
Then, it appears, a new dimension was added to the evolution of RuBPcase,
because the small subunit made its appearance. It is curious that there seems to
have been no progressive small subunit evolution of the type observed for the
large subunit. That is, no intermediate aggregates below the eight-large- + -eight-
small-subunit combination have been reported. Evidently, the evolutionary im-
perative that generated the need for small subunits did not occur until after the
structure of RuBPcase had advanced from two to eight large subunits.

Efficiency of RuBPCase in Relation to Evolution of Quaternary Structure

As far as is presently known, the RuBPcases of all organisms have the same
specificity: to catalyze the combination of CO_2 or O_2 with RuBP. We can presume
that the specificity has remained unchanged in spite of vast changes in the
quaternary structure of the enzyme. The vast changes appear to have occurred
very early in the evolution of RuBPcase and perhaps, to have ceased by the time
that a macromolecule of the eight large + eight small subunit type had evolved
in organisms as old as the purple sulfur bacteria. In this connection, it is worth
noting that the two-large-subunit type of RuBPcase found in *R. rubrum* has the
capacity to combine RuBP with either CO_2 or O_2. This is interesting since the
ancestors of *R. rubrum* are thought to have made their appearance before O_2

became part of the earth's atmosphere, yet their RuBPcase had already anticipated this event. The anticipation was most likely fortuitous with respect to the evolution of RuBPcase. As pointed out by Lorimer and Andrews (1973), the unique nature of the RuBP molecule made it inevitable that either CO_2 or O_2 could combine with it in the presence of a suitable catalyst.

Were the ancient changes in quaternary structure induced by a need for a more efficient enzyme, that is, an RuBPcase with higher specific activity? Apparently not, because McFadden states that *R. rubrum* RuBPcase consisting of only two large subunits has a turnover number very close to that of the eight-large- + - eight-small-subunit type of enzyme found in spinach, an angiosperm. I suspect that the changes in quaternary structure were in response to a need to fashion a crucial enzyme capable of coping with extreme changes in the environment that the ancient organisms encountered, such as fluctuations in temperature. F-1 protein in Angiosperms has the interesting property of reversibly changing its configuration in response to a change from high (50°C) to low (0°C) temperature and vice versa (Singh and Wildman, 1974). The change in configuration is accompanied by a reversible 30% change in RuBPcase activity. The enzymatic activity is greatest when F-1 protein has been preincubated at 50°C. The enzyme can thus survive temperatures that would be traumatic for terrestrial plants if exposure were prolonged.

The more than 90 sulfhydryl (SH) groups in a F-1 protein molecule seem to be protected against attack by atmospheric O_2. Crystalline F-1 protein has been kept without change in RuBPcase activity or physical properties for more than a year either as a suspension of crystals or as a solution of the protein. It might be illuminating to compare RuBPcase of procaryotes with angiosperm RuBPcase in regard to thermal stability and other physical properties. Differences, if found, might cast light on why the quaternary structure changed, and whether stability in regard to temperature and SH groups came before or after the small subunits made their appearance.

Amount of F-1 Protein in Relation to Evolution in Structure of Chloroplasts

F-1 protein of eukaryotes is localized within chloroplasts. In angiosperms, F-1 protein may amount to 25% of the total protein in the leaves of these plants. While I am unaware of quantitative estimates of the physical quantity of RuBPcase in prokaryotes, my impression is that the amount may be much less in relation to the total protein of blue-green algae or photosynthetic bacteria. A similar condition seems to prevail for unicellular eukaryotes. I wonder, therefore, whether a great increase in the amount of F-1 protein accompanied a change which occurred in the structure of the thylakoid system in chloroplasts. Unicellular eukaryotic chloroplasts have no grana to be seen by light and/or fluorescence microscopy, whereas angiosperms that contain large quantities of F-1 protein have chloroplasts with conspicuous grana separated from an equally obvious stroma. F-1 protein is the predominant component of the stroma of these chloroplasts, and I have arrived at the view that a reason for its great abundance is because F-1 protein serves a dual function. On the one hand, it is a vital enzyme for photosynthesis; on the other, F-1 protein also constitutes the matrix for the structure of the stroma of chloroplasts. I would argue, therefore, that less F-1 protein is found in chloroplasts without grana because the chloroplasts tend to be smaller in size and have less stroma in proportion to thylakoids than in chloroplasts with grana.

Angiosperms have evolved two different pathways for carbon assimilation during photosynthesis. Most angiosperms utilize the Calvin-Benson C-3 pathway, as do unicellular eukaryotes. The more recently evolved system combines the C-3 pathway, with RuBPcase at its core, with the Hatch-Slack C-4 pathway, the latter first assimilating CO_2 into organic acids without use of RuBPcase, and then reforming CO_2 so that it can be passed through the C-3 pathway with the aid of RuBPcase. The corn plant *(Zea mays)* is one example among many kinds of plants which utilize the C-4 pathway. To do this, the corn plant makes two kinds of chloroplasts. Strangely, corn mesophyll cells contain chloroplasts *with* grana but these chloroplasts have no F-1 protein, and if a stroma exists, it is different in character from the stroma in chloroplasts contained in C-3-type plants. The other kind of corn chloroplasts are located in bundle sheath cells. These chloroplasts are devoid of grana but contain a stroma and F-1 protein. They have thylakoids organized in a manner reminiscent of the structure of chloroplasts in unicellular eukaryotes. It would seem that the corn plant may have found it advantageous to reevolve a chloroplast whose structure mimics that of more primitive eukaryotes. Reversion to the structure without grana may have reduced the amount of stroma required by these chloroplasts, and perhaps on this account the amount of F-1 protein in corn plants is only about 10% of the amount found in angiosperms of the spinach or tobacco type.

Effect of Evolution on Amino Acid Composition

Akazawa et al. (1978) have made a statistical analysis of available data on the amino acid composition of RuBPcases from various organisms ranging from *R. rubrum,* at the bottom of the evolutionary scale, to angiosperm F-1 proteins, at the top. The two-large-subunit or *R. rubrum* type of RuBPcase seems to bear little or no resemblance in amino acid composition to a variety of RuBPcases of the eight-large- + -eight-small-subunit type. Thus, the previously mentioned lack of immunological similarity of *R. rubrum* RuBPcase to that in other organisms is also reflected in a similar absence of resemblance in amino acid composition. The analyses further suggested that extreme conservation of the amino acid composition of the large subunit ensued as eukaryotes evolved from prokaryotes. In marked contrast, the amino acid composition of the small subunit had been subject to very extensive change.

In studies made in my laboratory on angiosperms, the large subunit of spinach F-1 protein shared 17 out of 22 tryptic peptides in common with the F-1 protein large subunit of tobacco (Kawashima and Wildman, 1971). Spinach and tobacco are considered to be quite distantly related by authorities on higher-plant phylogeny. In comparison to the large subunit, 60% of the small subunit tryptic peptides of spinach F-1 protein were different from those of tobacco F-1 protein. The extreme difference in the degree of preservation of mutations in the primary sequences of the large and small subunits is seen even when F-1 proteins of closely related plants are compared. Among five species belonging to the genus *Nicotiana*, a χ^2 analysis indicated the large subunit to be more stable toward mutation than hemoglobin α-chains or cytochromes, whereas mutation of the small subunit had occurred to a much greater degree (Kwok and Wildman, 1974). Ten differences in amino acid composition were noted in comparing the small subunits of *N. glauca* and *N. tabacum* F-1 proteins. The sequence of the small subunit is composed of about 100 residues, while the large subunit consists of about 450 residues, yet only two differences in amino acid composition separated the large subunits of the two species.

Effect of Evolution on the Polypeptide Composition of the Large Subunit

When F-1 protein is S-carboxymethylated and electrofocused in 8 *M* urea, the large subunit resolves into three polypeptides whose isoelectric points differ by about 0.1 of a pH unit from each other. The three polypeptides are suspected of being posttranscriptional products of a single chloroplast DNA gene, because no differences were detected in amino acid or tryptic peptide composition between the most acidic and the least acidic of the three polypeptides obtained from the large subunit of *N. tabacum* F-1 protein (Gray et al., 1978). Large subunits composed of three polypeptides have been observed for many other species of plants representing a broad sampling of the plant kingdom. In a particular genus, a given species of plant could have a F-1 protein displaying three large-subunit polypeptides with isoelectric points of pH 6.0, 6.1, and 6.2, compared to F-1 protein from another species of plant in the same genus with large-subunit isoelectric points of 6.1, 6.2, and 6.3. The difference in isoelectric points of the cluster of three polypeptides is characteristic of the F-1 proteins from the two plant species. Furthermore, the genetic information controlling the isoelectric point differences separating the clusters of three polypeptides is inherited exclusively through the maternal line in hybrid plants produced by reciprocal, interspecific hybridization. Isoelectric point differences among clusters of large subunit polypeptides are referred to as "kinds" of polypeptides in Table 1, in which results obtained from electrofocusing F-1 proteins representative of the major phyla of the plant kingdom are summarized.

The large-subunit polypeptide composition of the 11 members of the Lemnaceae, 19 species belonging to the genus *Gossypium*, and 63 species of the genus *Nicotiana* are of particular interest. Fossils of members of the Lemnaceae dating back about 50 million years to the Upper Cretaceous period have been identified. While no fossils which could be used to date the *Nicotianas* and *Gossypiums* have been discovered, the existing species of both of these genera

Table 1. Fraction 1 Protein Polypeptide Composition Among Plants Representing Several of the Major Phyla of the Plant Kingdom[a]

Genus	Number of species analyzed	Kinds of polypeptides LS	SS	Range in number of kinds of small subunit polypeptides	References
Angiosperms					
Nicotiana	63	4	13	1–4	Chen and Wildman, 1979a
Lycopersicon	8	1	3	1–3	Gatenby and Cocking, 1978[b]; Uchimiya et al., 1979
Solanum	7	2	3	1–3	Gatenby and Cocking, 1978a
Brassica Sinapsis Rhaphanus	8	2	4	1–2	Uchimiya and Wildman, 1978
Beta Spinacia	2	2	2	1–2	Chen, K. et al. 1976
Oenothera	12	1	1	1	Chen, K., unpublished
Gossypium	19	4	8	2–4	Chen and Wildman, 1979a
Zea	3	1	2	1–2	Uchimiya et al., 1978
Sorghum	7	1	1	1	Chen, K., et al. 1976
Hordeum	4	1	1	1	Chen, K., et al. 1976
Triticum Aegilops	8	2	1	1	Chen et al., 1975
Avena	7	3	1	1	Steer and Kernoghan, 1977
Oryza	14	2	6	1–4	Uchimiya, unpublished
Lemna Spirodela Wolffiella Wolffia	11	4	8	1–4	Chen and Wildman, 1979a
Other Tracheophytes					
Gingko	1	1	1	1	Chen, K., et al. 1976
Selaginella	1	1	1	1	Chen, K. et al. 1976
Equisetum	1	1	1	1	Chen, K. et al. 1976
Algae					
Chlamydomonas	1	1	1	1	Chen, K. et al. 1976

[a]LS = large subunits; SS = small subunits.

occupy a unique geographical distribution. This suggests that ancestors of these plants could have been in existence at the time when continental drift began the separation of Africa, South America, and Australia 70-100 million years ago. Four differences affecting the isoelectric points of F-1 protein large subunits are all that exist among the members of the Lemnaceae, *Gossypiums,* or *Nicotianas,* and on this account, these angiosperms have been judged to be of similar antiquity

(Chen and Wildman, 1979a). Thus, in a time span extending beyond 50 million years, only two mutations ($2^x = 4$) affecting isoelectric points have been preserved in the primary structures of the F-1 protein large subunits.

Effect of Evolution on the Polypeptide Composition of the Small Subunit

As prokaryotes evolved into eukaryotes, the coding information for the F-1 protein large subunit became encapsulated within the DNA of the chloroplast, while that for the small subunit became a part of the DNA contained in the nucleus. We can, therefore, ponder the condition of the DNA coding for F-1 protein in prokaryotes. In the absence of chloroplasts and a nucleus, is all of the genetic information for an eight large + eight small subunit type of RuBPcase in a prokaryotic cell contained in a circular genome organized as in *E. coli?* Or has subdivision of the information already evolved to the extent that some coding information is located on a plasmid, separated from other information contained on the bacterial-like genome?

Pursuing the effects of evolution on the F-1 protein of eukaryotes, we see a remarkable further fragmentation of the coding information for the small subunit which has been discovered in angiosperm species. Angiosperms have been in existence for about 100 million years and have therefore occupied only a minute fraction of the total time that biological creatures have been evolving. Yet, in this small fraction of time, angiosperms have come to occupy a preeminent position in relation to other phyla of the plant kingdom. They appear to have undergone the most rapid evolution in terms of numbers of different species and this is responsible for the now ubiquitous and dominant distribution of angiosperm species on the solid surface of the globe. The rate of F-1 protein evolution, particularly of the small-subunit, seems also to have reached a zenith among angiosperms.

Electrofocusing F-1 protein resolves the small subunit into one, two, three, or four kinds of polypeptides of different isoelectric points, depending on the plant species from which the F-1 protein was isolated. As shown in Table 1, having more than one kind of polypeptide composing the small subunit of angiosperm F-1 protein is more the rule than the exception, whereas no instance has been uncovered in which polypeptides of different isoelectric points make up the small subunit of F-1 proteins isolated from plants more primitive than the angiosperms. Having acquired the ability to manufacture F-1 proteins with more than one kind of small subunit polypeptide also provided a mode of evolution of F-1 protein amino acid composition that could be independent of the simultaneous mutation of the genetic code.

F-1 protein of tobacco has a small subunit composed of two kinds of polypeptides. The more acidic polypeptide was acquired from nuclear coding information provided by *N. tomentosiformis,* whose pollen fertilized the egg cell of *N. sylvestris* to produce a hybrid plant. *N. sylvestris* provided the coding information for the isoelectric points of the large subunit and also for the least acidic of the two small subunit polypeptides in the hybrid (Gray et al., 1974). The hybrid was infertile because the chromosomes of the two species of *Nicotiana* were too unlike to permit proper pairing as the prelude to successful meiosis. Infertility was overcome by the phenomenon of *amphidiploidy,* wherein somatic doubling of both of the nuclear genomes occurred. Doubling of the number of chromosomes allowed the *N. tomentosiformis* chromosomes to pair with their partners and, likewise, the *N. sylvestris* chromosomes to pair with themselves; now meiosis could be completed and the hybrid plant was capable of self-fertilization and self-perpetuation by seeds (Goodspeed, 1954). The self-fertile hybrid, having phenotypic characters different from either of its two parents and twice as many chromosomes, was declared a new species and called *N. tabacum.* But creation of *N. tabacum* also evolved a new species of F-1 protein containing a small subunit composed of two different kinds of polypeptides in equal amounts. Self-perpetuation of *N. tabacum* by alternation of generations also provided self-perpetuation of the new species of F-1 protein, whose small subunit composition is different in the amounts of six different amino acids compared with the F-1 protein composition of the progenitor species.

Partial sequencing of the polypeptides shows isoleucine to occupy residue 7 in the small subunit polypeptide of *N. tomentosiformis* F-1 protein, while tyrosine is the occupant in *N. sylvestris* F-1 protein (Strobaeck et al., 1976). Substitution of tyrosine for isoleucine requires more than one base change in a coding triplet. Thus, genesis of *N. tabacum* F-1 protein of different composition from the F-1 proteins of either of the parents could not be attributed to a simple point mutation of the genetic code. In fact, evolution by amphiploidy resulted in a F-1 protein which differed also in the amounts of aspartic acid, glycine, valine, and histidine.

Another neat example of a change in small-subunit composition which probably arose by amphiploidy rather than point mutation in the genetic code concerns spinach F-1 protein. As shown in Table 1, the small subunit of spinach F-1 protein is composed of two polypeptides of different isoelectric points. Martin (1979) has determined the entire sequence of 120 amino acids of the small subunit. A tyrosine-proline substitution appears at residue 91. Again, more than one base change would be required to change the coding triplet for tyrosine into the code for proline.

McFadden and Purohit (1978) suggest that heterogeneity in small-subunit polypeptide composition might be an artifact induced by proteolysis during isolation of F-1 protein. This could not be the case for the two examples just described for the two kinds of polypeptides composing the small subunit of spinach and tobacco F-1 protein. Furthermore, single genes code for each kind of small-subunit polypeptide in tobacco F-1 protein with respect to amino acid sequence as well as isoelectric point. Another example (Kung et al., 1975) concerns the four kinds of small-subunit polypeptides of *N. digluta,* this plant having arisen from *N. glutinosa* ♀ hybridizing with *N. tabacum* ♂. The two kinds of small-subunit polypeptides of *N. glutinosa* F-1 protein have less acidic isoelectric points than the two kinds of polypeptides of *N. tabacum*. The two kinds of *N. glutinosa* polypeptides have kept their identity apart from the two kinds of *N. tabacum* polypeptides in F_2, F_3, and F_4 progeny of self-fertilized *N. digluta*. Genetic analysis of still other F-1 proteins composed of one, two, three, or four kinds of small-subunit polypeptides has shown that the isoelectric point of each kind of polypeptide is controlled by a single structural gene acting without dominance (Chen and Wildman, 1979).

Fraction-1 Protein Isozymes, A Consequence of Evolution by Amphiploidy

By opting for evolution by amphiploidy, F-1 protein acquired a means for an accelerated rate of evolution of its amino acid composition compared to previous dependence on point mutations in the genetic code and their very infrequent survival. The accelerated rate due to amphiploidy is illustrated by the fact that eight F-1 proteins of different composition have evolved along with 19 species of *Gossypium* (Chen and Wildman, 1979a). All of the *Gossypium* F-1 proteins have small subunits composed of more than one kind of polypeptide. Among 63 species of *Nicotiana*, 29 different kinds of F-1 proteins have evolved, and 60% have small subunits with more than one kind of polypeptide (Chen and Wildman, 1979). But evolution by amphiploidy engendered a further complication of a curious nature in regard to the stability of the composition of a newly evolved F-1 protein.

When two kinds of polypeptides compose the small subunit of F-1 protein, Hirai (1977) has shown that nine isozymes of F-1 protein are formed, which differ very slightly from each other in electrophoretic mobility. Considering the example of the two polypeptides of the small subunit of *N. tabacum* F-1 protein, the isozyme mixture consists of the two parental types having small subunits with single polypeptides, but together constituting less than 1% of the total isozyme mixture. The remaining seven isozymes are composed of all possible combinations of the two kinds of polypeptides, the most frequent combination (27%) being that of a F-1 protein with a small subunit which is composed of equal amounts of the two kinds of polypeptides. With three polypeptides of different charge, 45 isozymes could form, with only three structural genes providing coding information.

The remarkable feature of F-1 protein isozymes is their constancy in composition, compared to the inconstancy of other plant isozymes. The electrofocusing composition of F-1 protein remains the same regardless of the developmental stage of the plant at the time of isolation of protein. In contrast, Sheen et al. (1974) have shown that the ratio of 1 peroxidase isozyme in relation to 12 others changes appreciably as the tobacco plant develops. However, the composition of the F-1 protein isozyme mixture remains constant under the same circumstances of plant development.

Genetic analysis shows that constancy in composition of a mixture of F-1 protein isozymes is maintained because the coding information which determines the charge on small-subunit polypeptides is sequestered on heterologous chromosomes (Chen and Wildman, 1979b). That the genetic information for each kind of small-subunit polypeptide in the F-1 protein of *N. tabacum* is contained on a separate chromosome is known from the knowledge described earlier in this chapter of how the *N. tabacum* plant originated in the first place. The chromosomes in the *N. tabacum* genome derived from *N. tomentosiformis* do not pair with the chromosomes derived from *N. sylvestris,* so no opportunity exists for segregation of the two separate genes coding for the two kinds of *N. tabacum* small-subunit polypeptides during alternation of generations. Genetic studies involving *N. langsdorffii* + *N. glauca* parasexual hybrids, *N. otophora* × *N. tomentosiformis* hybrids, and other hybrids yield the same picture as seen in tobacco. When more than one kind of polypeptide composes the F-1 protein small subunit, the genetic information for each kind of polypeptide is located on a chromosome which does not exchange information with a chromosome bearing a dissimilar code during meiosis.

Evolution of F-1 protein by amphiploidy, together with concomitant generation of isozymes, poses an important problem to be solved in regard to regulation of F-1 protein biosynthesis. The large subunit is synthesized on chloroplast ribosomes; the small subunit on cytoplasmic ribosomes. But in the case of the small subunit, there may be as many as four different sites of transcription of the coding DNA, the sites being located on heterologous chromosomes physically separated from each other. The problem to be solved is therefore: What is the nature of the signal being transmitted to the different sites, and from what source does the signal arise? From the chloroplast? And once the signal has arrived, how is such exact coordination of the pace of syntheses maintained such that the same mixture of F-1 protein isozymes is invariably formed, irrespective of changes in metabolism that alter the composition of other kinds of isozyme mixtures?

References

Akazawa, T., Takabe, T., Asami, S., and Kobayashi, H. 1978. Ribulose bisphosphate carboxylases from *Chromatium vinosum* and *Rhodospirillum rubrum* and their role in photosynthetic carbon assimilation. *In* Photosynthetic Carbon Assimilation. H. Siegelman, and G. Hind (eds.). New York: Plenum Press, pp. 209–223.

Baker, T. S., Eisenberg, D., and Eiserling, F. 1977. Ribulose bisphosphate carboxylase: a two-layered, square-shaped molecule of symmetry 422. Science 196:293–295.

Chen, K., Gray, J. C., and Wildman, S. G. 1975. Fraction 1 protein and the origin of polyploid wheats. Science 190:1304–1306.

Chen, K., Kung, S. D., Gray, J. C., and Wildman, S. G. 1976. Subunit polypeptide composition of Fraction 1 protein from various plant species. Plant Sci. Lett. 7:429–434.

Chen, K., and Wildman, S. G. 1979a. Composition of Fraction 1 protein in relation to age of angiosperms. Plant System Evol.

Chen, K., and Wildman, S. G. 1979b. Inheritance behavior of information coding for small subunit polypeptides of fraction 1 protein. Biochem Gen. (in press).

Gatenby, A. A., and Cocking, E. C. 1978a. Fraction 1 protein and the origin of the European potato. Plant Sci. Lett. 12:177–181.

Gatenby, A. A., and Cocking, E. C. 1978b. The polypeptide composition of the subunits of Fraction 1 protein in the genus *Lycopersicon*. Plant Sci. Lett. 13:171–176.

Goodspeed, T. H. 1954. *The Genus Nicotiana*. Waltham, Massachusetts: Chronica Botanica Press.

Gray, J. C., Kung, S. D., and Wildman, S. G. 1978. Polypeptide chains of the large and small subunits of Fraction 1 protein from tobacco. Arch. Biochem. Biophys. 185:272–281.

Gray, J. C., Kung, S. D., Wildman, S. G., and Sheen, S. J. 1974. Origin of *Nicotiana tabacum* detected by polypeptide composition of Fraction 1 protein. Nature 252:226–227.

Hirai, A. 1977. Random assembly of different kinds of small subunit polypeptides during formation of Fraction 1 protein macromolecules. Proc. Nat. Acad. Sci. USA 74:3443–3445.

Kawashima, N., and Wildman, S. G. 1971. Studies on Fraction 1 protein. II. Comparison of physical, chemical, immunological and enzymatic properties between spinach and tobacco fraction 1 protein. Biochim. Biophys. Acta 229:749–760.

Kung, S. D., Sakano, K., Gray, J. C., and Wildman, S. G. 1975. The evolution of Fraction 1 protein during the origin of a new species of *Nicotiana*. J. Mol. Evol. 7:59–64.

Kwok, S. Y., and Wildman, S. G. 1974. Evolutionary divergence in the two kinds of subunits of ribulose diphosphate carboxylase isolated from different species of *Nicotiana*. J. Mol. Evol. 3:103–108.

Lorimer, G. H., and Andrews, T. J. 1973. Plant photorespiration—an inevitable consequence of the existence of atmospheric oxygen. Nature 243:359–360.

McFadden, B. A., and Purohit, K. 1978. Chemosynthetic, photosynthetic and cyanobacterial ribulose bisphosphate carboxylase. *In Photosynthetic Carbon Assimilation*. H. Siegelman and G. Hind (eds.). New York: Plenum Press, pp. 179–207.

Martin, P. G. 1975. Amino acid sequence of the small subunit of ribulose-1,5-bisphosphate carboxylase from spinach. Aust. J. Plant Physiol. 6:401–408.

Sheen, S. J. 1970. Peroxidases in the genus *Nicotiana*. Theo. Appl. Genet. 40:18–25.

Singh, S., and Wildman, S. G. 1974. Kinetics of cold inactivation of the ribulose diphosphate carboxylase activity of crystalline Fraction I proteins isolated from different species and hybrids of *Nicotiana*. Plant Cell Physiol. 15:373–379.

Steer, M. W., and Kernoghan, D. 1977. Nuclear and cytoplasmic genome relationships in the genus *Avena:* analysis by isoelectric focussing of ribulose bisphosphate carboxylase subunits. Biochem. Gen. 15:273–286.

Strobaeck, S., Gibbons, G. C., Haslett, G., Boulter, D., and Wildman, S. G. 1976. On the nature of the polymorphism of the small subunit of ribulose-1,5-diphosphate carboxylase in the amphidiploid *Nicotiana tabacum*. Carlsberg Res. Commun. 41:335–343.

Uchimiya, H., Chen, K., and Wildman, S. G. 1978. Polypeptide composition of Fraction 1 protein as an aid in the study of plant evolution. In Stadler Symposium, Vol. 9. G. P. Redei (ed.). University of Missouri Press, pp. 83–100.

Uchimiya, H., and Wildman, S. G. 1978. Evolution of Fraction 1 protein in relation to origin of amphidiploid *Brassica* species and other members of the Cruciferae. J. Hered. 69:299–303.

Uchimiya, H., Chen, K., and Wildman, S. G. 1979. Evolution of Fraction 1 protein in the genus *Lycopersicon*. Biochem. Gen. 17:333–341.
Wildman, S. G. 1979. Aspects of Fraction 1 protein evolution. Arch. Biochem. Biophys. 196:598–610.

Discussion of Presentation by Dr. Wildman

MARGULIES: Ribulose bisphosphate carboxylase is not synthesized throughout the life of a leaf, but is synthesized during growth of a leaf, and is then stable (Brady et al. 1971).

CASTELFRANCO: In parasexual hybrids is it true that the large subunits are of one or the other type but not a mixture? Did I hear you correctly? Then, how do you know which is the paternal and which is the maternal parent?

WILDMAN: Yes, they are of one or the other but not both types. Maternal and paternal types are terms not strictly applicable to parasexual hybrids which are (in the cases I discussed) amphidiploids created by fusion of protoplasts containing different nuclear genomes.

HALLICK: With respect to isoelectric variants of the large subunit, can one tell whether the polypeptides have different amino acid sequences, or the same sequence with some post translational modification?

WILDMAN: The individual polypeptides have been separated but display no differences in tryptic peptide maps, or amino acid composition. My prejudice is that they are posttranslational modifications, or possibly, a problem of incomplete carboxymethylation.

SCHIFF: Hagemann thinks that maternal inheritance, paternal inheritance, etc. of plastids are due to assortment of plastids at the last division of pollen formation. But the maternal inheritance in *Chlamydomonas* indicates this must be more complex. Perhaps below the gross level of pollen assortment other restriction events occur at a biochemical or informational level which determine whether maternal or paternal plastid genes persist in the offspring.

VON WETTSTEIN: In *Oenothera*, biparental inheritance of chloroplasts has been established by O. Renner. The paternal chloroplasts transmitted through the pollen tube can coexist with the maternally inherited chloroplasts, or they can be competed out or they can compete the maternally inherited chloroplasts out. This depends on their own genome (plastome) as well as on the genome present in the nucleus. Recombination among chloroplast genomes in *Oenothera* has not been observed. In *Chlamydomonas*, the chloroplast DNA of both the + mating-type (mt$^+$) and the − mating-type (mt$^-$) parent is degraded, but there is more extensive degradation of the mating-type − DNA. The genes located in *Chlamydomonas* chloroplast DNA are predominately inherited from the mating-type + parent. Fusion of the chloroplasts is observed, however, and recombination between the mt^+ and the mt^- chloroplast DNA can take place. The details of the mechanics of uniparental inheritance and chloroplast genome recombination are not known.

Reference to Discussion

Brady, C. J., Patterson, O. D., Jung, H. J., and Smillie, R. M. 1971. Protein and RNA synthesis during aging of chloroplasts in wheat leaves. *In* Autonomy and Biogenesis of Mitochondria and Chloroplasts. N. K. Boardman, A. W. Linnane, and R. M. Smillie, (eds.). Amsterdam: North Holland, pp. 453–465.

Mutants in the Analysis
of the Photosynthetic Membrane Polypeptides

Diter von Wettstein, Birger Lindberg Møller,
Gunilla Høyer-Hansen, and David Simpson

Gel electrophoresis of sodium dodecyl sulfate (SDS)-solubilized purified thylakoid membranes from greening barley seedlings has established the presence of a minimum of 43 polypeptides (Høyer-Hansen and Simpson, 1977). A review of the literature reveals that at least 30 different polypeptides are implicated in the photosynthetic function of these membranes. Other properties of thylakoids, such as chlorophyll biosynthesis, require additional membrane polypeptides.

Of the polypeptides resolvable by one-dimensional SDS polyacrylamide gel electrophoresis (PAGE), 14 can now be assigned a role in photosynthesis (Figure 1). Using SDS-PAGE at an ambient temperature of 5°C, five chlorophyll proteins can be seen (Machold et al., 1979). The 110-kD band contains the reaction center (P700) for photosystem I and the 29-kD chlorophyll protein is the light-harvesting chlorophyll a/b-protein 2. The 46-kD chlorophyll a-protein 3 is thought to contain the reaction center (P680) of photosystem II (Simpson et al., 1977; Machold et al., 1979). The 50-kD chlorophyll a-protein 2 and the 32-kd chlorophyll a/b-protein 1 may mediate exciton transfer between light-harvesting and reaction-center chlorophyll. The five subunits of the chloroplast coupling factor (CF_1) have been identified in the total polypeptide pattern by comparison with the pattern obtained from purified CF_1 by crossed immunoelectrophoresis, as illustrated in Figure 2 (Høyer-Hansen et al., 1979). Isolation was also used to locate the position of the dicyclohexyl carbodiimide (DCCD) binding proton channel proteolipid which is a component of the chloroplast coupling factor (intrinsic components) (CF_o) entity (Sigrist-Nelson and Azzi, 1979). The location

Address reprint requests to Dr. Diter von Wettstein, Professor, Department of Physiology, Carlsberg Laboratory, Gl. Carlsberg Vej 10, DK-2500 Copenhagen Valby, Denmark.

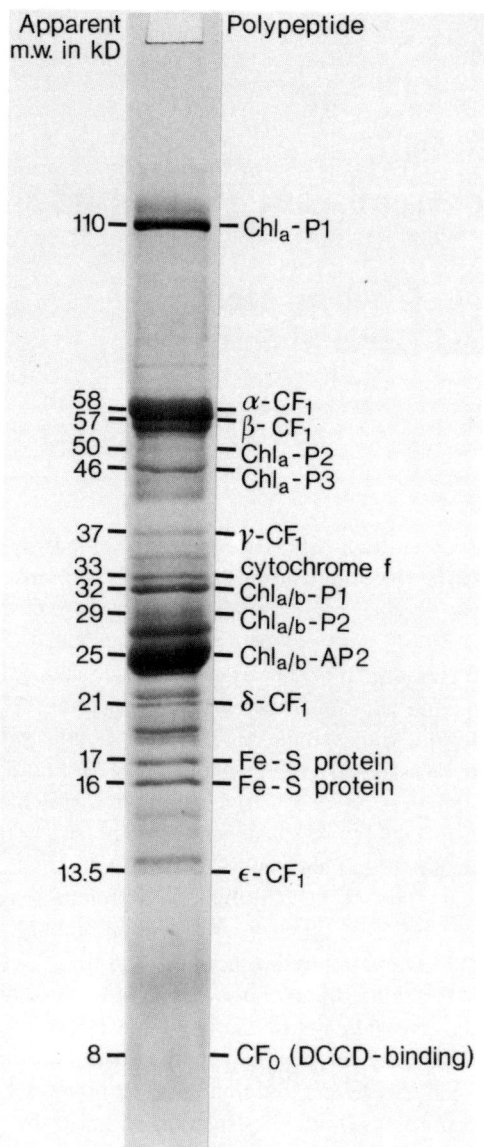

Figure 1. Polypeptide pattern obtained by SDS-polyacrylamide gel electrophoresis of thylakoid components from wild-type barley seedlings. CF_1 = extrinsic component of coupling factor comprising five different subunits; CF_0 = membrane component of coupling factor; Chl_a-P1, Chl_a-P2, Chl_a-P3 = chlorophyll a-proteins; $Chl_{a/b}$-P1, $Chl_{a/b}$-P2 = light-harvesting chlorophyll a/b-proteins; AP = apoprotein.

of cytochrome f was determined by using a specific staining procedure in which the haem covalently bound in cytochrome f catalyzes the oxidation of tetramethylbenzidine (Thomas et al., 1976). The 16- and 17-kD polypeptides are components of photosystem I particles, and based on their molecular weight (Bengis and Nelson, 1977), are tentatively identified as the iron-sulfur proteins on the reducing side of photosystem I.

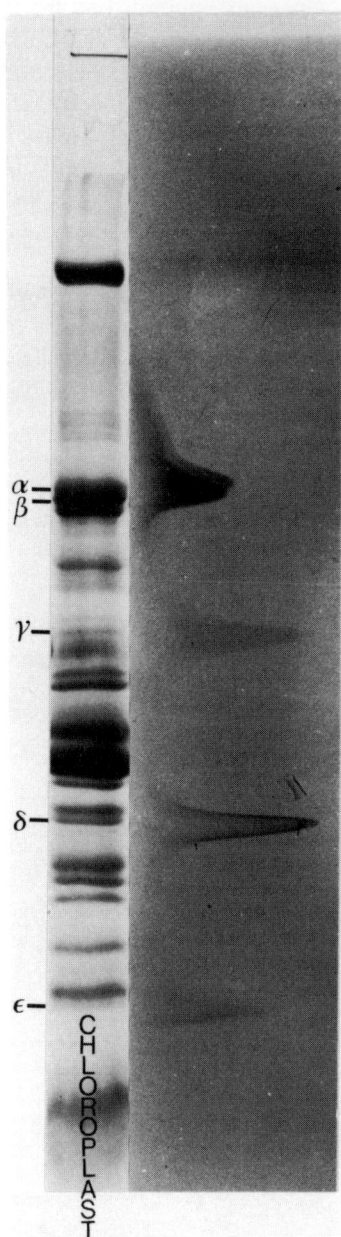

Figure 2. Crossed immunoelectrophoresis of wild-type chloroplast thylakoid polypeptides with antibodies to purified CF₁.

A likely arrangement of the 30 polypeptide components implicated so far in photosynthesis is presented in Figure 3. The positioning of the individual components within the membrane takes into consideration the sequence of electron transfer, information on neighbor relationships, accessibility to antibodies, artificial electron donors and acceptors, as well as structural features revealed by freeze-etching. The light energy absorbed by the light-harvesting chlorophyll a/b-protein 2 is transferred, possibly via the 32-kD chlorophyll a/b-protein 1, and the 50-kD chlorophyll a-protein 2 to the reaction center (P_{680}) of the photosystem II located in the 46-kD chlorophyll a-protein 3, which is expected to span the membrane to effect charge separation. From kinetic and inhibitor studies three components Z_1, Z_2, and M) are known to be involved on the oxidative side of photosystem II (water splitting). This organization is based primarily on the data in Burke et al., (1978); Conjeaud et al. (1979); Delepelaire and Chua (1979); Machold and Meister (1979); and Satoh (1979).

On the reducing side of photosystem II, electrons are transferred from the primary acceptor Q via the two-electron acceptor R to plastoquinone (PQ). From the latter, electrons are transferred sequentially to the Rieske iron-sulfur protein, cytochrome f, plastocyanin (PC), the hypothetical primary donor (PD) for pho-

Figure 3. A diagrammatic representation of known polypeptide components of photosynthesis arranged as required by the electron transport chain. Also considered is whether the polypeptide is accessible from one side or the other of the membrane, and whether the function of the polypeptide requires that it span the membrane.

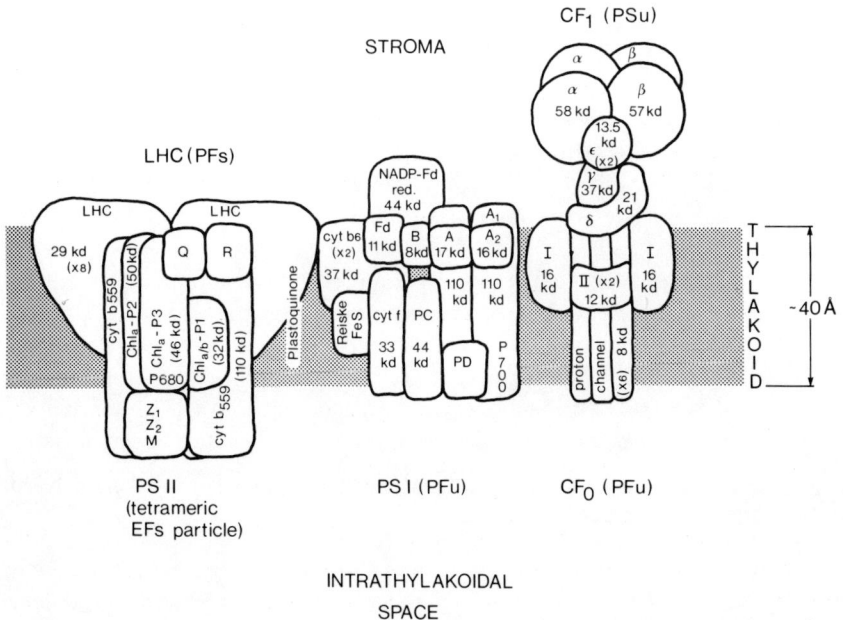

tosystem I, and finally to the reaction center itself (P_{700}) (Hauska et al., 1971; Nelson and Neumann 1972; Velthuys and Amesz, 1974; Siegelman et al., 1976; Bouges-Bocquet and Delosme, 1978; Govindjee and van Rensen, 1978; Malkin and Posner, 1978).

The 110-kD chlorophyll a-protein 1 contains the antenna chlorophylls which transfer the energy to the reaction center located on one of the two polypeptides which must span the membrane for charge speparation. From the primary electron acceptor A_1 located at or near the 110-kD polypeptide, the electron is transferred to the iron-sulfur protein A_2, and then to center B and/or center A. The transfer proceeds via ferredoxin (Fd) to ferredoxin-NADP reductase. Cytochromes b_{559} and b_6 have been placed in the diagram to allow cyclic photophosphorylation around photosystems II and I, respectively (Malkin et al., 1974; Ke et al., 1975; Cramer and Whitmarsh, 1977; Böhme 1977, 1978; Butler, 1978; Golbeck and Kok, 1978; Sauer et al., 1978).

The chloroplast coupling factor consists of extrinsic (CF_1) and intrinsic (CF_0) components responsible for converting the energy of the proton gradient established by the photosynthetic reaction center into ATP. The protons are conducted through the membrane by a channel consisting of six molecules of an 8-kD proteolipid to the catalytic site on the α- and β-subunits (Nelson, 1976; Baird and Hammes, 1979; Pick and Racker, 1979; Sigrist-Nelson and Azzi, 1979).

Data to support the localization and function of some of the polypeptides in the membrane can be obtained from studies with mutants. The lethal nuclear gene mutant $viridis$-n^{34} from barley has a chlorophyll content reaching 75% of that of the wild type. It has normal photosystem II activity, but lacks photosystem I activity, and there is no P_{700} signal, as measured by chemically oxidized minus-reduced spectra (Møller et al., 1980). Cytochromes f, b_6 and b_{559}, and ferredoxin-NADP reductase are present in normal amounts. As expected from the absence of P_{700}, chloroplasts of the mutant cannot generate a proton gradient (Figure 4). When the polypeptide pattern was examined the mutant was found to be partly deficient in the 110-kD P_{700} chlorophyll a-protein 1 and highly deficient in the 16- and 17-kD iron-sulfur proteins. The ultrastructure of the chloroplast membranes as analyzed by thin-sectioning and freeze-fracturing appears normal. The absence of P_{700} may prevent the incorporation of the iron-sulfur proteins into the membranes, or, alternatively, the absence of the iron-sulfur proteins prevents the formation of a functional P_{700} reaction center. This mutant reveals that failure to incorporate P_{700} and these two small polypeptides does not prevent the assembly of the other parts of the photosynthetic apparatus. The specific absence of these three components in an otherwise complete membrane offers unique possibilities to ascertain their function and to attempt reconstitution experiments.

Mutant $viridis$-m^{29}, a sublethal nuclear gene mutant, is specifically deficient in the 50-kD chlorophyll a-protein 2, which is close to the reaction center of photosystem II. This deficiency results in an increased fluorescence yield compared to the wild type. In thin sections, the lamellar system is organized into

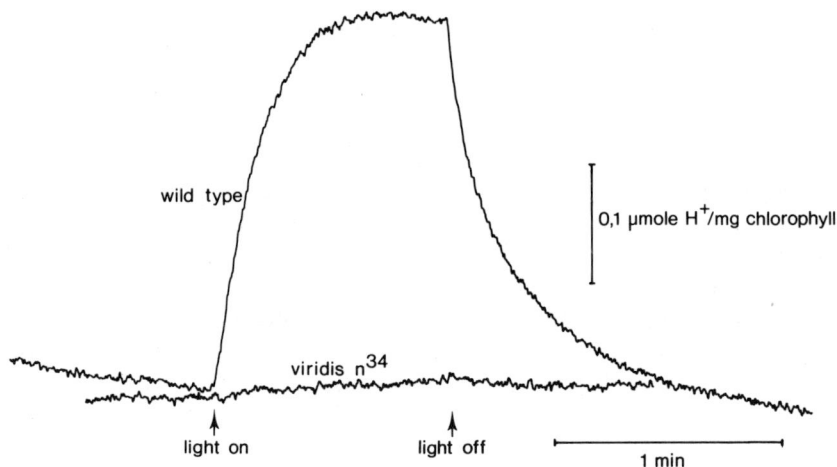

Figure 4. Light-induced changes in the pH of the medium containing chloroplasts of wild-type barley and of *viridis-n*[34]. The reaction mixture contains 81 μmol NaCl, 11 μmol $MgCl_2$, 1.35 μmol tricine (pH = 8.0), 0.543 μmol methylviologen, 81 μmol phenol red, 5.4 μmol NaN_3, and 60 μg chlorophyll, in a total volume of 1.5 ml. Red light of saturating intensity was used, and the pH changes were monitored by the absorbance changes of phenol red (A_{548}/A_{592}).

stroma and grana regions, the intrathylakoid spaces in the grana being somewhat reduced. Upon freeze-fracturing, the grana membranes have normal protoplasmic fracture faces (PFs), but are deficient in the number of particles on the endoplasmic fracture face (EFs) compared to the wild type. The 50-kD chlorophyll *a*-protein is thus required for the assembly of the large freeze-fracture particles found on the EF face of the grana.

Mutation in the *chlorina-f2* gene also results in a highly specific polypeptide deficiency in the composition of the thylakoid (Machold et al., 1977). The failure of mutants in this gene to synthesize chlorophyll *b* results in their inability to incorporate the 29-kD light-harvesting chlorophyll *a/b*-protein 2 into the thylakoids, irrespective of the fact that the polypeptide is probably synthesized on cytoplasmic ribosomes in the mutant (Apel and Kloppstech, 1978). In spite of the absence of this polypeptide from the membranes, stacking of thylakoids into grana is as distinctive as in the wild type. Freeze-fracture and freeze-etching analyses of these membranes using rotary shadowing are presented in Figure 5. The tetrameric particles on the ESs face are present in the mutant as in the wild type. Their spacing in arrays is closer in the mutant as previously shown (Miller, et al., 1976) and their size is 12% smaller. A small reduction in size is also found for the corresponding particles on the EFs face, and the particles are more densely packed in the mutant (See Figs. 5b, d). The most significant difference between the mutant and the wild type, however, is the reduction in the number of particles on the PFs face, from 6260/μm^2 in the wild type to 1120/μm^2 in the mutant (see Figs. 5b, d). The light-harvesting chlorophyll *a/b* protein is,

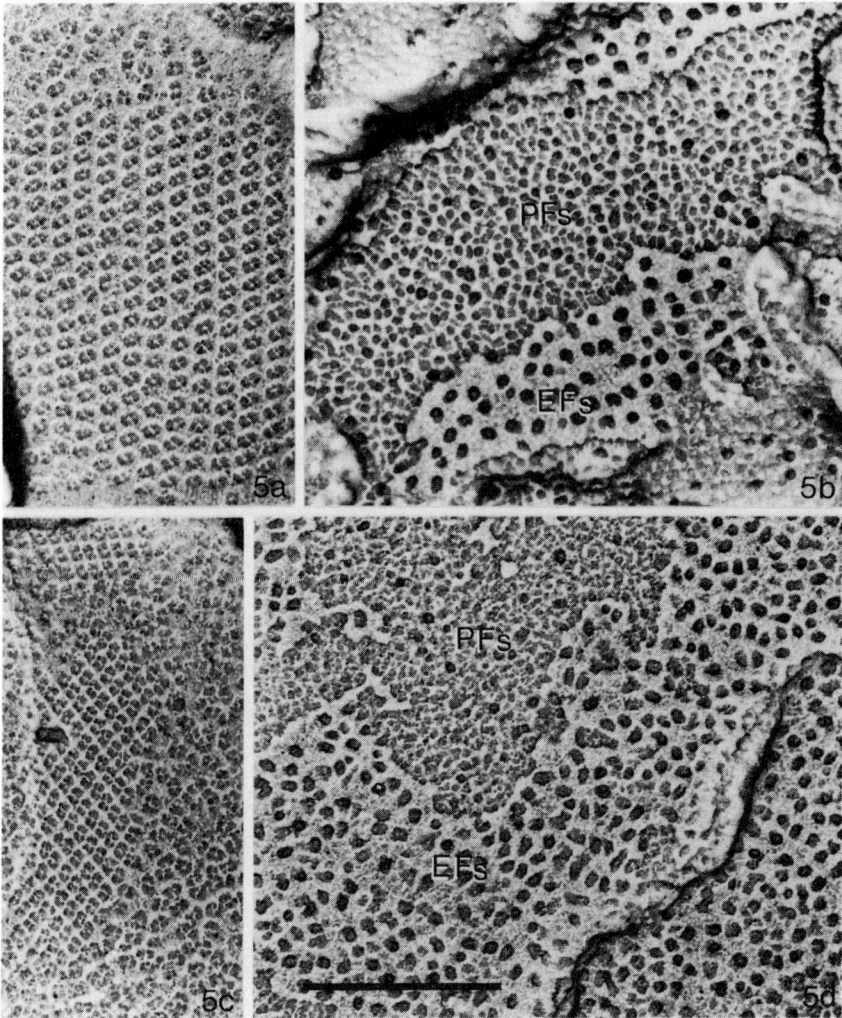

Figure 5. Comparison of the freeze-fracture ultrastructure of wild-type and *chlorina-f2* mutant of barley. EFs = endoplasmic fracture face of grana; PFs = protoplasmic fracture face of grana; ESs = endoplasmic surface of grana. × 200,000. (a) ESs particles forming a regular array in wild-type thylakoids; (b) appearance of the particles on the EFs and PFs faces of wild-type thylakoids; (c) ESs particles in an array in *chlorina-f2* thylakoids; (d) the EFs and PFs faces of *chlorina-f2* thylakoids. (Bar = 0.2 μm).

therefore, a major component of many of the particles on the PFs face (Simpson, 1979). This finding is the reason for placing the light-harvesting chlorophyll *a/b*-protein asymmetrically towards the stroma side of the membrane (Fig. 3). The freeze-fracture plane separates the light-harvesting complex from the photosystem II components of the EFs particle.

The purified light-harvesting chlorophyll *a/b* complex (Burke, et al., 1978) from barley contains large amounts of thylakoid lipids and self-assembles into membranous sheets. Upon storage in 30% glycerol at $-15°C$, these sheets spontaneously seal and form vesicles. When such a preparation is freeze-fractured, particles of approximately the same size are present, as on the PFs face of wild-type barley thylakoids (Figure 6).

The purified light-harvesting chlorophyll *a/b* complex can be separated by SDS-PAGE into a major apoprotein of 25 KD and a minor band of slightly lower molecular weight, in addition to the intact chlorophyll *a/b* protein 2 (29 KD). By chymotryptic peptide mapping the apoprotein is distinguishable from the minor band, but is identical to the monomeric chlorophyll protein and to the oligomeric forms induced by adding Mg^{2+} during SDS-PAGE. This can be seen by comparing the peptide fingerprints in Figures 7a, b.

Incorporation of the purified light-harvesting chlorophyll *a/b* complex into the *chlorina-f2* thylakoids lacking this complex has been attempted. Binding of the complex to the *chlorina-f2* thylakoids could be observed upon reisolation of the thylakoids by sucrose density gradient flotation in the presence of EDTA. How-

Figure 6. Freeze-fractured preparation of purified light-harvesting chlorophyll *a/b* complex (LHC) incorporated into large unilamellar vesicles (Bar = 0.5 μm). ×75,000.

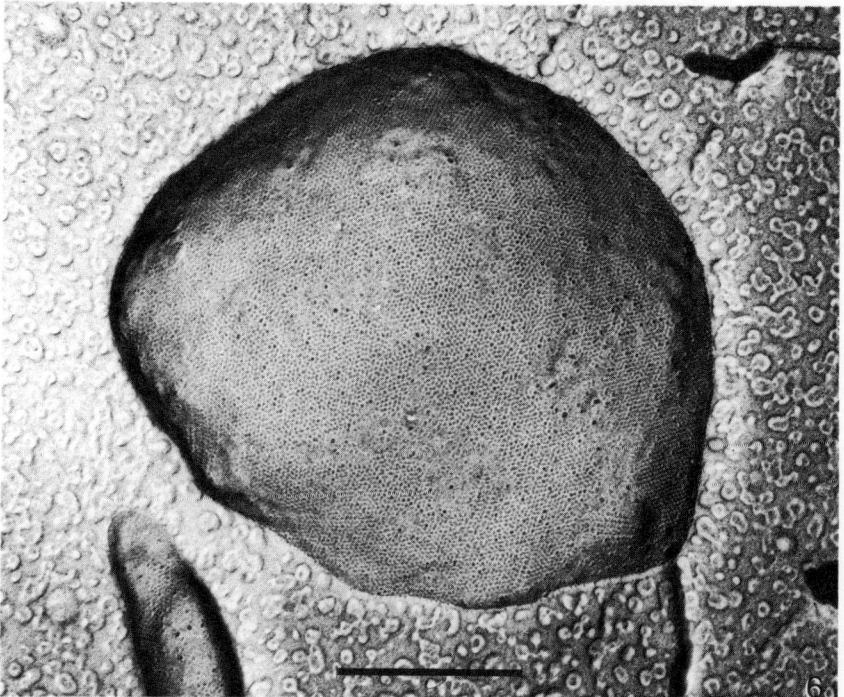

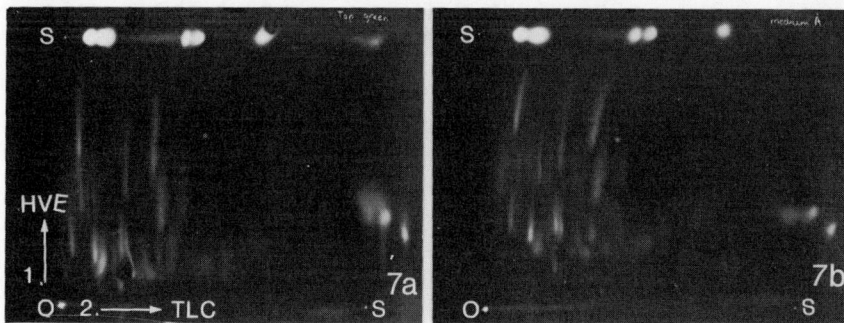

Figure 7. Two-dimensional separation of chymotryptic peptides from purified LHC on cellulose TLC plates. The sample was applied at the origin (O) and separated in the first dimension by high voltage electrophoresis (HVE) with pyridine: acetic acid: acetone: water (20:40:150:790). Thin-layer chromatography (TLC) in pyridine: acetic acid: butanol: water (40:12:90:48) was used for the second dimension separation. S = amino acid standards. (a) Oligomeric chlorophyll a/b-protein 2 from LHC separated by SDS-polyacrylamide gel electrophoresis (PAGE); (b) the major apoprotein after SDS-PAGE.

ever, we found that the conditions needed to maintain the light-harvesting complex in monomeric form caused the irreversible fragmentation of the *chlorina-f2* thylakoids. On the other hand, conditions preserving the stability of the mutant thylakoids caused the light-harvesting complex to aggregate.

The photosynthetic properties of mutants lacking certain thylakoid polypeptides have been reinvestigated employing fluorescence induction kinetics (Simpson and Arntzen, unpublished). The constant fluorescence (F_0) is a measure of the amount of chlorophyll not transferring energy to reaction centers. The amount of functional photosystem II reaction center is directly related to the variable fluorescence (F_v), and the maximum fluorescence (F_m) is the sum of the constant plus variable fluorescence. The increase in variable fluorescence in the presence of 5 mMMg^{2+} indicates that cation regulation of energy distribution is present.

The data are presented in Table 1. Mutant *vir-e*[64] has a high F_0 and no F_v, which is consistent with the absence of photosystem II activity and the deficiency in the 46-kD chlorophyll a-protein 2 (Simpson et al., 1978). The low amount of light-harvesting chlorophyll a/b-protein 2 in mutants *vir-k*[23] and *xan-l*[35] is reflected by the low variable fluorescence in the presence of Mg^{2+} and the absence of cation regulation of energy distribution between photosystem II and photosystem I.

The high F_v in the presence of Mg^{2+} in mutant *vir-n*[34] is consistent with the normal level of photosystem II activity. The absence of P$_{700}$ in the mutant is reflected in the absence of cation regulated energy distribution, and may be responsible for the high F_0.

The size of the photosynthetic unit can be rapidly determined by measuring the half-rise time of the fluorescence yield in the presence of 10^{-5} M DCMU

Table 1. Fluorescence Characteristics of Thylakoids Isolated from Wild-type and Mutant Barley Seedlings (mV 5 µg chlorophyll^{-1} ml^{-1})[a]

Mutant	Without Mg^{2+}			5 mM Mg^{2+}		
	F_0	F_v	F_m	F_0	F_v	F_m
vir-e[64]	2866	0	2862	3834	0	3828
vir-zd[69]	3210	440	3640	3616	1024	4640
xan-d[49]	1278	594	1872	1452	1738	3187
xan-b[12]	1898	530	2428	1563	941	2504
vir-n[34]	1394	1804	3198	1392	1840	3232
Wild type	323	331	654	355	1607	1962
vir-k[23]	158	370	528	160	296	456
xan-l[35]	228	133	361	186	134	320

[a]F_0 = constant fluorescence; F_v = variable fluorescence; F_m = maximum fluorescence.

(Table 2). The size of the photosynthetic unit is determined by the amount of light-harvesting chlorophyll a/b-protein 2, and is therefore inversely proportional to the chlorophyll a : b ratio. The half-rise time is also inversely related to the size of the photosynthetic unit.

It can be seen from Table 2 that the half-rise time increases with increasing chlorophyll a : b ratios, when different mutants are compared. In the presence of Mg^{2+} the photosynthetic unit size increases, except in mutants deficient in light-harvesting chlorophyll a/b-protein 2 (vir-k[23]; xan-l[35]). The absence of this increase in photosynthetic unit size in mutant vir-zd[69] may be due to the small amount of stroma membranes.

For future work on the organization of the photosynthetic membrane, more mutants are needed in which a single thylakoid polypeptide is either missing, or is present with an altered primary structure. A comparison of a given thylakoid polypeptide from different species (see Simpson et al., 1977) has revealed large

Table 2. Half-Rise Time of Fluorescence Yield in the Presence of 10^{-5} M DCMU

Mutant	Chl a/b	$t_{1/2}$ (Without Mg^{2+}) (msec)	$t_{1/2}$ (5 mM Mg^{2+}) (msec)
vir-e[64]	1.9	-	-
vir-zd[69]	2.2	4.0	4.1
xan-d[49]	2.5	12.8	7.6
xan-b[12]	2.6	13.2	7.8
vir-n[34]	2.8	10.4	5.5
Wild type	3.4	22.5	15.0
vir-k[23]	10	77.4	59.9
xan-l[35]	13	78.2	69.6

differences in electrophoretic mobility indicating evolutionary divergence in their primary structure. Induced alterations in the primary structure of individual membrane polypeptides should thus be obtainable in mutants of a single species.

References

Apel, K., and Kloppstech, K. 1978. The plastid membranes of barley *(Hordeum vulgare)*. Light-induced appearance of m-RNA coding for the apoprotein of the light-harvesting chlorophyll a/b protein. Eur. J. Biochem. 85:581–588.

Baird, B. A., and Hammes, G. G. 1979. Structure of oxidative- and photo-phosphorylation coupling factor complexes. Biochim. Biophys. Acta 549:31–53.

Bengis, C. and Nelson, N. 1977. Subunit structure of chloroplast photosystem I reaction center. J. Biol. Chem. 252:4564–4569.

Böhme, H. 1977. On the role of ferredoxin and ferredoxin-NADP$^+$ reductase in cyclic electron transport of spinach chloroplasts. Eur. J. Biochem. 72:283–289.

Böhme, H. 1978. Quantitative determination of ferredoxin, ferredoxin-NADP$^+$ reductase and plastocyanin in spinach chloroplasts. Eur. J. Biochem. 83:137–141.

Bouges-Bocquet, B. and Delosme, R. 1978. Evidence for a new electron donor to P-700 in *Chlorella pyrenoidosa*. FEBS Lett. 94:100–104.

Burke, J. J., Ditto, C. L., and Arntzen, C. J. 1978. Involvement of the light-harvesting complex in cation regulation of excitation energy distribution in chloroplasts. Arch. Biochem. Biophys. 187:252–263.

Butler, W. L. 1978. On the role of cytochrome b_{559} in oxygen evolution in photosynthesis. FEBS Lett. 95:19–25.

Conjeaud, H., Mathis, P., and Paillotin, G. 1979. Primary and secondary electron donors in photosystem II of chloroplasts. Rates of electron transfer and location in the membrane. Biochim. Biophys. Acta 546:280–291.

Cramer, W. A., and Whitmarsh, J. 1977. Photosynthetic cytochromes. Annu. Rev. Plant Physiol. 28:133–172.

Delepelaire, P. and Chua, N.-H. 1979. Lithium dodecyl sulphate/polyacrylamide gel electrophoresis of thylakoid membranes at 4°C: characterization of two additional chlorophyll *a*-protein complexes. Proc. Nat. Acad. Sci. USA 76:111–115.

Golbeck, J. H., and Kok, B. 1978. Further studies of the membrane-bound iron-sulfur proteins and P_{700} in a photosystem I subchloroplast particle. Arch. Biochem. Biophys. 188:233–242.

Govindjee, and van Rensen, J. J. S. 1978. Bicarbonate effects on the electron flow in isolated broken chloroplasts. Biochim. Biophys. Acta 505:183–213.

Hauska, G. A., McCarty, R. E., Berzborn, R. J., and Racker, E. 1971. Partial resolution of the enzymes catalyzing photophosphorylation VII. The function of plastocyanin and its interaction with a specific antibody. J. Biol. Chem. 246:3524–3531.

Høyer-Hansen, G., Møller, B. L., and Pan, L. C. 1979. Identification of coupling factor subunits in thylakoid polypeptide patterns of wild type and mutant barley thylakoids using crossed immunoelectrophoresis. Carlsberg Res. Commun. 44:337–351.

Høyer-Hansen, G., and Simpson, D. J. 1977. Changes in the polypeptide composition of internal membranes of barley plastids during greening. Carlsberg Res. Commun. 42:379–389.

Ke, B., Sugahara, K., and Shaw, E. R. 1975. Further purification of "triton subchloroplast fraction 1" (TSF-I particles). Isolation of a cytochrome-free high-P-700 particle and a complex containing cytochromes f and b_6, plastocyanin and iron-sulfur proteins. Biochim. Biophys. Acta 408:12–25.

Machold, O., and Meister, A. 1979. Resolution of the light-harvesting chlorophyll a/b protein of *Vicia faba* chloroplasts into two different chlorophyll-protein complexes. Biochim. Biophys. Acta 546:472–480.

Machold, O., Simpson, D. J., and Møller, B. L. 1979. Chlorophyll-proteins of thylakoids from wild-type and mutants of barley (*Hordeum vulgare* L.). Carlsberg Res. Commun. 44:235–254.

Machold, O., Meister, A., Sagromsky, H., Høyer-Hansen, G., and von Wettstein, D. 1977. Composition of photosynthetic membranes of wild-type barley and chlorophyll b-less mutants. Photosynthetica 11:200–206.

Malkin, R., Aparicho, P. J., and Arnon, D. I. 1974. The isolation and characterization of a new iron-sulfur protein from photosynthetic membranes. Proc. Nat. Acad. Sci. USA 71:2362–2366.

Malkin, R., and Posner, H. B. 1978. On the site of function of the Rieske iron-sulphur center in the chloroplast electron transport chain. Biochim. Biophys. Acta 501:552–554.

Miller, K. R., Miller, G. J., and McIntyre, K. R. 1976. The light-harvesting chlorophyll-protein complex of photosystem II. J. Cell Biol. 71:624–638.

Møller, B. L., Smillie, R. M., and Høyer-Hansen, G. 1980. A photosystem I mutant in barley (*Hordeum vulgare* L.) Carlsberg Res. Commun. 45:87–99.

Nelson, N. 1976. Structure and function of chloroplast ATPase. Biochim. Biophys. Acta 456:314–338.

Nelson, N., and Neumann, J. 1972. Isolation of a cytochrome b_6-f particle from chloroplasts. J. Biol. Chem. 247:1817–1824.

Pick, U., and Racker, E. 1979. Purification and reconstitution of the N,N'-dicyclohexyl-carbodiimide-sensitive ATPase complex from spinach chloroplasts. J. Biol. Chem. 254:2793–2799.

Satoh, K. 1979. Properties of light-harvesting chlorophyll a/b-protein, and photosystem I chlorophyll a-protein, purified from digitonin extracts of spinach chloroplasts by isoelectrofocusing. Plant Cell Physiol. 20:499–512.

Sauer, K., Mathis, P., Acker, S., and van Best, J. A. 1978. Electron acceptors associated with P-700 in triton solubilised photosystem I particles from spinach chloroplasts. Biochim. Biophys. Acta 503:120–134.

Siegelman, M. H., Rasched, I. R., Kunert, K. J., Kroneck, P., and Böger, P. 1976. Plastocyanin: possible significance of quaternary structure. Eur. J. Biochem. 64:131–140.

Sigrist-Nelson, K., and Azzi, A. 1979. The proteolipid of chloroplast adenosine triphosphatase complex. Mobility, accessibility, and interactions studied by a spin label technique. J. Biol. Chem. 254:4470–4474.

Simpson, D. 1979. Freeze-fracture studies on barley plastid membranes. III. Location of the light-harvesting chlorophyll-protein. Carlsberg Res. Commun. 44:305–336.

Simpson, D., Høyer-Hansen, G., Chua, N.-H., and von Wettstein, D. 1977. The use of single gene mutants in barley to correlate thylakoid polypeptide composition with the structure of the photosynthetic membrane. Proceedings of the 4th International Congress on Photosynthesis Reading: pp. 537–548.

Simpson, D., Møller, B. L., and Høyer-Hansen, G. 1978. Freeze-fracture structure and polypeptide composition of thylakoids of wild-type and mutant barley plastids. *In* Chloroplast Development. G. Akoyunoglou and J. Argyroudi-Akoyunoglou (eds.). Amsterdam: Elsevier/North-Holland Biomedical Press, pp. 507–512.

Thomas, P. E., Ryan, D., and Levin, W. 1976. An improved staining procedure for the detection of the peroxidase activity of cytochrome P-450 on sodium dodecyl sulfate polyacrylamide gels. Anal. Biochem. 75:168–176.

Velthuys, B. R., and Amesz, J. 1974. Charge accumulation at the reducing side of system 2 of photosynthesis. Biochim. Biophys. Acta 333:85–94.

Discussion of Presentation by Dr. von Wettstein et al.

JOLIOT: I think everybody agrees on the localization of the water-splitting enzyme(s) on the inner face of the thylakoid membrane. There are very convincing biophysical and biochemical arguments which favor this hypothesis. Nevertheless, I would be astonished if these enzymes were completely exposed to the water phase inside the thylakoids, as shown in the figure presented by Dr. von Wettstein. The water-splitting enzymes appear to be very well-shielded, and I would think it more probable that these enzymes are, at least partially, embedded in the membrane.

VON WETTSTEIN: A detailed model of the location of the water-splitting enzymes cannot at present be given. It is likely that the catalytic sites of the enzymes are embedded in the membrane. The prominent tetrameric structure exposed by freeze-etching is a logical candidate for the localization of the protein components of the water-splitting apparatus.

Brief Critique of Speculations on Microbial Evolution as Inspired by Protein Homology

Martin D. Kamen

It is appropriate to begin with a quote from Roger Stanier (1970) whose distinguished career has spanned four decades and whom I have been privileged to know as a friend and scientific colleague for nearly the same length of time: "Evolutionary speculation constitutes a kind of metascience, which has the same intellectual fascination for some biologists that metaphysical speculation possessed for some medieval scholastics. It can be considered a relatively harmless habit, like eating peanuts, unless it assumes the form of an obsession; then it becomes a vice." It may be added that it is not only biologists who play these games!

Of course, speculations about origins are a natural concomitant of the human condition, whenever data appear that may shed light on—rather than merely encourage belief in—the eventual solution of the riddle of evolution. It is natural to seize upon them to devise systems of phylogeny. This has become an activity of renewed fervor with the rise of molecular biology, which provides, for the first time, direct data on properties of macromolecules relevant to the elaboration of structure-function relationships. Of particular interest to us on this occasion are schemes proposed for phylogenetic connections in microbial evolution based on consideration of protein homology. (In this connection the fossil record has been and probably will remain patently inadequate.) I wish to comment briefly

Research that provided data relevant to this text was supported by grants from the National Science Foundation (BMS-75-13608 and PCM 76-81648); from the National Institutes of Health (GM 18528-18); and from the Department of Energy (486220-27143).

Address reprint requests to Dr. Martin D. Kamen, Professor Emeritus, Department of Chemistry, Box A-002, University of California, San Diego, La Jolla, California 92093.

on the strengths and limitations of this approach, especially as based on recent data defining primary structures of bacterial cytochromes c. Such schemes have excited some interest in our laboratory and among a few others associated with our efforts, because much of these data have been provided or made possible by our researches over the past two decades establishing a comparative biochemistry of cytochromes c. (Kamen, 1973; Kamen, 1978; Kamen et al, 1978).

Cytochromes of the c-type are distributed among a wide range of prokaryotes and exhibit a diversity of functions and structures which establish that these heme proteins are not a well-defined homogeneous group such as is the case for the well-known mitochondrial type cytochromes c. Whereas cytochrome c in mitochondria is invariably a relatively small single-chain polypeptide with the spectrochemical properties of a low-spin iron complex ("hemochrome" or "hemichrome") functioning as the substrate for the terminal portion of the uniquely eukaryotic electron transport system comprising complexes III and IV of the redox chain in mitochondria, no analogous generalization applies to the cytochromes c found in prokaryotes. I cannot expand on this subject on this occasion as a full discussion would require a separate symposium, but I may say here simply that on the basis of demonstrated properties exhibited by purified, isolated samples adequately characterized in terms of function and molecular structure, prokaryotic cytochromes c require classification into at least half a dozen categories, few of which include properties that resemble those of the mitochondrial subgroup (Horio and Kamen, 1970; Kamen and Horio, 1970). This fact encourages considerable hope for the extension of comparative studies on structure and enzymatic activity beyond those that can be provided by consideration only of the more restricted relationship found in mitochondrial cytochromes c. But, by the same token, they also discourage the notion that bacterial cytochromes c are proper molecules for construction of phylogenies in biochemical evolution, because the creation of such evolutionary schemes requires that function be rigorously defined and that it serve as the unifying principle so that the occurrence of mutations with structural consequences can lead to an ordered progression in time from one phenotype to another. This is far from being the case in the general group of prokaryotic cytochromes c, although there is one subgroup among them—the cytochromes c_2—which might provide the desired data for comparison with mitochondrial eukaryotic cytochromes c. These c-type cytochromes overlap with the mitochondrial types not only in structural homology, but in functional ability to react with the classical eukaryotic redox systems (Errede and Kamen, 1978).

In Table 1, the considerable overlap in numbers of identical residues, plus those derived by mostly conservative substitutions, between mitochondrial cytochromes c and cytochromes c_2 is shown. The homology is dramatically displayed in the typical case of a cytochrome c_2 from the budding, mycelial photosynthetic microorganism, *Rhodomicrobium vannielii*, and from horse heart cytochrome c (Figure 1). It becomes even more striking when comparing the tertiary structures determined for typical cytochromes of these two subgroups

Table 1. Homologies Between Cytochromes c and c_2[a]

Source c		c_2	I	S	Σ
Tuna	vs.	R. rubrum	36	42	78
Tuna	vs.	Rps. capsulata	36	34	70
Tuna	vs.	P. denitrificans	43	34	77
Tuna	vs.	Rsp. palustris	34	34	68
Tuna	vs.	Rps. sphaeroides	31	43	74
Tuna	vs.	Rm. vannielii	51	35	86
Tuna	vs.	R. molischianum	39	40	79
Horse	vs.	Rm. vannielii	51	35	86
Horse	vs.	R. viridis	55	29	84
Horse	vs.	R. rubrum	42	39	81
Horse	vs.	E. gracilis	56	34	90

[a]Numbers shown are identities (I), identities obtainable by one-base substitutions in codons (S), and summations of both (Σ). Abbreviations: *R, Rhodospirillum; Rsp, Rhodopseudomonas; Rm, Rhodomicrobium; E, Euglena; C, Chlorobium; P, Paracoccus.*

(Salemme et al., 1973; Dickerson and Timkovich, 1975). The conviction is overwhelming that these cytochromes originated from the same, or closely similar, prototypes. And, indeed, a number of evolutionary trees displaying the successive rise of present-day prokaryotes and eukaryotes in an ordered time sequence based on the use of these homologies have been proposed (Dickerson and Timkovich, 1975; Dickerson et al., 1976; Schwartz and Dayhoff, 1978). These schemes all require many assumptions, for example, that cytochromes c are evolutionarily representative of the whole genome for the electron transport pathways of which they are components; that rates of evolution be, on the average, the same for different cytochrome c subgroups; and that the number of mutations corresponding to large sequence differences can be predicted.

Examination of our data for cytochromes c_2 and those available from studies on sequences of cytochromes c reveals that the basis for belief that these assumptions are tenable is *under*whelming, to say the least. I will not presume to take the space here to go into an exhaustive discussion of these data. In any case they are available largely owing to the efforts of Dr. R. F. Ambler and his associates on sequence determinations (Ambler, 1979) and the work of Drs. T. E. Meyer and R. G. Bartsch in our laboratory, on isolation and characterization. At the time of writing, Dr. Meyer is preparing a paper on this subject which will go into this matter in some detail. What I may say here, having brought this subject up, is that the data concerned are amino acid sequences of the 12 most divergent mitochondrial cytochromes c compared to each other, and all those presently known for cytochrome c_2 (some 16 to 18, depending on the time it takes for this article to appear in print). The percentage sequence differences and their frequencies have been examined. Whether one compares cytochromes c_2 among themselves or cytochromes c_2 with mitochondrial cytochromes c, there

HORSE HEART CYTO. C vs. VANNIELI I CYTO. C
SEQUENCES

```
HH  ac  G D V E  K G  K K I  F  V Q K  C  A Q  C H   T   V  E K G G  K  H K T  G P  N L H G L
        1                    10                      20                       30
RV  A   G D A V  K G  E Q V  F  - K Q  C  K I  C H   Q   V  G P T A  K  N G V  G P  E Q N D V
```

```
HH  F G  R  K  T  G  Q A  P G F  T  Y  T  D A  N  K N  K G  I  T W  K  E  E  T L  M E  Y L E N P
             40                     50                    60                       70
RV  F G  Q  K  A  G  A R  P G F  N  Y  S  D A  M  K N  S G  L  T W  D  E  A  T L  D K  Y L E N P
```

```
HH  K  K Y I  P G T K M  I  F  A  G  I  K K K T E  R  E  D  I  I  A Y L K  K  A T N E
              80                    90                       100
RV  K  A V V  P G T K M  V  F  V  G  L  K N P Q D  R  A  D  V  I  A Y L K  Q L S G K
```

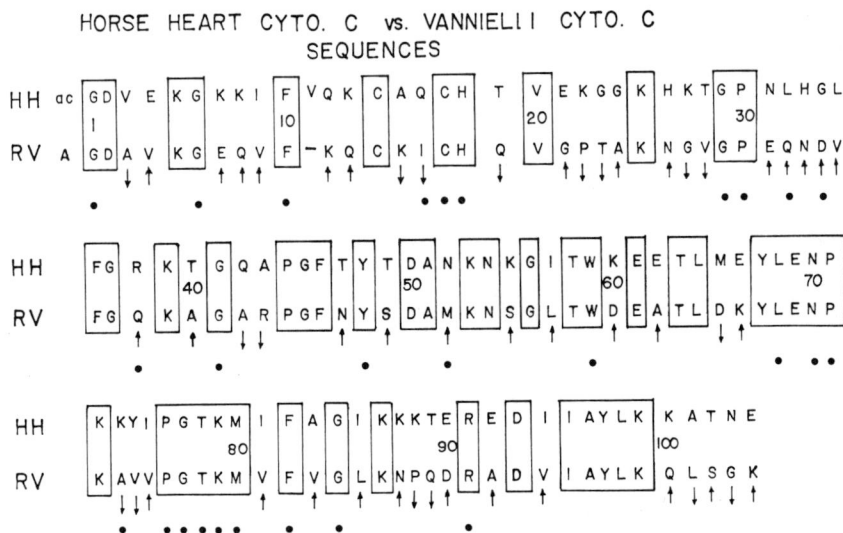

Figure 1 An example of homology between the subgroups mitochondrial cytochrome c and cytochromes c_2. The sequences shown are those from horse heart muscle (HH), and from the photosynthetic microorganism, *Rhodomicrobium vanielii* (RV), using the single-letter sequence code. The homology shown is obtained simply by aligning the invariant cysteines at positions 14 and 17, and the histidine at position 18, using only a single deletion (at position 11) in the RV sequence. Identical residues are boxed and number 51. The 35 changes which require only a single-base substitution in the trinucleotide codon for the amino acids in question are indicated by a downward arrow (↓). Those which require two such substitutions, 17 in number, and are less likely are indicated by upward arrows (↑). Circles under the positions shown indicate residues which are invariant (28 in all) in eukaryotic cytochromes c. The "ac" preceding the N-terminal glycine in HH refers to an acyl group, the presence of which does not affect consideration of the homology.

is a clear indication that the sequence differences of the most divergent of the latter subgroup and of the cytochromes c_2 both cluster around the same high value (~58-60%) rather than showing the expected increase in differences when cross-comparison is made between the two different subgroups. Cytochromes c_2 among themselves, are as divergent as they are when compared to cytochromes c.

They diverge to the point where the conclusion is inescapable they have reached saturation, the extreme beyond which further mutations are either neutral or lethal. They do not allow the construction of a scale linear in time based on a constant mutation rate, such as appears to be the case for the cytochromes c of mammals and other animals, which have been under the unique functional constraint of interdigitation with the cytochrome c oxidase complex (cytochrome aa_3) for a time perhaps no longer than one aeon. Those proteins of the eukaryotic subgroup which cluster around smaller percentage differences—that is, the larger majority from the higher animals which are the less divergent cytochromes *c*—are those from which we would expect that it would be possible to construct credible trees on the basis of their relatively small percentage differences. Use of current tree-building methodologies in the case of these cytochromes *c* gives phylogenies consistent with those derived on the basis of the sparse data relating to other molecules as well as with those from the fossil record (Dayhoff, 1970).

The extrapolation of these tree-building procedures to prokaryotic cytochromes *c* which have apparently approached a saturation value for mutation numbers seems untenable, at least in so far as cytochrome *c* sequences are concerned. Moreover, all the objections to the notions that cytochromes *c* are an adequate representation of the whole electron transport genome remain unanswered, as do worries about intergenic transfers which could prevent construction of trees for even closely related prokaryotic species, such as the Rhodospirillaceae, from which most sequence data are available. Finally, there are the difficulties arising from taxonomic anomalies within this group (Ambler, 1979).

The use of cytochromes *c* nevertheless can be rendered more plausible if the same dendrograms are derived from molecules more likely to have histories in accord with the above-mentioned assumptions, such as DNA or ribosomal RNA. A beginning has been made using some of these macromolecules (Schwartz and Dayhoff, 1978, Schwartz and Kossel, 1979; Gibson and Woese, 1980) but data are still much too sparse to indicate whether evolutionary schemes including both prokaryotes and eukaryotes can be deduced.

We are left with the realization that, while biochemical evidence is the only hope for elaboration of the evolution of life beyond what is contained in the fossil record, there is the caveat that, in the words of the poet Joyce Kilmer: "Poems are made by fools like me/But only God can make a tree."

References

Ambler, R. F. 1979. Cytochromes *c* from photosynthetic prokaryotes, new results and interpretations. *In* International Symposium on Photosynthetic Prokaryotes. Oxford (in press).

Dayhoff, M. O. 1970. Atlas of Protein Sequence and Structure, Vol. 5, Suppl. 2, pp. 26–27.

Dickerson, R. E., and Timkovich, R. 1975. Cytochromes *c*. *In* Enzymes, Vol. II. P. Boyer (ed.). New York: Academic Press, pp. 395–547.

Dickerson, R. E., Timkovich, R., and Almassy, R. J. 1976. The cytochrome fold and the evolution of bacterial energy metabolism. J. Mol. Biol. 100:473–491.

Errede, B., and Kamen, M. D. 1978. Comparative kinetic studies of cytochromes *c* in reactions with mitochondrial cytochrome *c* oxidase and reductase. Biochemistry 17:1015–1027.

Gibson, J., and Woese, C. R. 1980. Phylogenetic analysis of photosynthetic bacteria based on 16S ribosomal RNA. Catalogues, private communication.

Horio, T., and Kamen, M. D. 1970. Bacterial cytochromes II. Functional aspects. Annu. Rev. Microbiol. 24:399–428.

Kamen, M. D. 1973. Towards a comparative biochemistry of the cytochromes. *In* Proteins, Nucleic Acids and Enzymes, Vol. 18. Japan: pp. 753–773.

Kamen, M. D. 1978. The natural history of an ancient protein. Proc. Am. Phil. Soc. 122:214–221.

Kamen, M. D., Errede, B., and Meyer, T. E. 1978. Comparative studies of cytochromes c. *In* Symposium on Molecular Evolution, Kobe, Japan. H. Matsubara, and T. Yamanaka, (eds.). Tokyo: Japanese Scientific Society Press, pp. 373–387.

Kamen, M. D., and Horio, T. 1970. Bacterial cytochromes I. Structural aspects. Annu. Rev. Biochem. 39:673–700.

Meyer, T., Kamen, M. D., and Ambler, R. T. 1980. Can only God make a tree? Proc. Nat. Acad. Sci. USA (in preparation).

Salemme, F. R., Kraut, J., and Kamen, M. D. 1973. Structural bases for function in cytochromes c. An interpretation of comparative x-rays and biochemical data. J. Biol. Chem. 248:7701–7716.

Schwartz, R. M., and Dayhoff, M. O. 1978. Origins of prokaryotes, eukaryotes mitochondria and chloroplasts. Science 199:395–403.

Schwarz, Z. S., and Kossel, H. 1979. Sequencing of the 3′-terminal region of a 16S rRNA gene from *Zea mays* chloroplast reveals homology with *E. coli* 16S rRNA. Nature 279:520–522.

Stanier, R. Y. 1970. Some aspects of the biology of cells and their possible evolutionary significance. *In* Organization and Control in Prokaryotic and Eukaryotic Cells. H. P. Charles and B. C. J. G. Knight (eds.). Symp. Soc. Gen. Microbiol. 20:1–38.

Discussion of Presentation by Dr. Kamen

SCHIFF: We should keep in mind something Roger Stanier has said repeatedly: that even in molecular evolution we should be prepared to find gaps in the record—that is, molecules that have become extinct and are, therefore, missing links in the sequence of molecules available for testing.

The problem discussed by Dr. Kamen seems to be that the investigator is interested in genotype as reflected in amino acid sequence while evolutionary adaptation responds to phenotype—the function of the protein.

Rapporteur's Summary:
Origin and Evolution of Plastid Proteins

Diter von Wettstein

The chapter by Sam G. Wildman on ribulose-1,5-bisphosphate carboxylase and my own on the thylakoid polypeptides of higher plants—have brought into focus the close interdependence of the expression of the genes in the nucleus and the expression of the genes in the chloroplasts. This interdependence can be seen and studied at three levels.

An enzyme like ribulose-1,5-bisphosphate carboxylase contains one polypeptide chain coded for by a nuclear gene and another polypeptide coded for by a gene in chloroplast DNA. The nuclear-coded polypeptide is translated on cytoplasmic ribosomes and is then transported into the chloroplast where it combines with the polypeptide synthesized on chloroplast ribosomes directed by messenger RNA transcribed from chloroplast DNA. Besides having a coordinated transcription and coordinated translation in different compartments, the two polypeptides have to fit carefully together to carry out their enzymatic function.

One might think that this complicated interaction at three levels (coding, synthesis, and function) exerts a strong functional constraint on the primary structure of the plastid proteins. So far the data on this point are limited, but those obtained hint that, within eukaryotes, a considerable evolution of the primary structure of plastid proteins has taken place. The distribution of the genes between the nucleus and the chloroplast has, however, been carefully conserved, as have the distinctive features of the cytosol and organelle protein synthesizing machineries. This is the more remarkable since it is experimentally

Address reprint requests to Dr. Diter von Wettstein, Professor, Department of Physiology, Carlsberg Laboratory, Gl. Carlsberg Vej 10, DK-2500 Copenhagen Valby, Denmark.

feasible in cells like yeast to derive prokaryotic plasmid DNA from nuclear chromosomes or to insert prokaryotic plasmid DNA into nuclear chromosomes (see, for example, Beggs, 1978; Hinnen et al., 1978; and Kielland-Brandt et al., 1979).

We may briefly review to what extent polymorphism in the primary structure of chloroplast proteins coded for by nuclear and chloroplast genes are encountered. In Table 1 the N-terminal sequences of the nuclear coded small subunit of ribulose-1,5-bisphosphate carboxylase/oxygenase up to amino acid residue 25 are presented for seven plant species (for *Chlamydomonas* only the sequence up to position 11 is known). The amount of evolutionary change that has occurred in this polypeptide is quite large; only 3 of the first 11 residues show no variation. Considerable polymorphism is also known in the rest of the chain, comprising 120-130 amino acids. Analysis of 63 species within the genus *Nicotiana* (Chen et al., 1976) by determining the isoelectric focusing patterns of the small subunit point to a considerable evolution of the primary structure during a span of 10^8 years. As observed with amino acid polymorphism in other proteins, many of these variations have probably arisen by genetic drift, rather than being due to permanent selective advantages.

Table 1. N-Terminal Amino Acid Sequences of the Small Subunit of Ribulose-1,5-bisphosphate Carboxylase/Oxygenase in Seven Plant Species[a]

		5		10	
Oenothera biennis:		Phe-Asn-	Val-Trp-Pro-Pro-	Glu-Gly-Leu-	Lys-Lys-Phe-Glu-
Barley	:	Met-Gln-	-Ile-	-Ile-	
Pea	:	Met-Gln-	-Ile-	-Lys-	
Broad bean	:	Met-Gln-	-Ile-	-Lys-	
Spinach	:	Met-Gln-	-Leu-		
			-Ile-Asn-		
Tobacco	:	Met-Gln-		Lys-	-Tyr-
			-Tyr-Gly-		
Chlamydomonas	:	Met-Met-	-Thr-	-Val-Asn-Asn-	-Met

		15	20	25
Oenothera biennis:		-Thr-Leu -Ser-Tyr -Leu-Pro	-Pro-Leu-Thr	-Arg-Glu-Gln
Barley	:		-Ser-Thr-	-Ala
Pea	:	-Trp-	-Pro-Asp-	
Broad bean	:		-Gln-Asp-	
Spinach	:		-Thr-	
Tobacco	:	-Asp-	-Ser -Gln-Gln	

Source: Adapted from von Wettstein et al., 1978; Martin, 1979; and Schmidt et al., 1979.

[a]Only residues differing from those of *Oenothera biennis* are indicated.

On the other hand, the polymorphism in residues seven and eight of the amphidiploid (allotetraploid) *Nicotiana tabacum* ($n = 2x = 24$) is the result of the permanent heterozygosity established for the nuclear gene coding for the small subunit when this species arose by chromosome doubling and hybridization of *Nicotiana sylvestris* ($n = x = 12$) and *Nicotiana tomentosiformis* ($n = x = 12$). The new species inherited from its mother *(sylvestris)* the allele specifying the polypeptide chain containing Ile-Asn, and from its father *(tomentosiformis)* the allele coding for the chain containing Tyr-Gly (Strøbæk et al., 1976). This heterozygosity has been preserved for several hundred generations, maybe as long as 2×10^3 years. It is generally considered that the selective advantage of self-fertilizing polyploid species lies in the possibility of harboring two or more alleles at a gene locus in order to form a variety of heterooligomeric enzyme molecules; these may provide the plant with an improved fitness in a wider range of environments (von Wettstein et al., 1978). The precise physiological advantages or disadvantages of different homomeric and heteromeric forms of chloroplast proteins remain to be explored in future studies.

With regard to the gene in chloroplast DNA coding for the amino acid sequence of the large subunit of ribulose-1,5-bisphosphate carboxylase, information is available for 210 of the 475 residues of the barley enzyme and for several fragments of the spinach protein (Poulsen, 1979; Poulsen et al., 1979). Of the 37 residues that can be compared between barley and spinach, four differences have been encountered. Three of these can be derived by single nucleotide substitutions, whereas the fourth requires two nucleotide substitutions. One of the detected amino acid changes is in the vicinity of the catalytic site of the enzyme.

As the evening primrose *Oenothera* spread across the American continent, extensive speciation took place. Evolution of both the nuclear genome and the chloroplastic genome occurred as the species in this genus formed (Cleland, 1972; Stubbe, 1964). As first recognized by Otto Renner (1937) the evolution of the nuclear and chloroplast genomes of *Oenothera* resulted in the present-day incompatibility of certain nuclear genome classes with certain chloroplast genomes. This comes to light as aberrant pigment and plastid properties, seedling lethality or embryo abortion in certain species hybrids within the subgenus *Oenothera* (or *Euoenothera*). Analysis of these incompatibilities has led to the recognition of five chloroplast genomes (plastomes), each adapted to cooperate with one or several of the five known nuclear gene complexes (Stubbe, 1959; Schötz, 1969; Kutzelnigg and Stubbe, 1974).

Chymotryptic peptide mapping of the large subunit of ribulose-1,5-bisphosphate carboxylase from representatives of the five chloroplast genomes in *Oenothera* revealed a striking overall similarity in the patterns of the peptides but also distinct and reproducible differences among them (Holder, 1978; von Wettstein et al., 1978). It is likely that the peptide differences detected among members of different chloroplast genomes, as well as among members of the same chloroplast genome, are the result of single amino acid substitutions. These

differences indicate the evolutionary divergence of the chloroplast genes within
the genus *Oenothera* in agreement with the expected polymorphism of its chloroplast DNA. Mapping of restriction endonuclease cleavage sites in the DNA
of representatives of the five chloroplast genomes of *Oenothera* have revealed
small but significant changes in the patterns of restriction fragments. Differences
in electrophoretic mobilities of thylakoid polypeptides known in other organisms
to be coded in chloroplast DNA have also been observed when comparing plants
containing different *Oenothera* chloroplast genomes (Gordon et al., 1980, and
personal communication). The molecular bases for the nuclear-chloroplastic genome incompatibilities in *Oenothera* are not yet understood. The determination
of the precise amino acid sequence differences now becoming apparent, and
their effect upon the assembly and function of oligomeric enzymes requiring the
cooperation of the nuclear and organellar genetic systems, will tell how this type
of cooperation evolves and might have arisen.

Of the 30 known thylakoid polypeptides functioning in photosynthesis (see
von Wettstein, Møller, Høyer-Hansen, Simpson, Chapter 15 of this volume),
9 are so far considered to be coded in chloroplast DNA and synthesized on
chloroplast ribosomes, whereas 7 are considered to be coded for by nuclear
genes, synthesized on cytoplasmic ribosomes and transported into the chloroplast
(see Gillham, 1978, and Kirk and Tilney-Basset, 1978 for references). Evidence
exists in one or several species that the following polypeptides are coded in
chloroplast DNA: chlorophyll *a*-protein 2 (Chua, 1976); chlorophyll *a*-protein
3 (Chua and Gillham, 1977); cytochrome b_{559} (Zielinski and Price, 1977); cytochrome f (Doherty and Gray, 1979); coupling factor (CF_1) subunits α, β, and
ε (Mendiola-Morgenthaler et al., 1976); proton channel polypeptide (CF_0) DCCD
binding protein (Doherty and Gray, 1979); and subunit I of CF_0 (Nelson, personal
communication). Evidence exists in one or several species that the following
polypeptides are coded in nuclear DNA: chlorophyll *a/b*-protein 2 (Apel and
Kloppstech, 1978); plastocyanin (Haslett and Cammack, 1974); ferredoxin (Armstrong et al., 1971); ferredoxin-NADP$^+$ reductase (Armstrong et al., 1971);
coupling factor (CF_1) subunits γ and δ (Mendiola-Morgenthaler et al., 1976);
subunit II of CF_0 (Nelson, personal communication).

A physical map of *Chlamydomonas* chloroplast DNA has been constructed
by J.-D. Rochais using the restriction endonucleases EcoRI and Bam H 1, and
many of the fragments have been cloned (Figure 1). With the aid of antibodies
against individual polypeptides prepared by N.-H. Chua, it was possible to place
the genes for the large subunit of ribulose-1,5-bisphosphate carboxylase on the
map (Malnoë et al., 1979) and, tentatively, those for the membrane polypeptides
chlorophyll *a*-apoprotein 2 (= *Chlamydomonas* polypeptide 5), chlorophyll *a*-
apoprotein 3 (= *Chlamydomonas* polypeptide 6), the α- and β-subunits of
coupling factor (CF_1), and two peptides D_1 and D_2 of unknown function (Rochaix,
personal communication), as indicated in Fig. 1 and Table 2. Assignment of the
polypeptides to cloned restriction fragments was done by coupled transcription
and translation of the fragments and characterization of the products by immu-

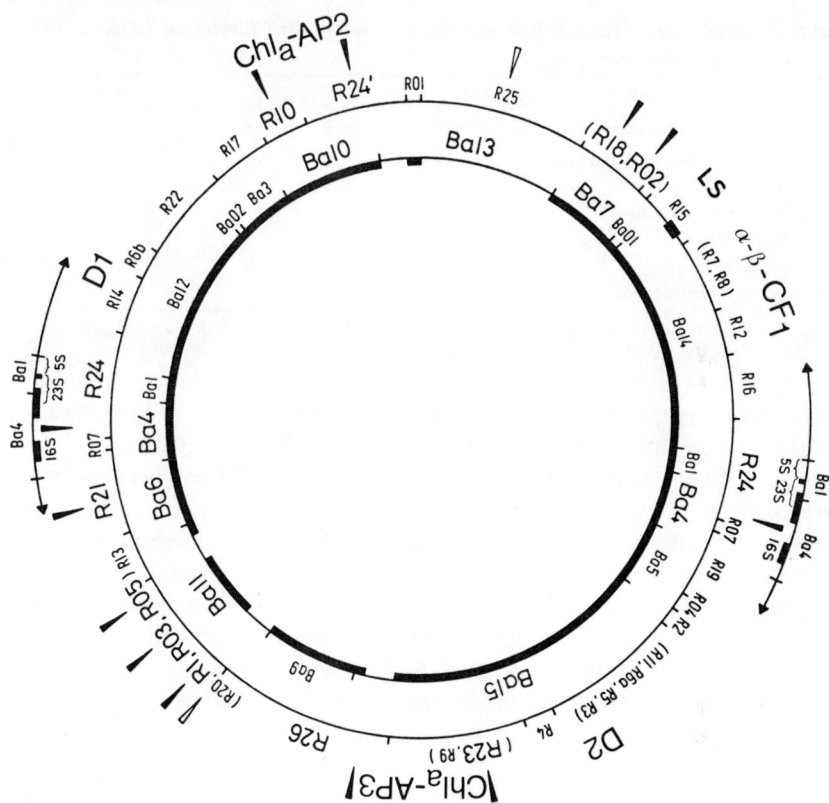

Figure 1. Physical map of the chloroplast DNA of *Chlamydomonas reinhardtii*. The EcoRI and Bam H I restriction endonuclease fragments are denoted by *R* and *Ba*, and are shown on the outer and inner circles, respectively. The cloned regions of the genome are indicated by a thick line on the inner circle. The two ribosomal RNA gene units are shown on the outside and the inverted repeats are indicated. Arrows indicate the location of 4S RNA genes and fragments labeled with large letters hybridize to 4S RNA. The location of the gene for the large subunit of ribulose-1,5-bisphosphate carboxylase is indicated on fragment R15. According to personal information from J.-D. Rochaix, the tentative location of the genes for the α- and β-subunits of CF$_1$ is on fragment R 7, the gene for chlorophyll *a*-apoprotein 2 on fragment Ba 10, the gene for chlorophyll *a*-protein 3 on fragment R 9, and the genes for thylakoid polypeptides D$_1$ and D$_2$ on fragments R 14 and R 3, respectively. *Source:* From Malnoë et al., 1979.

noprecipitation. Alternatively, the cloned DNA fragments were melted and hybridized to messenger RNA prior to translation in vitro; arrested translation then identifies the restriction fragment coding for the polypeptide in question. Nonpolyadenylated messenger RNA molecules were isolated from cells by sucrose gradient centrifugation and were separated by electrophoresis. By hybridization to the labeled restriction fragments the size of the messenger RNA for the

Table 2. Genes for Thylakoid Polypeptides Located in the Chloroplast DNA of
Chlamydomonas

	β-CF$_1$	α-CF$_1$	Chl$_a$AP2	Chl$_a$-AP3	D$_1$	D$_2$
Coupled transcription-translation (IgG)	R7	R7	Ba10	R9	—	R3
Hybrid-arrested translation	—	—	—	—	R14	R3
Messenger RNA (kb)	2.14	1.82	1.93	1.73	1.12	1.10
Coding capacity (kD)	71	61	65	58	37.5	37
Mol wt of polypeptide (kD)	53	53	50	47	35	30

Source: From tentative results by J.-D. Rochaix and P. Malnoë (personal communication).

individual polypeptides could be estimated as given in Table 2. In all cases, a coding capacity sufficent for the molecular weight of the polypeptides was found.

Primary sequence data of these chloroplast DNA coded polypeptides of the thylakoids are not yet available and we cannot, therefore, judge to what extent evolution of these polypeptides has occurred. However, comparison of electrophoretic patterns from six different plant species in Figure 2 reveals quite dramatic differences in the electrophoretic mobilities of the α- and β-subunits of coupling factor, of the chlorophyll *a*-protein 3 and cytochrome *f*. This indicates extensive evolution of the primary structures of these proteins. Using antibodies against individual polypeptides, Chua and Blomberg (1979) could identify equally large differences in apparent molecular weight between spinach and *Chlamydomonas* polypeptides.

Interspecific variation in the apparent molecular weight is also recognizable for the nuclear coded polypeptides (such as chlorophyll *a/b*-apoprotein 2 in Fig. 2). Complete amino acid sequences for the 95–100 residues of prokaryotic and eukaryotic ferredoxins have been determined; they reveal a strong conservation of the primary structure (Hall and Rao, 1977). The sequences of the 99 amino acid residues of plastocyanin among nine higher plants, one alga and one cyanobacterium revealed 30 invariant positions (Ramshaw et al., 1976). The amino acid differences between *Anabaena* and *Chlorella* amounted to 50, whereas between 42 and 49 differences are observable when *Chlorella* is compared with various higher-plant species.

The chloroplast polypeptides synthesized on cytoplasmic ribosomes are made in a precursor form of larger molecular weight, the extension being 44 amino acids at the N-terminus (Figure 3) for the small subunit of *Chlamydomonas* ribulose bisphosphate carboxylase (Schmidt et al., 1979). The precursor sequence is cleaved off by an endoprotease in connection with its transfer across the chloroplast envelope (Dobberstein et al., 1977; Highfield and Ellis, 1978). The translation of a precursor molecule from polyadenylated messenger RNA on cytoplasmic ribosomes, its posttranslational transport into intact chloroplasts, and its processing by proteolytic cleavage in the organelle has been also estab-

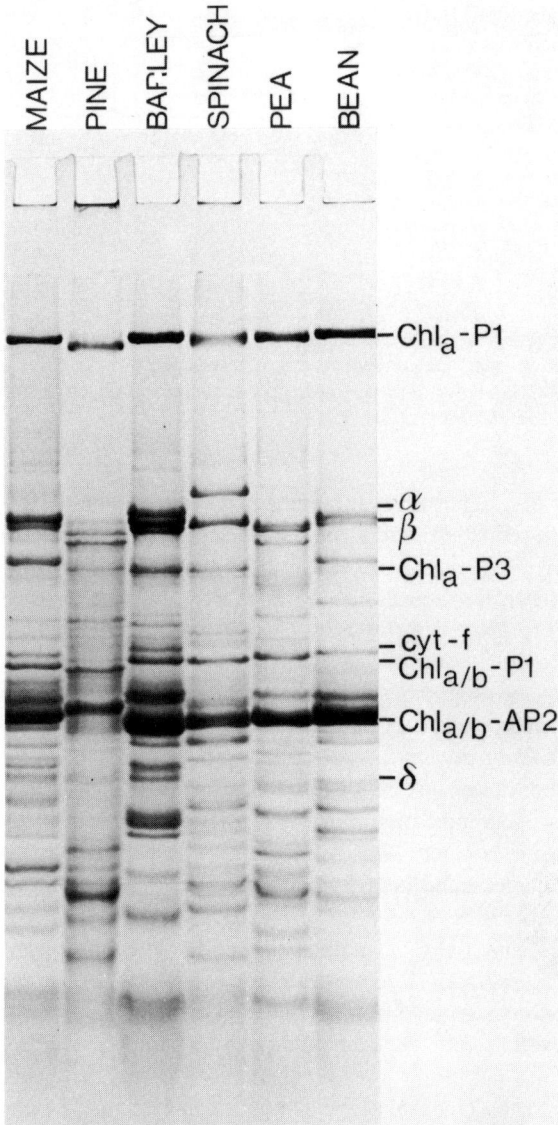

Figure 2. Electrophoretic separation of sodium dodecyl sulfate solubilized thylakoid polypeptides of six different plant species. A polyacrylamide gradient gel of 11–15% was employed for the separation (see Høyer-Hansen and Simpson, 1977). Designation of some polypeptide bands refers to the barley thylakoids. Polypeptides coded for by chloroplast DNA are chlorophyll a-apoprotein 3 (Chl$_a$-P3), cytochrome f (cyt-f), and the subunits α and β of coupling factor (CF$_1$). Polypeptides coded for by nuclear genes are chlorophyll a/b-apoprotein 2 (Chl$_{a/b}$-AP2), and the subunit δ of coupling factor (CF$_1$). Chl$_a$-P1 = chlorophyll a-protein 1. *Source:* Courtesy of G. Høyer-Hansen.

270   D. von Wettstein

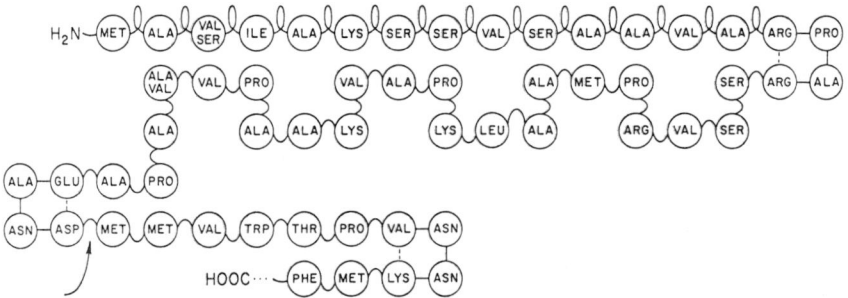

Figure 3. The N-terminal sequence of the precursor molecule for the small subunit of ribulose-bisphosphate carboxylase from *Chlamydomonas* synthesized on cytoplasmic ribosomes. The N-terminal 44 amino acids are cleaved off upon transport of the molecule into the chloroplast. *Source:* From Schmidt et al., 1979.

lished for the apoprotein of the light-harvesting chlorophyll *a*/*b*-protein 2, ferredoxin, ferredoxin-NADP+, reductase, and a considerable number of thylakoid polypeptides with as yet unknown functions (Huisman et al., 1978; Bartlett et al., 1979; Schmidt et al., 1979). Reciprocal heterologous reconstitution of large and small subunits of ribulose bisphosphate carboxylase from pea and spinach has been achieved by translation in vitro of polyadenylated messenger RNA for the small subunit of one species and transport of the translation product into isolated chloroplasts of the other species (Chua and Schmidt, 1978). Transport of the small subunit from *Chlamydomonas* into higher-plant chloroplasts after translation in vitro was not possible, and a reconstitution of the algal and higher-plant suunits has so far failed. A comparative analysis of the precursor sequences and the transport mechanism is awaited with anticipation, as it may inform us about the origin of organellar compartmentation of the genes for chloroplast polypeptides.

There are two hypotheses which seek to explain the modes of origin of eukaryotic cells containing chloroplasts (Bogorad, 1975). One is the origin by endosymbiosis whereby genomes from different prokaryotes have become established in the same (eukaryotic?) cell. The other alternative is called the *clusterclone route,* and envisions the partitioning of the genome of a single cell into three organelles: the nucleus, the chloroplast, and the mitochondrion. The endosymbiotic hypothesis is supported by discoveries of the same proteins, enzyme complexes, and pathways in prokaryotic cells and in the chloroplast. A good example is the phycobilisome, the light-harvesting pigment protein complex of the prokaryotic cyanobacteria and the plastids of the eukaryotic *Rhodophyta.* Close sequence homologies have been established for N-terminal portions of the polypeptide chains in phycocyanins and allophycocyanins from the cyanobacterial and rhodophytan proteins (Glazer et al., 1976). Chlorophyll *b* as a light-harvesting pigment has been found in the prokaryotic *Prochloron,* which in cellular organization is similar to a cyanobacterium (Whatley et al., 1979), and may contain a light-harvesting chlorophyll *a/b* protein comparable to that in green algal and higher-plant chloroplasts.

Roger Stanier (1977) has characterized the prokaryotic cyanobacteria "as microorganisms that harbor, within a typically prokaryotic cell, a photosynthetic apparatus similar in structure and function to that located in the chloroplast of phototrophic eukaryotes." He has evaluated for us how this group of organisms harbors alternative capacities for oxygenic and anoxygenic photosynthesis; how they can combine fixation of atmospheric nitrogen with oxygenic photosynthesis; and, together with Germaine Cohen-Bazire, he has presented the structural and functional stages that lead from the photosynthetic cell membrane in the purple bacterium via the chlorobium vesicle to the thylakoids of the red and green algae. Apparently these evolutionary stages are preserved in various members of the cyanobacteria.

For students who would like to learn more about stages in endosymbiosis, a prokaryote with 80S, that is, cytoplasmic ribosomes, would be a welcome discovery, as would be a cell with only nucleosomic chromosomes and prokaryotic 70S ribosomes. The cluster-clone route could perhaps be explored by knocking out a gene in chloroplast DNA and inserting a corresponding wild-type gene in the nucleus, or, vice versa, by eliminating a gene specifying a chloroplast protein in the nucleus and inserting it into chloroplast DNA.

References

Apel, K., and Kloppstech, K. 1978. The plastid membranes of barley (*Hordeum vulgare*). Light-induced appearance of m-RNA coding for the apoprotein of the light-harvesting chlorophyll a/b protein. Eur. J. Biochem. 85:581–588.

Armstrong, J. J., Surzycki, S. J., Moll, B., and Levine, R. P. 1971. Genetic transcription and translation specifying chloroplast components in *Chlamydomonas reinhardtii*. Biochemistry 10:692–701.

Bartlett, S. G., Grossman, A. R., and Chua, N.-H. 1979. *In vitro* synthesis of ferredoxin and its transport into intact spinach chloroplasts. J. Cell Biol. 83:369a.

Beggs, J. D. 1978. Transformation of yeast by a replicating hybrid plasmid. Nature 275:104–109.

Bogorad, L. 1975. Evolution of organelles and eukaryotic genomes. Science 188:891.

Chen, K., Johal, S., and Wildman, S. G. 1976. Role of chloroplast and nuclear DNA genes during evolution of fraction I protein. *In* Genetics and Biogenesis of Chloroplasts and Mitochondria. Th. Bücher W. Neupert, W. Sebald and S. Werner (eds). Amsterdam: Elsevier/North-Holland Biomedical Press, pp. 3–12.

Chua, N.-H. 1976. A uniparental mutant of *Chlamydomonas reinhardtii* with a variant thylakoid membrane polypeptide. *In* Genetics and Biogenesis of Chloroplasts and Mitochondria. Th. Bücher et al. (eds.). Amsterdam: Elsevier/North-Holland Biomedical Press, pp. 323–330.

Chua, N.-H., and N. W. Gillham, 1977. The sites of synthesis of the principal thylakoid membrane polypeptides in *Chlamydomonas reinhardtii*. J. Cell Biol. 74:441–452.

Chua, N.-H., and Schmidt, G. W. 1978. Post-translational transport into intact chloroplasts of a precursor to the small subunit of ribulose-1,5-bisphosphate carboxylase. Proc. Nat. Acad. Sci. USA 75:6110–6114.

Chua, N.-H., and F. Blomberg. 1979. Immunochemical studies of thylakoid membrane polypeptides from spinach and Chlamydomonas reinhardtii. J. Biol. Chem. 254:215–223.

Cleland, R. E. 1972. Oenothera Cytogenetics and Evolution. London: Academic Press.

Dobberstein, B., Blobel, G., and Chua, N.-H. 1977. In vitro synthesis and processing of a putative precursor for the small subunit of ribulose-1,5-bisphosphate carboxylase of Chlamydomonas reinhardtii. Proc. Nat. Acad. Sci. USA 74:1082–1085.

Doherty, A., and Gray, J. C. 1979a. Synthesis of cytochrome f by isolated pea chloroplasts. Eur. J. Biochem. 98:87–92.

Doherty, A., and Gray, J. C. 1979b. Synthesis of a dicyclohexylcarbodiimide-binding proteolipid by isolated pea chloroplasts. Biochem. Soc. Trans. 7:1114–1115.

Gillham, N. W. 1978. Organelle Heredity. New York: Raven Press.

Glazer, A. N., Apell, G. S., Hixson, C. S., Bryant, D. A., Rimon, S., and Brown, D. M. 1976. Biliproteins of cyanobacteria and Rhodophyta: Homologous family of photosynthetic accessory pigments. Proc. Nat. Acad. Sci. USA 73:428–431.

Gordon, K. H. J., Hildebrandt, J. W., Bohnert, H. J., Herrmann, R. G. and Schmitt, J. S. 1980. Analysis of plastid DNA in an Oenothera plastome mutant deficient in ribulose bisphosphate carboxylase. Theor. Appl. Genet. 57:203–207.

Hall, D. O., and Rao, K. K. 1977. Ferredoxin. In Encyclopedia of Plant Physiology (n.s.). Vol. V, Photosynthesis I, Photosynthetic Electron Transport and Photophosphorylation. A. Trebst and M. Avron (eds.). Berlin: Springer Verlag, pp. 206–216.

Haslett, B. G., and Cammack, R. 1974. The development of plastocyanin in greening bean leaves. Biochem. J. 144:567–572.

Highfield, P. E., and Ellis, R. J. 1978. Synthesis and transport of the small subunit of chloroplast ribulose bisphosphate carboxylase. Nature 271:420–424.

Hinnen, A., Hicks, J. B., and Fink, G. R. 1978. Transformation of yeast. Proc. Nat. Acad. Sci. USA 75:1929–1933.

Holder, A. A. 1978. Peptide mapping of the ribulose bisphosphate carboxylase large subunit from the genus Oenothera. Carlsberg Res. Commun. 43:391–399.

Høyer-Hansen, G. and Simpson, D. J. 1977. Changes in the polypeptide composition of internal membranes of barley plastids during greening. Carlsberg Res. Commun. 42:379–389.

Huisman, J. G., Moorman, A. F. H., and Verkley, F. N. 1978. In vitro synthesis of chloroplast ferredoxin as a high molecular weight precursor in a cell-free protein synthesizing system from wheat germs. Biochem. Biophys. Res. Commun. 82:1121–1131.

Kielland-Brandt, M. C., Nilsson-Tillgren, T., Holmberg, S., Petersen, J. G. L., and Svenningsen, B. A. 1979. Transformation of yeast without the use of foreign DNA. Carlsberg Res. Commun. 44:77–87.

Kirk, J. T. and Tilney-Bassett, R. A. E. 1978. The Plastids. Amsterdam: Elsevier-North-Holland.

Kutzelnigg, H., and Stubbe, W. 1974. Investigations on plastome mutants in Oenothera. 1. General considerations. Subcell Biochem. 3:73–89.

Malnoë, P., Rochais, J.-D., Chua, N.-H., and Spahr, P.-F. 1979. Characterization of the gene and messenger RNA of the large subunit of ribulose-1,5-diphosphate carboxylase in Chlamydomonas reinhardtii. J. Mol. Biol. 133:417–434.

Martin, P. G. 1979. Amino acid sequence of the small subunit of ribulose-1,5-bisphosphate carboxylase from spinach. Aust. J. Plant Physiol. 6:401–408.

Mendiola-Morgenthaler, L. R., Morgenthaler, T. T., and Price, C. A. 1976. Synthesis of coupling factor CF₁ protein by isolated spinach chloroplasts. FEBS Lett. 62:96–100.

Poulsen, C. 1979. The cyanogen bromide fragments of the large subunit of ribulose bisphosphate carboxylase from barley. Carlsberg Res. Commun. 44:163–189.

Poulsen, C., Martin, B., and Svendsen, I. 1979. Partial amino acid sequence of the large subunit of ribulose bisphosphate carboxylase from barley. Carlsberg Res. Commun. 44:191–199.

Ramshaw, J. A. M., Scawen, M. D., Jones, E. A., Brown, R. H., and Boulter, D. 1976. The amino acid sequence of plastocyanin from *Lactuca sativa* (lettuce). Phytochemistry 15:1199–1202.

Renner, O. 1937. Zur Kenntnis der Plastiden- und Plasmavererbung. Cytologia (Tokyo) Fujii Jubilee Vol.: 644–653.

Schmidt, G. W., Bartlett, S., Grossman, A. R., Cashmore, A. R., and Chua, N.-H. 1979. *In vitro* synthesis, transport, and assembly of the constituent polypeptides of the light-harvesting chlorophyll a/b protein complex. *In* Genome Organization and Expression in Plants. C. J. Leaver (ed.) Plenum Press, New York.

Schmidt, G. W., Devillers-Thiery, A., Desruisseaux, H., Blobel, G., and Chua, N.-H. 1979. NH₂-terminal amino acid sequences of precursor and mature forms of the ribulose-1,5-bisphosphate carboxylase small subunit from *Chlamydomonas reinhardtii*. J. Cell Biol. 83:615–622.

Schötz, F. 1969. Effects of the disharmony between genome and plastome on the differentiation of the thylakoid system in *Oenothera*. Symp. Soc. Exp. Biol. 24:39–54.

Stanier, R. Y. 1977. The position of cyanobacteria in the world of prototrophs. Carlsberg Res. Commun. 42:77–98.

Strøbæk, S., Gibbons, G. C., Haslett, B., Boulter, D., and Wildman, S. G. 1976. On the nature of the polymorphism of the small subunit of ribulose-1,5-diphosphate carboxylase in the amphidiploid *Nicotiana tabacum*. Carlsberg Res. Commun. 41:335–343.

Stubbe, W. 1959. Genetische Analyse des Zusammenwirkens von Genom und Plastom bei *Oenothera*. Z. Vererb. 90:288–298.

Stubbe, W. 1964. The role of the plastome in evolution of the genus *Oenothera*. Genetica 35:28–33.

Whatley, J. M., John, P., and Whatley, F. R. 1979. From extracellular to intracellular: The establishment of mitochondria and chloroplasts. Proc. R. Soc. B 204:165–187.

von Wettstein, D., Poulsen, C., and Holder, A. A. 1978. Ribulose-1,5-bisphosphate carboxylase as a nuclear and chloroplast marker. Theor. Appl. Genet. 53:193–197.

Zieliński, R. E., and Price, C. A. 1977. Synthesis of cytochrome b₅₅₉ by isolated spinach chloroplasts. Plant Physiol. 59:8 (suppl).

PART IV
MOLECULAR BIOLOGY
AND CONTROL OF
PLASTID DEVELOPMENT

Science . . . warns me to be careful how I adopt a view
which jumps with my preconceptions, and to require stronger
evidence for such belief than for one to which I was
previously hostile. My business is to teach my aspirations to
conform themselves to fact, not to try and make facts
harmonize with my aspirations.

—Thomas Henry Huxley

Regulation of Intracellular Gene Flow in the Evolution of Eukaryotic Genomes

Lawrence Bogorad

In the early and mid-1960s unequivocal evidence for the presence of DNA and unique information-processing systems in chloroplasts revived interest in the long-standing proposal that these organelles are derived from endosymbionts and have retained their autonomy. In contrast to this view that chloroplasts and mitochondria are derived from free-living organisms that brought new complete genomes into a nucleated cell, the cluster-clone hypothesis proposes that eukaryotic cells arose by the subdivision and compartmentalization of a single genome (Bogorad, 1975). The objectives of this paper are, first, to review current knowledge regarding genes for chloroplast components and the organization of chloroplast genomes and, second, to sift through this information for clues to the origin and evolution of eukaryotic genomes and chloroplasts.

Sizes of Chloroplast Chromosomes

Closed circular cpDNA (chloroplast DNA) molecules have been found in plastids of almost every species examined. Supercoiled molecules have been observed among the circles (Kolodner and Tewari, 1972, Manning and Richards, 1972). Table 1, adapted from Bedbrook and Kolodner (1979), shows that the molecular

The research from my laboratory has been supported by grants from the National Institute of General Medical Sciences, The Competitive Research Grants Office of the U.S. Department of Agriculture, and the National Science Foundation. It has also been supported in part by the Maria Moors Cabot Foundation of Harvard University.

Address reprint requests to Dr. Lawrence Bogorad, Professor Biology, Harvard University, Biological Laboratories, 16 Divinity Avenue, Cambridge, Massachusetts 02138.

L. Bogorad

Table 1. Sizes of Chloroplast DNAs and of Large Repeated Sequences in
Some Species[a,b]

	Mol wt × 10⁶	Kilobase pairs	Repeats (KBP)
Sphaerocarpos castellani	85.2	-	-
Zea mays	91.1	138	22.5 × 2 (I)
Avena sativa	92.2	-	-
Pisum sativa	95.0	-	-
Euglena gracilis	97.1	140	5.6 × 3 (T)
Beta vulgaris	99.6	-	-
Spinacia oleracea	101.4	153	24.5 × 2 (I)
Anthirrhinum majus	101.4	-	-
Lactuca sativa	103.2	156	24.5 × 2 (I)
Chlamydomonas reinhardii	143.0	200	19.0 × 2 (I)
Cyanobacterial genomes:	1600–7600		

[a]Repeats: I = inverted; T = tandem.
Source: Adapted from Bedbrook and Kolodner (1979).

weights of cpDNAs range from 85.2 to 143 megadaltons (mD). (These values
are different from earlier published ones because of more accurate information
now available on the size of ϕX 174 RF DNA molecules which were used as
size standards.) The amount of DNA per chloroplast ranges from about 10 to
60 times the size of a single circle, yet the kinetic complexities which have been
measured are in the ranges of 100 mD, indicating that all the cpDNA circles in
a plastid are entirely or very largely of one type. This conclusion is supported
by the observation in maize, for example, that the sum of fragments generated
by digestion of cpDNA with the restriction endonuclease Sal I (Bedbrook and
Bogorad, 1976) is approximately equal to the size estimated by electron microscopy of cpDNA molecules (Kolodner and Tewari, 1975).

Repeated sequences have been observed in cpDNA molecules by both electron
microscopy (Kolodner and Tewari, 1979) and by restriction endonuclease mapping (Bedbrook and Bogorad, 1976; Hobom et al., 1977; Rochaix, 1978; Whitfeld et al., 1978). In maize and spinach the repeated sequences are 22.5–24.5
kilobase pairs (KBP) in length, and are in inverted orientation to one another.
The shortest distances between the ends of the two inverted repeats range from
12.6 to 19.5 KBP. On the other hand, pea cpDNA contains a pair of repeated
sequences in the 20 KBP size range, but they are not inverted with respect to
one another (Kolodner and Tewari, 1979). *Euglena gracilis* cpDNA has three
much smaller tandem repeats close to one another (Gray and Hallick, 1978).

In summary, cpDNA almost always, perhaps universally, occurs as supercoiled
circles. Almost all other features are variable—the sizes of the circles so far
measured ranges from about 85 to 143 md; the number of circles per chloroplast
varies from about 10 to 60; some cpDNAs have inverted repeats although their
distances from one another and their sizes are not constant, but other cpDNAs

contain tandem repeats of either large or small size; and the space between inverted repeats varies from one species to another.

Genes on Chloroplast Chromosomes

Chloroplast Ribosomal Proteins, rRNAs, and tRNAs

One structural gene for a chloroplast component has been mapped to the chloroplast chromosome by transmission genetics. This is the structural gene for the *Chlamydomonas* chloroplast ribosomal protein LC 4 (Mets and Bogorad, 1972; Hanson et al., 1974; Boynton et al., 1976).

Genes for rRNAs, some tRNAs, and a few proteins have been mapped physically with precision on chloroplast chromosomes. Genes for 23, 16, and 5S rRNAs were mapped on maize cpDNA (Bedbrook, et al., 1977) and subsequently on other cpDNAS (see review by Bedbrook and Kolodner, 1979). In these experiments chloroplast rRNAs were labeled with ^{32}P in vitro and hybridized to fragments of cpDNA. The fragments produced by digestion of cpDNA with a restriction endonuclease were separated according to size by agarose gel electrophoresis and transferred to nitrocellulose filters by the method of Southern (1975). Maize 16, 23, and 5S rRNAs were all found to hybridize to fragment *a* produced by digesting maize cpDNA with the enzyme Eco RI. Fragment *a*, the largest segment produced by digestion of maize cpDNA with Eco RI, is present twice on the chromosome, once in each of the two inverted repeats.

The location of the 16, 23, and 5S rDNAs within fragment *a* were established by two methods. The first method involved determining to which subfragments of *a* each of the rRNAs hybridized. Fragment *a* was isolated from a chimeric plasmid (into which it had been incorporated) that was cloned in *Escherichia coli*. It was digested with some restriction enzymes other than Eco RI, and the orientations of these subfragments with respect to one another were determined, that is, restriction endonuclease recognition sites were mapped on *a*. Fragments of *a* were separated electrophoretically and transferred to nitrocellulose sheets (as described above) for hybridization of cpDNA fragments with ^{32}P-labeled 5S, 16S, or 23S rRNA. In this way each of the rRNAs could be assigned a location on *a*, but the precision of the location by this method depends on the sizes of the subfragments of *a* used for hybridization. The second method involved heteroduplex mapping. Fragment *a* was isolated from the chimeric plasmid, denatured, and hybridized to a mixture of 16 and 23S chloroplast rRNAs. The hybrids were examined by electron microscopy. By measuring the lengths of hybridized portions of fragment *a*, the locations of the genes for these rRNAs could be established with considerable precision (Bedbrook et al., 1977).

In every case studied to date the rRNAs occur in repeated cpDNA sequences. *Euglena gracilis* cpDNA contains three tandemly repeated sequences, each with a set of genes for 16S and 23S rRNAs (Gray and Hallick, 1978; Jenni and Stutz, 1978 Rawson et al., 1978). A fourth partially repeated sequence has a gene for 16S RNA without that for 23S rRNA (Jenni and Stutz, 1979). In all other cases there is one complete set of genes per repeat in each of the two repeats per chromosome. As indicated above, the repeats are either in inverted (maize, spinach, *Chlamydomonas*) or tandem orientation to one another. Within each cistron, the order appears to be: 16S rDNA; spacer, which is 1.68 KBP in *Chlamydomonas reinhardii* (Rochaix and Malnoe, 1978), 2.1 KBP in *Zea mays* (Bedbrook et al., 1977), and 2.2 KBP in *Spinacea oleracea* (Whitfeld et al., 1978); 23S rDNA; and 5S rRNA. There is evidence from *S. oleracea* (Hartley and Ellis, 1973) that the 23S and 16S rRNAs are transcribed together as a single strand which is presumably processed later.

The order of the genes for rRNAs as well as their transcription as a single unit is similar to the situation in *E. coli*. These observations are sure to be pointed to with pleasure by bacteria watchers and endosymbiont lovers, who will also call attention to the homology in the base sequences of 16S rDNAs of maize and *E. coli* (Schwarz and Kossel, 1980). But differences among cpDNAs should be noted as well. First, each 23S rRNA gene in *C. reinhardii* is interrupted by a 0.94-KBP intervening sequence 270 base pairs from the 5' end of the RNA sequence (Rochaix and Malnoe, 1978). Inserts of this sort have been found in *Neurospora crassa* DNA for 24S RNA (Raj Bhandary et al., 1977) and in the comparable gene in some yeast mitochondrial DNAs (Bos et al., 1978), but not in HeLa mtDNA (Attardi et al., 1979), maize or spinach cpDNA, or in *E. coli* rDNAs. A gene for tRNA^{Val} is located about 300 nucleotides upstream from the start of the maize plastid gene for 16S rRNA (Schwartz et al., 1981)—there is nothing similar in *E. coli*. Furthermore, a gene for 4.5S rRNA has been found between the genes for 23 and 5S rRNA in *Z. mays* (Dyer and Bedbrook, 1980). The gene for this small rRNA appears to be absent from cpDNA of *C. reinhardii* (Rochaix and Malnoe, 1978). *E. coli* lack 4.5S rRNA.

Twenty-five tRNAs have been mapped to restriction fragments of spinach cpDNA (Driesel et al., 1979) and a smaller number have been localized on fragments of maize cpDNA (G. Burkhard, L. McIntosh, Z. Schwarz, R. Selden, A. Steinmetz, J. Weil, and L. Bogorad, unpublished). A few regions of these two chromosomes have a higher than average density of tRNA genes, but the locations of these regions with respect to the inverted repeats are not the same.

Ribulose Bisphosphate Carboxylase

The gene for the large subunit of the enzyme ribulose bisphosphate carboxylase (RuBPCase) has been localized to a 1.6-KBP-long uninterrupted stretch of maize cpDNA (Coen et al., 1977; Bedbrook et al., 1979; Bogorad et al., 1979; Link and Bogorad, 1980). In the course of this research the direction of transcription of the gene for the large subunit (LS) of RuBPCase was established. It was also found that another gene is transcribed in the opposite direction (i.e., from the other DNA strand), starting about 0.33 KBP from the 3' end of the RuBPCase gene. The function of the product of the 2.2-KBP gene is not known at this time

(Link and Bogorad, 1980). Structural genes are relatively closely packed in at least this segment of maize cpDNA. In leaves of C-4 plants like maize, chloroplasts of mesophyll cells lack RuBPCase and fix CO_2 into four carbon acids, but plastids of adjacent bundle sheath cells contain RuBPCase. We have used cpDNA sequences from the interior of the 2.2 KBP structural gene and from the interior of the LS RuBPCase as probes to determine whether RNAs complementary to these two genes are present in bundle sheath and mesophyll chloroplasts. RuBPCase LS mRNA was found in extracts of bundle sheath cells, but not of mesophyll cells (Link et al., 1978). RNA complementary to the 2.2-KPB gene was found in extracts of both cell types (Link and Bogorad, unpublished).

Regardless of the origin of cpDNA, that is, whether by subdivision of a genome already present, or through the addition of a genome from outside of the cell, some elements for the control of expression have evolved around the LS RuBPCase gene in C-4 plants.

Photogene 32

Photogene 32 is also associated with control elements which probably evolved long after the origin of eukaryotic cells. The processed product of this photogene is a 32-kd chloroplast membrane polypeptide (Grebanier et al., 1978). Plastids of dark-grown maize seedlings lack mRNA transcribed from photogene 32, but this species of mRNA is abundant during light-induced plastid development (Bedbrook et al., 1978). This mRNA is colinear with a 1.3-KBP DNA sequence which has been cloned in *E. coli* (L. McIntosh, G. Link, and L. Bogorad, unpublished).

Summary

To summarize the properties of chloroplast genomes and genes which have been studied in detail to date:

1. The two identified plastid structural genes studied in detail so far, the gene for LS RuBPCase and photogene 32, are colinear with their messages in maize. Regardless of their origin, elements controlling expression of these genes must have evolved long after the origin of the organelle and very likely in concert with evolution of nuclear elements. For at least part of its length, and perhaps for its entire length, the 2.2-KBP gene is colinear with its message.

2. In the few cases analyzed to date, the equivalent of one complete strand of cpDNA is transcribed. The absence of long untranscribed stretches is supported in a very small way by detailed analysis of the region between the gene for LS RuBPCase and the 2.2 KB gene.

3. Genes for chloroplast rRNAs are arranged more like those in *E. coli* than those in nuclear genomes and, at least in spinach, all the rDNAs are included in a single transcript. The gene sequence is 16S, spacer, 23S, and 5S. Chloroplasts of some species differ from bacteria in the presence of a gene for a 4.5S

rRNA between the genes for 23S and 5S rRNAs in their DNA. *C. reinhardii* cp rDNA also differs from that of *E. coli* in the presence of an intervening sequence within the 23S rRNA gene. Chloroplast genomes differ from one another with respect to whether rDNAs are on inverted or tandemly repeated sequences, as well as whether the repeats are close together (e.g., *E. gracilis*) or far from one another.

Plastid chromosomes most resemble bacterial ones in the way rDNAs are oriented with respect to one another (with the several exceptions noted) and the similarities in 16S rDNA sequences.

Before adherents of the endosymbiont hypothesis for the origin of all organelles take great comfort from the similarity of the arrangement of rDNAs in *E. coli* and some plant chloroplasts, the arrangement of rRNA genes on mitochondrial chromosomes should be recalled. Genes for the large and small mitochondrial rRNAs are widely separated from one another and are present in one copy per chromosome in mitochondrial genomes examined so far. Mitochondria in some yeast contain an approximately 1-KPB insert in the gene for the large rRNA. In all, there is little reason to use rDNA arrangements as the sole or major basis for a hypothesis on the origin of organelles in eukaryotes; patterns are very diverse.

Transcriptional Apparatus

A great deal is known about the properties of the DNA-dependent RNA polymerase of *E. coli*. Much less is know about the polymerases of cyanobacteria (blue-green algae) and chloroplasts. Still, some comparative studies have been carried out.

RNA polymerase from *E. coli,* the filamentous blue-green algae *Fremyella diplosiphon,* and the chloroplasts of *Zea mays* are all multimeric enzymes comprised of two large subunits (140–180 kd) plus a number of smaller subunits. The precise sizes of the larger subunits and of the smaller subunits vary among the enzymes from these organisms. However, each of the three types of DNA-dependent RNA polymerases found in nuclei is also comprised of two large and several smaller polypeptide subunits although, again, these polymerases differ among themselves and from the trio already mentioned with regard to the sizes and numbers of components. The striking exceptions to this pattern are RNA polymerases of mitochondria; these enzymes consist of single subunits about 60 kd in size.

Initiation of transcription by blue-green algal and *E. coli* RNA polymerase has been compared using a restriction fragment of the bacteriophage lambda as a template in vitro. Both enzymes start transcribing at the same points but their sensitivity to heparin is different (Miller et al., 1979). So these RNA polymerases from these two organisms are not quite the same.

A 27-kd polypeptide isolated from maize chloroplasts, the *S factor,* has a striking effect on the action of the maize chloroplast polymerase. When cloned supercoiled chimeric plasmid DNA containing cpDNA sequences is used as a template, transcription of the cpDNA sequences is highly favored when the S factor is present. The S factor has no effect on the rate of transcription of circular plasmid DNA by *E. coli* RNA polymerase, nor does *E. coli* RNA polymerase sigma factor affect transcription by the chloroplast enzyme (Jolly and Bogorad, 1980).

Interest in the possible endosymbiotic origin of plastids and mitochondria was boosted by the observation that protein synthesis by these organelles is inhibited by antibiotics that block protein synthesis in bacteria but is insensitive to cycloheximide which inhibits protein production by cytoplasmic ribosomes in eukaryotes. This line of thought has also generated interest in the sensitivity or resistance of RNA polymerases to the antibiotic rifamycin. Both blue-green algal and bacterial RNA polymerases are sensitive to very small amounts of rifamycin SV in vitro. The maize chloroplast enzyme appears to be sensitive in vivo (Bogorad and Woodcock, 1970), but is only partially sensitive to quite high concentrations of the antibiotic under some conditions in vitro (Bogorad et al., 1973). Plastid RNA synthesis in vivo appears to be blocked in *C. reinhardii* (Surzycki, 1969) by a form of rifamycin that inhibits bacterial polymerases. The relative ease of obtaining rifamycin-resistant mutants of microorganisms suggests that sensitivity to this agent may not be a very significant characteristic for establishing the evolutionary lineage of an organelle polymerase. Even if the plastid originated as an endosymbiont, it is easily conceivable that all but one or a few subunits of its RNA polymerase might have been lost only to be replaced by nuclear-coded components.

Multimeric Plastid Components Are the Products of Dispersed Genes

Structural genes for chloroplast ribosomal proteins L4 and L6 have been mapped on the plastid and nuclear genomes of *C. reinhardii*, respectively (Mets and Bogorad, 1972; Hanson et al., 1974, Davidson et al, 1974; Boynton et al., 1976). Like genes for chloroplast rRNAs of other species, these genes are on the chloroplast chromosome in *C. reinhardii* (Rochaix and Malnoe, 1978). Thus, structural genes for components of plastid ribosomes are dispersed among the nuclear and chloroplast genomes.

The structural gene for the LS of RuBPcase has been mapped on maize cpDNA, as already described. The small subunit of RuBPCase has been shown by transmission genetics to be located in the nuclear genome in the case of tobacco (Kawashima and Wildman, 1972). Thus, the genes for the two components of this multimeric enzyme are dispersed.

The product of maize cpDNA photogene 32 is a 32-kd polypeptide of the chloroplast thylakoid membrane (Grebanier et al., 1978), but some other chloroplast membrane proteins, including those of the chlorophyll *a/b* light-harvesting complex, are nuclear gene products (Apel and Koppstech, 1978).

There are several well-documented cases of gene dispersal in yeast. A number of multimeric components of yeast mitochondria have been shown to be made up of products of nuclear and mitochondrial genomes.

Gene dispersal is clearly an important principle—perhaps the only principle to date—of organelle biology. It provides a realistic picture of the limits to the independence of organelles, and raises important questions for students of organelle origin and evolution (Bogorad, 1975; Bogorad et al., 1977). It also forces us to ask: Where did the organelle come from? Where did nuclear genes for organelle components come from? Which way did the genes go? Or, did they move?

The Origin of Chloroplasts and the Evolution of Eukaryotes

Two possibilities for the origin of chloroplasts and eukaryotic cells are described by the endosymbiont and cluster-clone hypotheses.

An implication of the endosymbiont hypothesis is that after anaerobic nonphotosynthetic prokaryotic cells became established, one line diverged to give rise to nonphotosynthetic anaerobic nucleated organisms, while other descendents of the original prokaryote evolved electron transport systems, phosphorylation, photosynthesis, and aerobic respiration. Later, the nucleated cell was invaded by one or two of the morphologically less-compartmentalized descendants of the common ancestor. Complete new genomes would have been added to a genome already present and well established if this was how the organelles originated. By contrast, the cluster-clone hypothesis proposes that the DNA-containing subcellular organelles evolved within a single cell as separate partially isolated compartments.

The cluster-clone hypothesis postulates that clusters of genes in a prokaryotic cell were cut off into separate compartments where they were (and continue to be) cloned and to evolve interdependently—one clone became mitochondria, another plastids, and a third nuclei. Numerous alternative sequences of events, for example, the evolution of nuclei before some genes separated into progenitors of plastids and mitochondria, have been suggested and reviewed (Bogorad, 1975), but such differences are conceptually minor.

Arguments favoring the endosymbiont hypothesis have rested mainly on what have been assumed to be biochemical relics of prokaryotes in eukaryotic organelles. The fact that similarities are indisputable does not settle the origin of the characteristics, nor does it tell us whether a feature arose by recombination of two lines of cells originally evolved from a single type, that is, one line in which relics have persisted and another more differently evolved line, or whether evolution occurred within a single cell following gene partition. One of the central features of eukaryotes and their organelles is dispersal of genes for multimeric components and the present location in the nuclear chromosome of genes for many unique single polypeptide enzymes which are transferred solely into the chloroplast where they function in carbon metabolism, porphyrin biosynthesis, amino acid metabolism, etc. Regardless of one's adherence to the endosymbiont hypothesis or the cluster-clone progression, the genes and their products were originally in the same compartment of whatever the progenitor of the modern eukaryotic cell was. What mechanisms can we imagine for gene dispersal, and would one mechanism be favored over others, depending upon how organelles originated?

The present state of gene dispersal could have been the first step in the evolution of organelles in one form of the cluster-clone progression (Bogorad, 1975). If each gene were present in one copy in the cell, initial compartmentalization according to the cluster-clone hypothesis could have resulted in the pattern of gene dispersal for multimeric organelle components we recognize in eukaryotic cells today. Any other mode of organelle origin I can imagine, including partition from a single genome with any or all genes present in more than one copy, would require some gene redistribution or new gene evolution to account for current gene dispersal patterns.

Two possible mechanisms of gene redistribution can be considered (Bogorad, 1975). One is gene transfer—the movement of a gene from one genome to another as one might visualize could result from the breaking up or excision of a piece of DNA from one chromosome and its integration, either, physically or functionally, into another genome. For example, pieces of DNA from an endosymbiont breaking up in a cell could become incorporated into the nuclear genome. The same gene might persist in the endosymbiont, as well as in the nucleus, until one or the other might be lost without detriment to the entire organism.

Another possibility for gene dispersal is protein and gene substitution. In such a series of events, if a gene in the organelle mutated to uselessness and if another protein in the cell, for example a nuclear gene product, could serve as a substitute, even if it were an inferior one, the function would transfer from the organelle to the nucleus. A nuclear gene and its product would substitute for the gene of the organelle chromosome. If genes for the same polypeptides were present in more than one genome a somewhat comparable scenario could be imagined easily.

Finally, gene dispersal could occur after the origin of organelles by evolution. For example, suppose a primitive monomeric enzyme were to take on another subunit and thus be converted into a multimeric form by evolution in the nucleus of a gene or genes whose products could enter the organelle and interact with the polypeptide coded by the organelle gene.

Gene transfer would be blocked or at least not favored: if common DNA sequences which could facilitate homologous recombination were absent; if the transcription and translation systems in the two compartments were highly incompatible, so that DNA moved from one compartment to another would not be transcribed and "foreign" RNA would not be translated; or if different modification-restriction mechanisms were established with each of the genomes so that "non-self" DNA would be destroyed. These situations would all be more likely in the previously genetically isolated evolving lines of cells visualized to be partners in endosymbiosis than in compartmentalization within a single cell following the cluster-clone path. But this is a discussion of probabilities, and what is improbable is not impossible.

The protein and gene substitution mechanism for gene dispersal would be favored by the presence of common metabolic paths in each of the compartments. These common paths would not need to be populated by identical enzymes— only by enzymes with similar functions. Such a situation would seem especially likely in the endosymbiont proposal. Although each partner might have paths with somewhat differently evolved enzymes, the biochemical systems essential for life would likely be retained by both partners from their common progenitor. From this starting point, various reassortments of functional elements could go on within the proeukaryote with the deletion of unnecessarily duplicated genes and gene products in the cell.

Protein and gene substitution would also be promoted by a membrane system undiscriminating enough to permit relatively free movement of proteins between cell compartments. Such common membranes would be more likely at the earliest stages of the cluster-clone progression.

Thus, if the choice as to the most likely mechanism for the origin of organelles and eukaryotic genomes is to be made between the endosymbiont hypothesis and the cluster-clone progression on the basis of ease of gene dispersal by either of the two kinds of mechanisms mentioned, the cluster-clone progression would seem to be favored overall. First, genes could be partitioned "correctly" at the initial clustering stage (although "correct" initial partitioning is not an essential element of the cluster-clone progression), and gene redistribution could occur more freely after the initial clustering. Second, the presence of common DNA sequences in all genomes and the initial compatibility of systems for DNA replication, transcription, and translation should be favorable for gene dispersal by transfer. Third, if protein and gene substitution mechanisms play a role in gene dispersal, the common membrane type to be expected at the beginning stages of the cluster-clone progression would seem to favor the possibility of proteins produced in one compartment travelling into another. Genome isolation would come later in this process, and specific membrane transport systems could evolve together with barriers to transport across the membrane (discussed in Bogorad, 1975).

Yet, it is difficult, if not impossible, to tell from the properties of today's eukaryotes whether evolution of molecules or mechanisms occurred within a single cell or in separate lines of cells which arose from a common progenitor. Endosymbionts do exist today, although it seems that we cannot say for certain whether the host's proteins go across the membranes into the endosymbiont. And, in the crown gall disease, a segment of the Ti plasmid from *Agrobacterium tumifaciens* is incorporated into a chromosome of the plant cell's nuclear genome (Schell and Van Montagu, 1980).

The End Point Problem

Mechanisms seem to be in place today to prevent jumbled genomes—to prevent unregulated and random shifting of genes among the several genomes of a eukaryotic cell. No case yet has been seen in which a gene is shifted from the nuclear to the mitochondrial to the chloroplast genome in various cells or from one generation to the next in the same organism. Modern creatures have barriers against gene shifting: membranes which limit the transfer of DNA or RNA; sets of incompatible machinery for replication, transcription, and translation in nuclear, mitochondrial, and chloroplast compartments; and perhaps modification-restriction systems similar to those in bacteria which could mop up loose pieces of DNA from one genome compartment which float into another (marginal evidence for the existence of such a system is the presence of 5-methyl-cytosine in plant nuclear DNA, but ordinarily not in chloroplast DNA). Perturbing these barriers and looking for subsequent gene transfer, or seeking mutants with impaired barriers, are enticing experimental goals in the study of the evolution of such barriers.

Another end point question is: Why are any genes left in the organelle at all? There are DNA-free organelles, such as the Golgi apparatus and lysosomes, which are probably specified by self-assembling selective membrane transport systems. A single gene inside of a selective membrane system is another relatively simple possibility for specifying an organelle. Yet, we see many genes retained within mitochondria and chloroplasts. Why? How?

We can consider first that some classes of gene products may be untransportable. The kinds of outer membranes needed for chloroplast and mitochondrial function may be incompatible with the inclusion of nucleic acid transport systems. Proteins for which transport devices evolve too slowly, even if the gene is transferred, provide another example of operationally nontransportable elements of multimeric components.

Or, it may be necessary to retain the gene for at least one element of each multimeric component to lock components of cytoplasmic origin into the organelle (Bogorad, 1975). Such lock-in devices would have been necessary if there was a time in the establishment of eukaryotic cells when the internal membranes, that is, the organelle limiting membranes, were relatively unselective. As selectivity increased through the elaboration of more highly differentiated membranes, the probability of the evolution of additional protein transport systems for those components still coded within the organelle might have been reduced drastically.

Gene dispersal may have an extraordinarily simple basis. Once all of the components of a self-assembling element such as a ribosome are together in the appropriate environment they self-assemble! So, keeping gene products apart until all are in the compartment where they need to assemble for proper function could account for continued separation of genes for these components in different compartments, that is, either all of the genes for all of the components would be retained within the organelle or some elements critical for the aggregation of the others would be retained within the organelle. As in much of the rest of this speculation, there is an alternative. Maintenance of different ionic or other environmental conditions in the two compartments could impede self-assembly in one and promote it in the other. It is amusing to consider that the simple necessity of keeping self-assembling components apart, as discussed here, could account for the existence of all organelles in eukaryotic cells!

Each of the barriers to gene transfer, such as selective membranes, incompatible information transducing systems, and distinctive defenses against foreign DNA in each compartment (such as sequence diversity and modification-restriction systems), could have come into place at different times in the evolution of each line of proeukaryotes. The distribution of genes among organelle and nuclear genomes could be affected by the time at which barriers were set into place. Thus far the distribution of genes for multimeric organelle components has been studied in only a few organisms of a smaller number of major phylogenetic lines. As the number of investigations increases, it should not be surprising to find different gene distribution patterns among divergent phylogenetic lines. This diversity, as well as the experimental possibility of raising the barriers, would provide experimental material for studying the evolutionary questions discussed here.

Summary

1. Chloroplast genomes which have been studied physically to date represent a modest diversity of sizes, but all include some repeated sequences carrying genes for rRNAs. These repeats are of four types: (a) the maize type with two large inverted repeats separated by a unique region (within this class there are differences in the sizes of the inverted repeats and the unique sequences between them); (b) the pea type with a pair of tandem, rather than inverted, repeats of about the same sizes and distance apart as those in the maize type; (c) the *Chlamydomonas* type, which is in general like the maize type but includes an intron in the gene for the large rRNA gene; and, finally, (d) the *Euglena* type, which contains three (and one-half) tandem repeats quite close to one another.

The examples available to date seem too small in number to expose the full range of variations in the physical organization of chloroplast genomes. The systematic study of this problem will probably expand rapidly in the next few years.

2. Control elements in the chloroplast DNA of maize, such as those for the expression of ribulose-bisphosphate carboxylase in mesophyll cells of maize and the photocontrol of genes, as in the case of maize photogene 32, are unlikely to have been present in the organelle progenitor. Here and elsewhere we see the

evolution of features unique to chloroplasts. Organelles, prokaryotes, and nuclear genomes each have their distinctive features, and no two are equivalent.

3. The endosymbiont hypothesis and the cluster-clone progression represent two possible patterns for the origin of organelles and eukaryotic genomes. Two lines of organisms are presumed to have evolved independently from a common progenitor and to have merged later according to the endosymbiont hypothesis. According to the alternative cluster-clone progression, the modern eukaryotic cell arose by the partitioning of a single genome. In all proposals each gene and its product would have been in a single compartment originally, but now genes for some organelle components are found in the nuclear genome. Genes for multimeric organelle components are dispersed in the nuclear and organelle genomes. This appears to be a principle of organelle biology. Nuclear genes code for numerous enzymes which function exclusively in plastids.

4. Some possible mechanisms of gene dispersal have been considered here: (a) initial separation of genes in the cluster-clone progression; (b) gene transfer; (c) protein and gene substitution; and (d) evolution in the nuclear-cytoplasmic system of components which become associated with organelle elements. The relative probabilities of gene dispersal by each of these mechanisms have been considered in relation to the two classes of possibilities for the origins of organelles and eukaryotic genomes.

5. Finally, we have considered end point problems. What are the barriers that prevent jumbled genomes in modern organisms? Why are any genes retained within an organelle? Linked to these two questions but not explored in this paper, although touched on elsewhere (Bogorad, 1975; Bogorad et al., 1977) is the possibility that these barriers could have come into place at different times in different phylogenetic lines in progression toward modern eukaryotic cells. As more different organelle genomes are studied we may find different gene dispersal patterns in different groups of organisms.

Added in proof: It now appears [Koller, B. and Delius, H. (1980) Molec. Gen. Genet. 178:261–269] that the *Vicia faba* chloroplast DNA contains a single set of rDNA genes and no large repeated segments. There is also evidence that *Pisum sativa* chloroplast chromosome, like that of *Vicia faba* has a single set of genes for rDNAs [Palmer, J. D. and Thompon, W. F. (1981) Proc. Nat. Acad. Sci. U.S.A., in press]. The "pea type" of chloroplast chromosome is thus one without any large repeated sequences.

This chapter, like other chapters in this volume, was presented at a symposium dedicated to Roger Stanier to recognize and honor his contributions, all of which are important, to an area of science in which our interests overlap. I thank the organizers of this symposium for permitting me to join in this celebration of Roger Stanier's extraordinary effect on our field. For me this seems to be an especially appropriate time to acknowledge my particular personal debt to C. B. van Niel for his enrichment of my intellectual life—I had the remarkable good fortune to be exposed and thus, of course, to be influenced by his personality and ideas early in my scientific career. That I have not yet recovered from the exposure is shown by both the intensity of the influence and my pleasure in it.

I am also indebted to the graduate students, postdoctoral fellows, and other colleagues with whom I have been associated in pursuit of the truth about chloroplasts (and sundry other matters) over the past years for the education I have had from them and for their usual tolerance of my ideas.

290   L. Bogorad

References

Apel, K., and Kloppstech, K. 1978. The plastid membranes of barley (Hordeum vulgare). Light-induced appearance of mRNA coding for the apoprotein of the light-harvesting chlorophyll a/b protein. Eur. J. Biochem. 85:581–588.

Attardi, G., Cantatore, P., Ching, E., Crews, S., Gelfand, R., Merkel, C. and Ojala, D. 1979. The organization of the genes in the human mitochondrial genome and their mode of transcription. 1979 In ICN-UCLA Symposia on Molecular and Cellular Biology, Vol. 15, *Extrachromosomal DNA*. D. J. Cummings, P. Borst, I. B. Dawid, S. M. Weissman, and C. F. Fox (eds.). New York: Academic Press, pp. 443–470.

Bedbrook, J. R., and Bogorad, L. 1976. Endonuclease recognition sites mapped on *Zea mays* chloroplast DNA. Proc. Nat. Acad. Sci. USA 73:4309–4313.

Bedbrook, J. R., Kolodner, R., and Bogorad, L. 1977. *Zea mays* chloroplast ribosomal RNA genes are part of a 22,000 base pair inverted repeat. Cell 11:739–749.

Bedbrook, J. R., Link, G., Coen, D. M., Bogorad, L., and Rich, A. 1978. Maize plastid gene expressed during photoregulated development. Proc. Nat. Acad. Sci. USA 75:3060–3064.

Bedbrook, J. R., Coen, D. M., Beaton, A. R., Bogorad, L., and Rich, A. 1979. Location of the single gene for the large subunit of ribulose bisphosphate carboxylase on the maize chloroplast chromosome. J. Biol. Chem. 254:905–910.

Bogorad, L. 1975. Evolution of organelles and eukaryotic genomes. Science 188:891–898.

Bogorad, L., and Woodcock, C. L. F. 1979. Rifamycins: The inhibition of plastid RNA synthesis *in vivo* and variable effects on chlorophyll formation in maize leaves. *In* Autonomy and Biogenesis of Mitochondria and Chloroplasts. N. K. Boardman, A. W. Linnane, and R. M. Smillie (eds.). Amsterdam: North-Holland Publishing Company, pp. 92–97.

Bogorad, L., Mets, L. J., Mullinix, K. P., Smith, H. J., and Strain, G. C. 1973. Possibilities for intracellular integration: The RNA polymerases of chloroplasts and nuclei; genes specifying ribosomal proteins. Biochem. Soc. Symp. 38:17–41.

Bogorad, L., Bedbrook, J. R., Davidson, N. J., Hanson, M., and Kolodner, R. 1977. Genes for plastid ribosomal proteins and RNAs. *In* Genetic Interaction and Gene Transfer. Brookhaven Symposia in Biology 29:1–15.

Bogorad, L., Link, G., McIntosh, L., and Jolly, S. O. 1979. Genes on the maize chloroplast chromosome. *In Extrachromosomal DNA*. D. J. Cummings, P. Borst, I. B. Dawid, S. M. Weissman, and C. F. Fox (eds.). New York: Academic Press, pp. 113–126.

Bohnert, H. J., Driesel, A. J. Crouse, E. J., Gordon, K., Hermann, R. G., Steinmetz, A., Mubumbilia, M., Keller, M., Burkard, G., and Weil, J. H. 1979. Presence of a transfer RNA gene in the spacer sequence between the 16S and 23S rRNA genes of spinach chloroplast DNA. FEBS Lett. 103:52–56.

Bos, J. L., Heyting, C., Borst, P., Arnberg, A. C., and Van Druggen, E. F. J. 1978. An insert in the single gene for the large ribosomal RNA in yeast mitochondrial DNA. Nature 275:336–338.

Boynton, J. E., Gillham, N. W., Harris, E. H., Tingle, C. L., Van Winkle Swift, K., and Adams, G. M. 1976. Transcription segregation and recombination of chloroplast genes in *Chlamydomonas*. *In* Genetics and Biogenesis of Chloroplasts and Mitochondria. T. H. Bücher, W. Neupert, and S. Werner (eds.) Amsterdam: North Holland Publishing Company, pp. 312–322.

Coen, D. M., Bedbrook, J. R., Bogorad, L., and Rich, A. 1977. Maize chloroplast DNA fragment encoding the large subunit of ribulose bisphosphate carboxylase. Proc. Nat. Acad. Sci. USA 74:5487–5491.

Davidson, J. N., Hanson, M. R., and Bogorad, L. 1974. An altered chloroplast ribosomal protein in ery-M1 mutants of *Chlamydomonas reinhardii*. Mol. Gen. Genet. 132:119–129.

Driesel, A. J., Crouse, E. J., Gordon, K., Bohnert, H. J., Herrmann, R. G., Steinmetz, A., Mubumbila, M., Keller, M., Burkard, G., and Weil, J. H. 1979. Fractionation and identification of spinach chloroplast transfer RNAs and mapping of their genes on the restriction map of chloroplast DNA. Gene 6:285–306.

Dyer, T. A., and Bedbrook, J. R. 1980. The organization in higher plants of the genes coding for chloroplast ribosomal RNA. *In Genome Organization and Expression in Plants*. C. J. Leaver (ed.). New York: Plenum Publishing Company, pp. 305–312.

Gray, P. W., and Hallick, R. B. 1978. Physical mapping of the *Euglena gracilis* chloroplast DNA and ribosomal RNA gene region. Biochemistry 17:284–289.

Grebanier, A. E., Coen, D. M., Rich, A. and Bogorad, L. 1978. Membrane proteins synthesized but not processed by isolated maize chloroplasts. J. Cell Biol. 78:734–746.

Hanson, M. R., Davidson, J. N., Mets, L. J., and Bogorad, L. 1974. Characterization of chloroplast and cytoplasmic ribosomal proteins of *Chlamydomonas reinhardii* by 2-dimensional gel electrophoresis. Mol. Gen. Genet. 132:105–118.

Hartley, M. R., and Ellis, R. J. 1973. Ribonucleic acid synthesis in chloroplasts. Biochem. J. 134:249–262.

Hobom, G., Bohnert, H. J., Driesel, A., and Herrmann, R. G. 1977. Restriction fragment map of the circular plastid DNA from *Spinacia oleracea*. *In* Acides Nucléiques et Synthèse des Proteines Chez Les Végétaux. Colloque International du CNRS No. 261. L. Bogorad and J. A. Weil (eds.). Pp. 195–212.

Jenni, B., and Stutz, E. 1978. Physical mapping of the ribosomal DNA region of *Euglena gracilis* chloroplast DNA. Eur. J. Biochem. 88:127–134.

Jenni, B., and Stutz, E. 1979. Analysis of *Euglena gracilis* chloroplast DNA, mapping of a DNA sequence complementary to 16S rRNA outside of the rRNA gene sets. FEBS Lett. 102:95–99.

Jolly, S. O., and Bogorad, L. 1980. Preferential transcription of cloned maize chloroplast DNA sequences by maize chloroplast RNA polymerase. Proc. Nat. Acad. Sci. USA 77:822–826.

Kawashima, N. and Wildman, S. G. 1972. Studies on fraction I protein IV. Mode of inheritance of primary structure in relation to whether chloroplast or nuclear DNA contains the code for a chloroplast protein. Biochim. Biophys. Acta 262:42–49.

Kolodner, R., and Tewari, K. K. 1972. Molecular size and conformation of chloroplast deoxyribonucleic acid from pea leaves. J. Biol. Chem. 247:6355–6364.

Kolodner, R., and Tewari, K. K. 1979. Inverted repeats in the chloroplast DNA from higher plants. Proc. Nat. Acad. Sci. USA 76:41–45.

Link, G., Coen, D. M., and Bogorad, L. 1978. Differential expression of the gene for the large subunit of ribulose bisphosphate carboxylase in maize leaf cell types. Cell 15:725–731.

Link, G., and Bogorad, L. 1980. Sizes, locations, and directions of transcription of two genes on a cloned maize chloroplast DNA sequence. Proc. Nat. Acad. Sci. USA 77:1832–1836.

Manning, J. E., and Richards, O. C. 1972. Isolation and molecular weight of circular chloroplast DNA from *Euglena gracilis*. Biochim. Biophys. Acta 259:285–296.

Mets, L. J., and Bogorad, L. 1972. Altered chloroplast ribosomal proteins associated with erythromycin-resistant mutants in two genetic systems of *Chlamydomonas reinhardii.* Proc. Nat. Acad. Sic. USA 69:3779–3783.

Miller, S. S., Ausubel, F. M. and Bogorad, L. 1979. Cyanobacterial RNA polymerases recognize lambda promoters. J. Bacteriol. 140:246–250.

RajBhandry, U. L. Heckman, J. E., Yin, S., and Alzner-Deweerd, B. 1979. Mitochondrial tRNAs and rRNAs of *Neurospora crassa:* Sequence studies, gene mapping and cloning. *In* ICN-UCLA Symposia on Molecular and Cellular Biology. Vol. 15, Extrachromosomal DNA. D. J. Cummings, P. Borst, I. B. Dawid, S. M. Weissman, and C. F. Fox (eds.) New York: Academic Press, pp. 379–394.

Rawson, J. R. Y., Kushner, S. R., Vapnek, D., Alton, N. K. and Boerma, C. L. 1978. Chloroplast ribosomal RNA genes in *Euglena gracilis* exist as three clustered tandem repeats. Gene 3:191–209.

Rochaix, J. D. 1978. Restriction endonuclease map of the chloroplast DNA of *Chlamydomonas reinhardii.* J. Mol. Biol. 126:597–617.

Rochaix, J. D., and Malnoe, P. 1978. Anatomy of the chloroplast ribosomal DNA of *Chlamydomonas reinhardii.* Cell 15:661–670.

Schell, J., and Van Montagu, M. 1980. The Ti-plasmids of *Agrobacterium tumefaciens* and their role in crown gall formation. *In* Genome Organization and Expression in Plants. C. J. Leaver. (ed.). New York: Plenum Press, pp. 453–470.

Schwarz, Zs. and Kossel, H. 1980. The primary structure of 16S rDNA from *Zea mays* chloroplast is homologous to *E. coli* 16S rRNA. Nature 283:739–742.

Schwarz, Zs., Kössel, H., Schwarz, E. and Bogorad, L. 1981. A gene coding for tRNA[Val] is located near the 5' terminus of the 16S rRNA gene in the *Zea mays* chloroplast genome. Proc. Nat. Acad. Sci. U.S.A. 78:in press.

Southern, E. M. 1975. Detection of specific sequences among DNA fragments separated by gel electrophoresis. J. Mol. Biol. 98:503–517.

Surzycki, S. J. 1969. Genetic functions of the chloroplast of *Chlamydomonas reinhardii:* Effects of rifampin on chloroplast DNA-dependent RNA polymerase. Proc. Nat. Acad. Sci. USA 63:1327–1334.

Whitfeld, P. R., Herrmann, R. G., Bottomley, W. 1978. Mapping of the ribosomal RNA genes on spinach chloroplast DNA. Nucleic Acid Res. 5:1741–1751.

Discussion of Dr. Bogorad's Presentation

VON WETTSTEIN: I think the 2-μm plasmid in yeast provides an interesting example of DNA entering organelles, for example, the nucleus. Transformation with this plasmid (Kielland-Brandt et al., 1979) produces either cells in which the plasmid carries the gene necessary for the cell's survival or in which the gene gets incorporated into the nuclear chromosome. The presence of plasmids in the cell is conserved strictly by selective pressure. A given organelle DNA being a plasmid, may likewise be perpetuated solely because of its selective advantage. Experiments with chloroplast plasmids may help us to get more of an understanding of these phenomena.

BOGORAD: My concurrence with this view is included in a paper I wrote (Bogorad, 1979).

VON WETTSTEIN: For the *Drosophila* X-chromosome the work of Dr. Lucchesi and co-workers (Lucchesi, 1977; Maroni and Lucchesi, 1980) has convincingly demonstrated that dosage compensation for many genes occurs by increased and decreased transcriptional activity. Thus for the same chromosome there is demonstrated amplification of the genetic information by local amplification of some genes (bobbed locus, ribosomal RNA genes), and by increasing the number of transcripts from other genes. By analogy, it should be possible to produce enough of the small subunits of ribulose-bisphosphate carboxylase from a single copy of the gene in the nucleus to match the number of large subunits produced from the many copies of the gene present in the multiple chloroplast DNA molecules.

BOGORAD: I agree that this would be a perfectly reasonable way for things to work. It would be nice to know how that happens. Of course, within the chloroplast not all messages are equally represented. The mRNA for the large subunit of ribulose bisphosphate carboxylase seems to be very abundant in chloroplasts of light-grown seedlings, compared with many other messages for which the same number of genes is probably present per chloroplast.

MARGULIES: Are there any data which show how many of the chloroplast DNA copies in a chloroplast are active in transcription?

BOGORAD: No.

SCHIFF: Since organelles lack the regularity of strict mitosis, it may be advantageous to have a large number of plastid genomes to insure that some complete copies are passed to the progeny despite random losses of pieces of DNA.

MULLINIX: Do you know if the photogenes are clustered in the chloroplast genome?

BOGORAD: We are very interested in this question and are now trying to explore it. We do know that maize chloroplast Eco RI fragment *l*, which is adjacent to the photogene and just inside the inverted repeat, contains a gene for tRNA^{His}, as well as a gene that codes for a 1.6 kilobase RNA (Schwarz, Steinmetz, and Bogorad, unpublished) but we have no clear information whether this is a photogene. We are studying this gene further and are also looking in the other direction from photogene 32 on the chromosome.

MULLINIX: Are the RuBP carboxylase gene and the adjacent gene that is transcribed in the opposite direction similarly influenced by light?

BOGORAD: Light seems to have a very strong influence on the production of ribulose bisphosphate carboxylase in peas, but there isn't much effect in maize. It's an interesting question in the former case, as well as in the case of photogene 32, as to whether the *primary effect* is on the expression of the chloroplast gene (i.e., the large-subunit gene) or on the expression of the other subunit in the nuclear-cytoplasmic compartment.

SCHIFF: You and Woodcock (Woodcock and Bogorad, 1970) showed that many chloroplasts in *Acetabularia* lack plastid DNA. Thus there may be a germ line which throws off functional but nonreplicative chloroplasts. Therefore it may not be appropriate to multiply genomes per chloroplast by numbers of chloroplasts in all cases.

BOGORAD: That may well be correct—it is one of the possibilities in *Acetabularia*—but in looking at cross sections of bean leaves, for example, there appears to be some DNA in each chloroplast. My guess is that the situation in *Acetabularia*—with only about 20–35% of the chloroplasts containing detectable amounts of DNA (Woodcock and Bogorad, 1970)—may be extreme but, as you suggest, the failure of every plastid to contain DNA in a plant may not be exceptional.

SCHIFF: Again, if organelles and cells are to live in harmony, one has to regulate the activities of the other. If the chloroplast is reduced to the point where it has no genome it could have no regulatory influence on the rest of the cell.

BOGORAD: I suppose the contrary argument would be that the influence of an organelle on a cell can be metabolic, as in the case of Golgi apparatus, but I agree that without information of your own you can't have very much of a regulatory influence—you can only be regulated.

BARNETT: Are there examples of regulation of chloroplast gene expression involving effectors other than light?

BOGORAD: Yes. Mesophyll-bundle sheath differentiation in leaves of C-4 plants is another case. The wise scientists of olden times discovered and named: *chromoplasts*—cartenoid-containing plastids such as in ripe tomato skins; *amyloplasts*—plastids differentiated to contain starch grains and no photosynthetic apparatus; *elaioplasts*—oil-containing plastids; and probably some others. The factor influencing each type of differentiation is more obscure than light. I should also add that plastid development in higher plants is sharply influenced by the state of nutrition, as was shown very elegantly a number of years ago by Professor von Wettstein and his colleagues (Eriksson et al., 1961) for the amino acids leucine and aspartate, and is well known in cases of mineral nutrient deficiencies.

WILDMAN: Are the DNAs in mesophyll and bundle sheath chloroplasts identical in sequence?

BOGORAD: We don't know whether the DNAs are identical in sequence. We do know that the gene for the large subunit of ribulose bisphosphate carboxylase is present in mesophyll chloroplasts (Link et al., 1978) but this experiment cannot tell us whether the sequences within the gene are identical. However, it is most likely that the gene for the large subunit which we have cloned comes from mesophyll chloroplast DNA. We grind leaves relatively gently and briefly to increase our yield of intact chloroplasts, but this has the effect of disrupting only a few of the bundle sheath cells. Thus, most or all of the

DNA we use in cloning experiments comes from mesophyll cells. The cloned chloroplast DNA, as I mentioned in my paper, can drive the linked transcription-translation system to produce the large subunit polypeptide in vitro.

WILDMAN: Do you have a notion to explain why the RuBP carboxylase large subunit genes in mesophyll chloroplast DNA are repressed?

BOGORAD: We do not know yet the mechanism of this repression, but we are optimistic about the future.

References to Discussion

Bogorad, L. 1979. The chloroplast, its genome and possibilities for genetically manipulating plants. *In* Genetic Engineering, Vol I. J. K. Setlow and A. Hollaender (eds.). Pp. 181–204.

Eriksson, G., Kahn, A., Walles, B., and von Wettstein, D. 1961. Zur makromolekularen Physiologie der Chloroplasten III. Ber. D. Tsch. Bot. Ges. 74:221–232.

Kielland-Brandt, M. C., Nilsson-Tillgren, T., Holmberg, S., Litske Petersen, J. G., and Svenningsen, B. A. 1979. Transformation of yeast without the use of foreign DNA. Carlsberg Res. Commun. 44:77–88.

Link, G., Coen, D. M., and Bogorad, L. 1978. Differential expression of the gene for the large subunit of ribulose bisphosphate carboxylase in maize leaf cell types. Cell 15:725–731.

Lucchesi, J. C. 1977. Dosage compensation: transcription-level regulation of X-linked genes in *Drosophila*. Amer. Zool. 17:685–693.

Maroni, G., and Lucchesi, J. C. 1980. X-chromosome transcription in *Drosophila*. Chromosoma 77:253–261.

Woodstock, C. L. F., and Bogorad, L. 1970. Evidence for variation in the quantity of DNA among plastids of *Acetabularia*. J. Cell. Biol. 44:361–375.

Organization and Expression
of the Chloroplast Genome of *Euglena gracilis*

Richard B. Hallick, Barry K. Chelm,
Emil M. Orozco, Jr., Keith E. Rushlow,
and Patrick W. Gray

Many clues leading toward an understanding of events in the origins of chloroplasts reside in the structure and properties of the plastid nucleic acids. In studies of chloroplast DNA structure, gene organization, and RNA synthesis, some of these clues may be revealed. We have been studying the organization and expression of genes of *Euglena* chloroplast DNA with the primary goal of understanding the mechanisms involved in selective transcription. Some of this work, which may be of interest and importance in a discussion of the origins of chloroplasts, is described below.

Chloroplast RNA Synthesis In Vivo

To gain an overall perspective on the problem of chloroplast RNA synthesis, we have attempted to determine the total extent of transcription of chloroplast DNA. For *Euglena gracilis,* it has also been possible to determine how the extent of transcription varies during light-induced chloroplast development. The experimental approach involves RNA hybridization in solution to specific restriction nuclease fragments of chloroplast DNA (ctDNA). From the fraction of ctDNA that forms a hybrid at the completion of a reaction, one can determine the extent of transcription of that DNA. From the kinetics of the hybridization reaction, the cellular abundance of chloroplast transcripts can be estimated. For the ex-

This work was supported by NIH Grant GM 21351. Richard B. Hallick is recipient of NIH Research Career Development Award KO4-GM 00372.

Address reprint requests to Dr. Richard B. Hallick, Associate Professor, Department of Chemistry, University of Colorado, Boulder, Colorado 80309.

periments to be described, three different cellular RNAs were used. "0-hr" refers to RNA from exponentially growing cells, dark-adapted in an organotrophic medium. "4-hr" and "72-hr" RNAs are from cells following 4 and 72 hr, respectively, of light-induced development in a phototrophic medium. Details of this experimental approach, and the treatment of the data have previously been reported (Chelm and Hallick, 1976; Chelm et al., 1978, 1979).

A summary of the results of cell RNA hybridization to the five Pst I restriction nuclease fragments of *Euglena* chloroplast DNA is presented in Table 1. Following 72 hr of light-induced development, the total extent of transcription is 67 kilobases (kb), or 67,000 nucleotides, from the 130–140 kilobase pairs (kbp) genome. In dark-adapted cells, only 54 kb of DNA is transcribed, and the 4-hr RNA represents a transcript of 50 kbp. It is evident that several types of regulatory patterns can be described for the expression of the chloroplast genome during chloroplast development. The classes of DNA transcripts for the five Pst I fragments are listed in Table 2. They may be summarized as follows:

1. Transcripts from 45 kb of the genome are present at all stages of development examined. This is the largest class of transcripts. These genes are scattered throughout the chloroplast genome. Major components of this class are the chloroplast rRNAs, as described below.

2. A second class of transcripts, totaling 7 kb, and located within the Pst A and Pst B fragments, are present in dark-adapted cells, disappear with the initial 4 hr of development, and reappear by 72 hr. The disappearance of these transcripts must involve RNA degradation.

3. At the same time that transcripts disappear from the Pst A and Pst B regions, during the initial 4 hr of greening, transcripts from Pst C and Pst E, totaling 3 kb, appear in the cell. These RNAs are absent in dark-adapted cells.

Table 1. Extent of Transcription of Pst I Fragments of *Euglena gracilis* Chloroplast DNA[a]

DNA fragment	Fragment size (kbp)[b]	Size of transcribed region (kb)[c]		
		0-hr	4-hr	72-hr
Pst A	53	26.3	22.3	29.8
Pst B	35	14.7	11.8	17.4
Pst C	25	8.8	11.1	15.0
Pst D	10	2.7	2.2	2.2
Pst E	7.9	1.0	2.2	2.3
Total	131	53.5	49.6	66.7

[a]Total cell RNA from dark-adapted cells, cells exposed to light for 4 hr (early development), or for 72 hr (mature chloroplasts) was hybridized in solution to [³H]Pst I restriction-nuclease fragments. The size of the transcribed region is calculated as the fraction of the [³H]DNA hybridized at the completion of the reaction times the single strand size of the fragment. The protocol for greening cultures has been described by Chelm et al. (1979).

[b]Kilobase pairs.

[c]Kilobases.

Table 2. Transcription Pattern for the Pst I fragments of *Euglena* Chloroplast DNA
During Light-induced Chloroplast Development[a]

Pattern of expression	Size of transcribed region (kb)					
	Pst A	Pst B	Pst C	Pst D	Pst E	Total
Continuously present	21	12	9	2	1	45
Present at 0 and 72 hr, absent at 4 hr	5	2	0	0	0	7
Absent at 0 hr, present at 4 and 72 hr	0	0	2	0	1	3
Present only at 72 hr	4	3	4	0	0	11

Source: Data summary from Table I, and from Chelm et al., 1979.

4. A final class of chloroplast genes gives transcripts which accumulate in the cell between 4 hr and 72 hr of development. The transcripts, totaling 11 kb, are located in the Pst A, Pst B, and Pst C regions.

The transcripts that appear after the onset of chloroplast development, that is, categories 3 and 4 above, represent 14 kb of RNA. These RNAs presumably arise from genes whose gene products are not required for maintenance of the proplastid, but are essential in the mature chloroplast. It would have been possible to determine more precisely the period during development that these RNAs appear if additional RNA samples had been taken. These transcripts are presumably located on multiple transcription units, each of which may be independently regulated. It is interesting that the largest fraction of these inducible RNAs is encoded in the Pst C region.

The location of the Pst fragments on the restriction nuclease map of the chloroplast DNA (Gray and Hallick, 1978), and a schematic representation of the size and location of the four classes of transcription units is shown in Figure 1. The various classes of transcripts are scattered throughout the genome. They are interspersed with regions which are transcriptionally inactive.

Expression of the Chloroplast Ribosomal RNA Genes

Euglena gracilis, strain Z chloroplast DNA contains three sets of 16S and 23S rRNA genes, arranged in a tandem array (Gray and Hallick, 1978; Jenni and Stutz, 1978; Rawson et al., 1978). Also encoded in the rRNA region are 5S rRNA (Gray and Hallick, 1979) and tRNAs (Hallick et al., 1978). An additional 16S rRNA gene sequence, either complete or nearly complete, is located outside of the tandem array (Jenni and Stutz, 1979). The arrangement of all these genes, which are located entirely within the Pst A fragment (Gray and Hallick, 1978) is shown in Figure 2. Assuming sizes for these RNAs of 1.5 kb for 16S rRNA, 2.8 kb for 23S rRNA, 0.12 kb for 5S rRNA, and 0.08 kb for each of two or

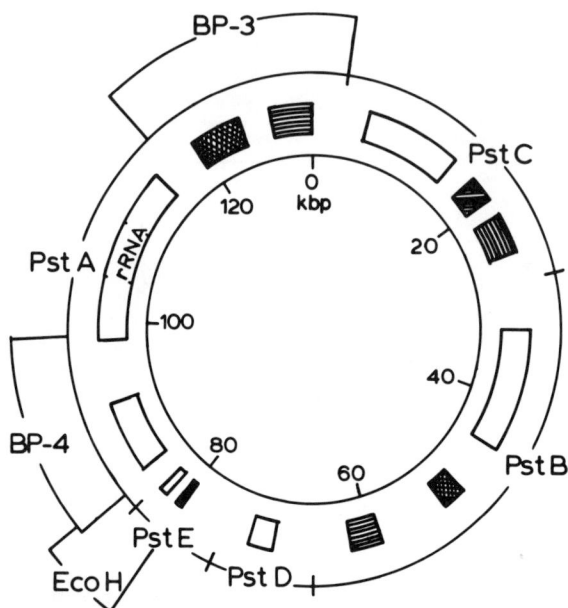

Figure 1. Physical map of *Euglena gracilis* chloroplast DNA, with the positions of the various classes of chloroplast genes illustrated. With the exception of the rRNA genes, the exact locations of genes within the Pst fragments are unknown. *BP-3* and *BP-4* refer to subfragments of Pst A, which were also analyzed (data not shown; see Chelm et al., 1978; 1979). Classes of transcripts, according to the categories described in the text and Table I are (a) [＿＿＿] , continuously present; (b) ▓▓▓▓ , present at 0 and 72 hr, absent at 4 hr; (c) ▨▨▨ , absent at 0 hr, present at 4 and 72 hr; and (d)▤▤▤ , present only at 72 hr.

three tRNAs per repeating unit, these stable RNAs would hybridize with 15 kb of Pst A DNA. These data can be directly related to the hybridization studies described above. The kinetics of the hybridization of 0-hr, 4-hr, and 72-hr RNA to [³H]Pst A DNA are shown in Figure 3. For each RNA sample the hybridization kinetics are consistent with at least two cellular abundance classes of RNA (Chelm et al., 1979). The high abundance components have measured complexities of 14.3, 14.8, and 13.8 kb for the 0-, 4-, and 72-hr RNAs, respectively. This is consistent with the assignment of these high abundance RNAs as rRNAs and tRNAs from the rRNA transcription units. The cellular levels of these transcripts as determined from the hybridization kinetics are in agreement with values previously determined using hybridization reactions specific for chloroplast rRNA (Chelm et al., 1977). In both sets of experiments, chloroplast rRNAs were found to be the major chloroplast DNA transcripts in *Euglena* accounting for from 2.3% of total cell RNA in dark-adapted cells to 26% of total cell RNA following 72 hr of light-induced development.

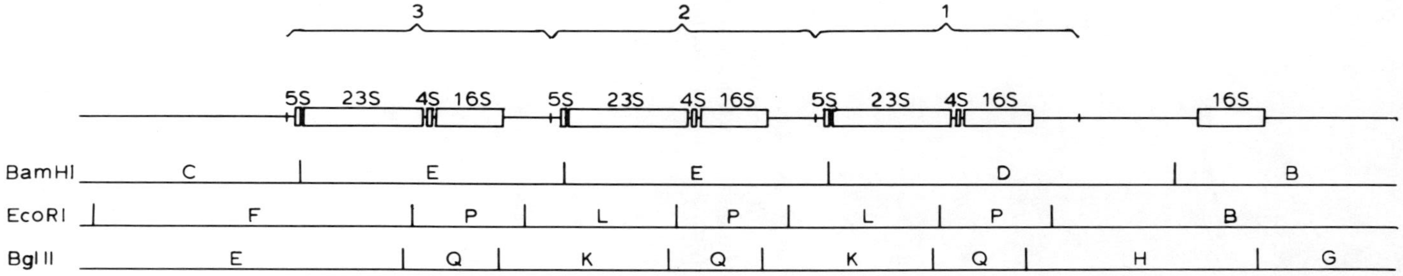

Figure 2. Relative locations of rRNA and tRNA genes within the Pst A fragment of *Euglena gracilis* Klebs., strain Z Pringsheim, chloroplast DNA. The numbers and corresponding brackets at the top of the figure refer to the three identical tandemly repeated rRNA transcription units. The size of each unit is approximately 6 kbp. The restriction nuclease cleavage sites shown are for the enzymes of Bam HI, Eco RI, and Bgl II. Approximately 30 kbp of the 130 kbp genome is shown. *Source:* From Rushlow et al., 1980.

301

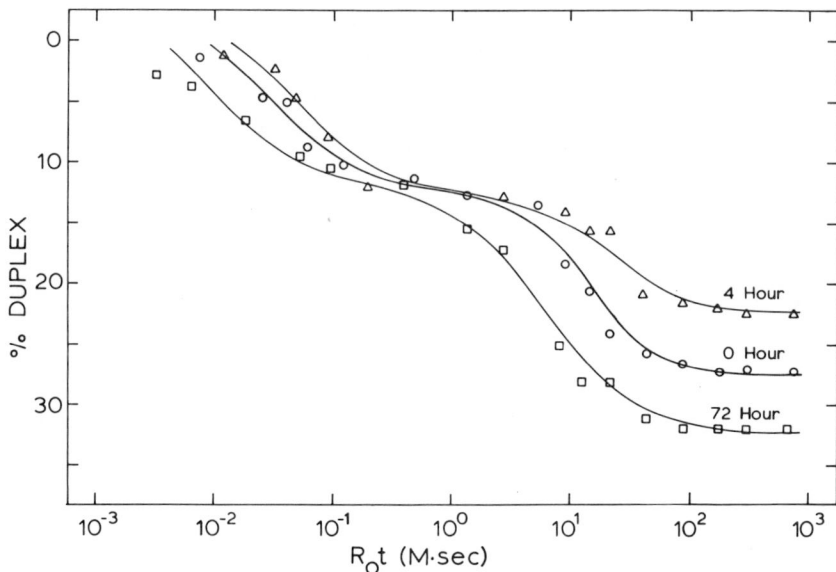

Figure 3. Kinetics of the hybridization of RNA from different stages of chloroplast development to [³H]Pst A DNA. A plot of R_0t, (the product of the initial RNA concentration in moles of nucleotides/liter and the reaction time in sec) against percentage duplex (the percentage of the input [³H]Pst A resistant to digestion from Sl-nuclease for 0-, 4-, and 72-hr RNA samples) is shown, as described in the text.

Coding Capacity of Chloroplast DNA

It is possible to make some estimates about the coding capacity of *Euglena* chloroplast DNA. The total extent of transcription is approximately 66 kb. Of this, approximately 45 kb is continuously present, and the remaining 21 kb may be present or absent in plastids, depending on the stage of development (Table 2). Contained within the continuous class are the rRNAs, which account for nearly 15 kb of transcripts. *Euglena* chloroplast DNA is reported to encode approximately 25 tRNAs (McCrea and Hershberger, 1976; Schwartzbach et al., 1976). These tRNAs would account for another 2 kb of transcript, presumably also part of the continuous class. The remaining RNA, with a complexity of approximately 50 kb, must represent the chloroplast messenger RNA population. This is sufficient to code for a maximum of approximately 50 polypeptides of average mol wt 30,000–35,000.

Chloroplast rRNA Operons

Many of the characteristics of ribosomes and of protein synthesis in chloroplasts are more similar to protein synthetic systems in prokaryotes, for which *E. coli* is a major example, than in eukaryotes. These would include the initiation of protein translation with formyl methionyl tRNA, the sensitivity of plastid ribosomes to streptomycin, erythromycin, and chloramphenicol, but not to cyclohexamide, and the 70S size of the ribosomes, as opposed to 80S ribosomes in the cytoplasm of eukaryotic cells. Nucleic acids involved in protein synthesis are also most similar to their prokaryotic counterparts. The 5S, 16S, and 23S rRNAs of chloroplasts are very nearly the same size as their bacterial counterparts. The ribonuclease T_1 oligonucleotides of *Euglena* chloroplast 16S rRNA have been characterized as most comparable to similar prokaryotic RNAs (Zablen et al., 1975). Plastid tRNAs are also thought to be most closely related to prokaryotic tRNAs, especially when secondary structures are compared (see the chapter by W. E. Barnett in this volume).

Since there are so many similarities between ribosomes of prokaryotes and plastids, it is reasonable to speculate that the chloroplast rRNA genes may be organized into a single transcription unit, analogous to the rRNA operons of prokaryotes. The *E. coli* genome has at least seven rRNA operons (Kenerley et al., 1977; Kiss et al., 1977), which have been designated rrn A, B, C, D, E, F, and X (reviewed by Nomura et al., 1977). Each operon contains genes for 16S, 23S, and 5S rRNA, as well as tRNA genes in the intergenic spacer between the 16S and 23S genes, and (in some operons) distal to the 5S rRNA gene. Each operon is transcribed into a single primary transcript containing sequences for all of the stable RNA products (for recent reviews, see deBoer et al., 1979; Young and Steitz, 1979).

As shown in Figure 2, the three sets of rRNA genes of *Euglena* chloroplast DNA have the same overall organization as the *E. coli* rRNA operons. The arrangement of each unit is 16S-23S-5S, with tRNA gene(s) in the intergenic spacer between the 16S and 23S gene. To gain a more detailed knowledge of the rRNA gene region, restriction nuclease mapping studies of the chloroplast Eco P fragment have been undertaken. As seen in Figure 2, this 2.5 kbp Eco RI fragment contains the entire 16S rRNA gene, the intergenic spacer, and part of the 23S rRNA gene. A detailed restriction endonuclease map of Eco P is shown in Figure 4. (E. M. Orozco, Jr., unpublished observations). To determine the precise location of the various coding regions, [^{32}P]tRNAs, [^{125}I]16S rRNA, and [^{125}I]23S rRNAs were individually hybridized to membrane-filter blots of various restriction-nuclease fragments from Eco P DNA. The locations of the coding regions as determined in these experiments are summarized in Figure 4. It was possible to position the spacer tRNA(s) to a region of approximately 300–500 base pairs (bp). This region contains two Taq I cleavage sites, separated by 75–80 bp, and two Hinf I cleavage sites, the same distance apart. These Taq I and Hinf I sites map either very close to one another, or actually overlap (Fig. 4). Overlapping Taq I and Hinf I cleavage sites within tRNA genes could readily be accounted for with available tRNA sequence data. tRNAs contain several invariant and semiinvariant residues located in the same relative positions in all tRNAs (Rich and RajBhandary, 1976). These include ψ_{55}, C_{56}, purine$_{57}$, A_{58}, pyrimidine$_{60}$, and C_{61}. The resulting DNA sequence coding for positions 55–61 within a tRNA gene would be 5' T-C-G or A-A-N-T or C-C. This sequence can

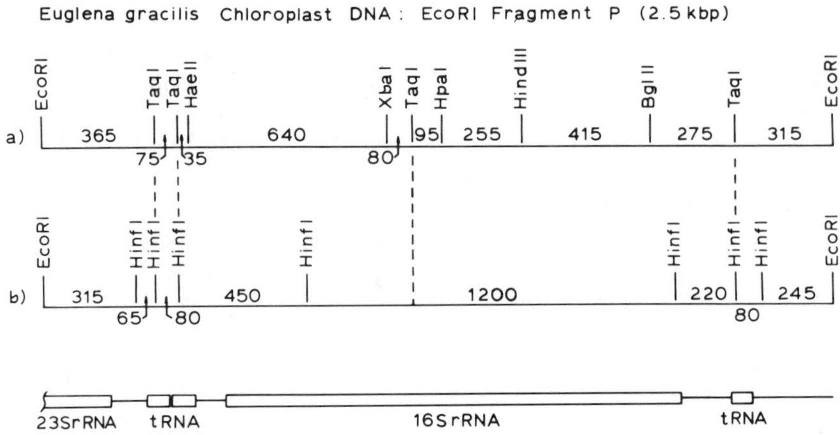

Euglena gracilis   Chloroplast   DNA :   EcoRI   Fragment   P   (2.5 kbp)

Figure 4. Restriction endonuclease cleavage map of chloroplast DNA fragment Eco P. Estimated distances between cleavage sites (in base pairs) and the location of coding regions for rRNAs and tRNAs are shown. The cleavage properties and nomenclature for the enzymes have been reviewed. *Source:* From Roberts, 1976.

contain both the Taq I and Hinf I recognition sequences, which are 5' T-C-G-A and 5' G-A-N-T-C, respectively. We would therefore predict that the Taq I and Hinf I cleavage sites in the spacer region between the 16S and 23S genes are located in the T-ψ-C regions of two adjacent tRNA genes. Note that there is an additional region of tRNA hybridization on Eco P (Fig. 4), which also contains Taq I and Hinf I cleavage sites in very close proximity.

A comparison of the rRNA gene organizations of *E. coli* rrn operons and the *Euglena* chloroplast rRNA coding regions is shown in Figure 5. The gene arrangements, as well as the sizes of the genes and spacer regions, are strikingly similar. A comparison of the tRNAs encoded in the spacer region is also possible. It has recently been reported (Burkard et al., 1979) that the Eco P fragment of *Euglena* chloroplast DNA contains the genes for tRNA[ala] and tRNA[ile]. If these

Figure 5. A comparison of the gene organization of the ribosomal RNA transcription units of *E. coli* DNA and *Euglena* chloroplast DNA.

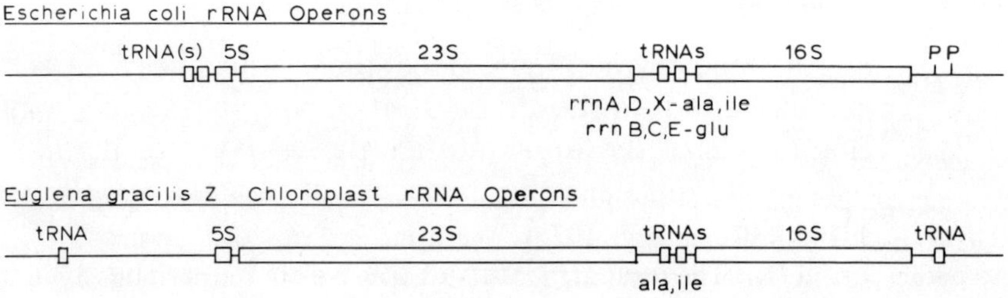

are assumed to be the spacer tRNAs, then the *Euglena* rRNA transcription unit is nearly identical in organization to the rrnA, D, and X operons of *E. coli,* which also have tRNA^{ala} and tRNA^{ile} genes in their spacer regions.

Results of the type described above should provide additional insights into the question of the origins of chloroplasts. It seems reasonable to conclude that the rRNA transcription units of *Euglena* chloroplast DNA are of a prokaryotic type, presumably inherited from a prokaryotic ancestor. This finding is consistent with previous conclusions about the nature of chloroplast ribosomes and ribosomal RNAs. Whether other chloroplast DNA transcription units have prokaryotic homologues remains an open question.

References

Burkard, G., Canaday, J., Crouse, E., Guillemaut, P., Imbault, P., Keith, G., Keller, M., Mubumbila, M., Osorio, L., Sarantoglou, V., Steinmetz, A., and Weil, J. H. 1980. Transfer RNAs and aminoacyl-tRNA synthetases in Plant organelles. *In* Genome Organization and Expression in Plants. C. J. Leaver (ed.). New York: Plenum, pp. 313–320.

Chelm, B. K., and Hallick, R. B. 1976. Changes in the expression of the chloroplast genome of *Euglena gracilis* during chloroplast development. Biochemistry 15:593–599.

Chelm, B. K., Gray, P. W., and Hallick, R. B. 1978. Mapping of transcribed regions of *Euglena gracilis* chloroplast DNA. Biochemistry 17:4239–4244.

Chelm, B. K., Hallick, R. B., and Gray, P. W. 1979. Transcription program of the chloroplast genome of *Euglena gracilis* during chloroplast development. Proc. Nat. Acad. Sci. USA 76:2258–2262.

Chelm, B. K., Hoben, P. J., and Hallick, R. B. 1977. Expression of the chloroplast ribosomal RNA genes of *Euglena gracilis* during chloroplast development. Biochemistry 16:776–781.

deBoer, H. A., Gilbert, S. F., and Nomura, M. 1979. DNA sequences of promoter regions for rRNA operons *rrnE* and *rrnA* in *E. coli*. Cell 17:201–209.

Gray, P. W., and Hallick, R. B. 1978. Physical mapping of the *Euglena gracilis* chloroplast DNA and ribosomal RNA gene region. Biochemistry 17:284–289.

Gray, P. W., and Hallick, R. B. 1979. Isolation of *Euglena gracilis* chloroplast 5S ribosomal RNA and mapping of the 5S rRNA gene on chloroplast DNA. Biochemistry 18:1820–1825.

Hallick, R. B., Gray, P. W., Chelm, B. K., Rushlow, K. E., and Orozco, E. M., Jr. 1978. *Euglena gracilis* chloroplast DNA structure, gene mapping, and RNA transcription. *In* Chloroplast Development. G. Akoyunoglou, and J. H. Argyroudi-Akoyunoglou (eds.). Amsterdam: Elsevier/North Holland; pp. 619–622.

Jenni, B., and Stutz, E. 1978. Physical mapping of the ribosomal DNA region of *Euglena gracilis* chloroplast DNA. Eur. J. Biochem. 88:127–134.

Jenni, B., and Stutz, E. 1979. Analysis of *Euglena gracilis* chloroplast DNA. Mapping of a DNA sequence complementary to 16S rRNA outside of the three rRNA gene sets. FEBS Lett. 102:95–99.

Kenerley, M. E., Morgan, E. A., Post, L., Lindahl, L., and Nomura, M. 1977. Characterization of hybrid plasmids carrying individual ribosomal ribonucleic acid transcription units of *Escherichia coli*. J. Bacteriol. 132:931–949.

Kiss, A., Sain, B., and Venetianer, P. 1977. The number of rRNA genes in *Escherichia coli*. FEBS Lett. 79:77–79.

McCrea, J. M., and Hershberger, C. L. 1976. Chloroplast DNA codes for transfer RNA. Nucleic Acid Res. 3:2005–2018.

Nomura, M., Morgan, E. A. and Jaskunas, S. R. 1977. Genetics of bacterial ribosomes. Annu. Rev. Genet. 11:297–347.

Rawson, J. R. Y., Kushner, S. R., Vapnek, D., Alton, N. K., and Boerma, C. L. 1978. Chloroplast ribosomal RNA genes in *Euglena gracilis* exist as three clustered tandem repeats. Gene 3:191–209.

Rich, A., and RajBhandary, U. L. 1976. Transfer RNA: molecular structure, sequence, and properties. Annu. Rev. Biochem. 45:805–860.

Roberts, R. J. 1976. Restriction Endonucleases. Crit. Rev. Biochem. 4:123–164.

Rushlow, K. E., Orozco, E. M. Jr., Lipper, C., and Hallick, R. B. 1980. J. Biol. Chem. 255:3786–3792.

Schwartzbach, S. D., Hecker, L. I., and Barnett, W. E. 1976. Transcriptional origin of *Euglena* chloroplast tRNAs. Proc. Nat. Acad. Sci. USA 73:1984–1988.

Young, R. A., and Steitz, J. A. 1979. Tandem promoters direct *E. coli* ribosomal RNA synthesis. Cell 17:225–234.

Zablen, L. B., Kissil, M. S., Woese, C. R., and Buetow, D. E. 1975. Phylogenetic origin of the chloroplast and prokaryotic nature of its ribosomal RNA. Proc. Nat. Acad. Sci. USA 72:2418–2422.

Discussion of Presentation by Drs. Hallick, Chelm, Orozco, Rushlow, and Gray

SCHIFF: Is the arrangement of genes you show to be the same for *Euglena* chloroplast DNA and for *E. coli* due to relatedness, or to an optimal functional arrangement to be expected in all prokaryotic situations?

HALLICK: One obvious advantage of having all three rRNAs on the same transcription unit is that they are produced in the correct stoichiometry. There is no obvious explanation at present for the clustering of tRNAs and rRNAs on the same operons, either in bacteria or in chloroplasts.

Note Added in Proof: The presence of tRNA^{Ala} and tRNA^{Ile} genes in the 16S–23S rRNA spacer has been confirmed by DNA sequence analysis (E. M. Orozco Jr, K. E. Rushlow, J. R. Dodd, and R. B. Hallick 1980. J. Biol. Chem. 255:10997–11003.)

The Evolution of Chloroplasts as Determined by Transfer RNA Sequence Analysis

John Delehanty, William G. Farmerie, Simon Chang, and W. Edgar Barnett

Chloroplasts, the semiautonomous organelles of plant cells, may represent the most cleverly disguised and perfectly adapted prokaryotic symbionts in nature. While perhaps taking unjustified freedom in describing these organelles as prokaryotic, our laboratory has proceeded from this hypothesis in providing evidence for the origin of chloroplasts. Clearly, we are not alone in accepting this view. Indeed, the preponderance of recent experimental evidence based on the analysis of ribosomes, membrane components, and mechanisms of electron flow and CO_2 fixation overwhelmingly suggests that chloroplasts and mitochondria are descendants of prokaryotes that established a symbiotic relationship with the host cell (Margulies, 1970). However, this is not to ignore the enormous genetic contribution made by the nuclear genome to organellar function. As has been shown to be characteristic of many eukaryotic genes, very recent evidence (Bos et al., 1978; Rochaix and Malnoe, 1978; Hahn et al., 1979; Heckman and RajBhandary, 1979) has also shown that several organellar genes contain sequence inserts. Clearly, neither the endosymbiotic theory nor the eukaryotic compartmentalization theory is currently sufficient to explain all facets of chloroplast evolution. Perhaps, as suggested by Bogorad (1975), we must examine more closely the rules and effects of gene dispersal in eukaryotes, regardless of whether those genes are eukaryotic or have originated from a prokaryotic invader.

Molecular biology has made a rather late entry into the study of evolution but its contribution has been extremely significant. The most rewarding aspect has been the development of techniques for analyzing proteins and nucleic acids

Address reprint requests to Dr. W. Edgar Barnett, Director, Oak Ridge Graduate School of Biomedical Science, Biology Division, Oak Ridge National Laboratory, Oak Ridge, Tennessee 37830.

which permit the use of sequencing as a phylogenetic tool. The evaluation of both chloroplastic (Ambler and Bartsch, 1975; Phillips and Carr, 1975; Zablen et al., 1975; Bonen and Doolittle, 1976; Schwartz and Dayhoff, 1978; Schwartz and Kossel, 1979) and mitochondrial components (Starki and Sinclair, 1974; Jakovcic et al., 1975; Bonen et al., 1977) by direct sequence comparison or by hybridization homology of their genomes and transcription products with eukaryotic and prokaryotic nucleic acids has indicated that organellar nucleic acids resemble those of bacteria.

My laboratory has been engaged in determining the transcriptional identity of chloroplast transfer RNA (tRNA) (Reger et al., 1970; Fairfield and Barnett, 1971; Hecker et al., 1974; Barnett et al., 1976; Schwartzbach et al., 1976). Subsequent to determining that these molecules are transcribed from chloroplast DNA (Schwartzbach et al., 1976), we have and are continuing to sequence chloroplast tRNAs in an attempt to discern phylogenetic influences in the evolution of both the molecule and the organelle (Barnett et al., 1978).

Transfer RNAs are poor monitors of evolutionary time as gauged by linear base substitutions. Jukes and Holmquist (1972) and Holmquist et al. (1973) have deduced that tRNAs as a group have diverged so widely that unambiguous taxonomic separation can no longer be detected by examining primary sequences. Differences in the primary sequences of contemporary tRNAs are the result of an interaction of selective and stochastic processes that favored moderate evo-

Figure 1. Sequence homology of *Euglena* and bean chloroplast phenylalanine tRNAs. The shaded areas indicate differences in bases between the two tRNAs.

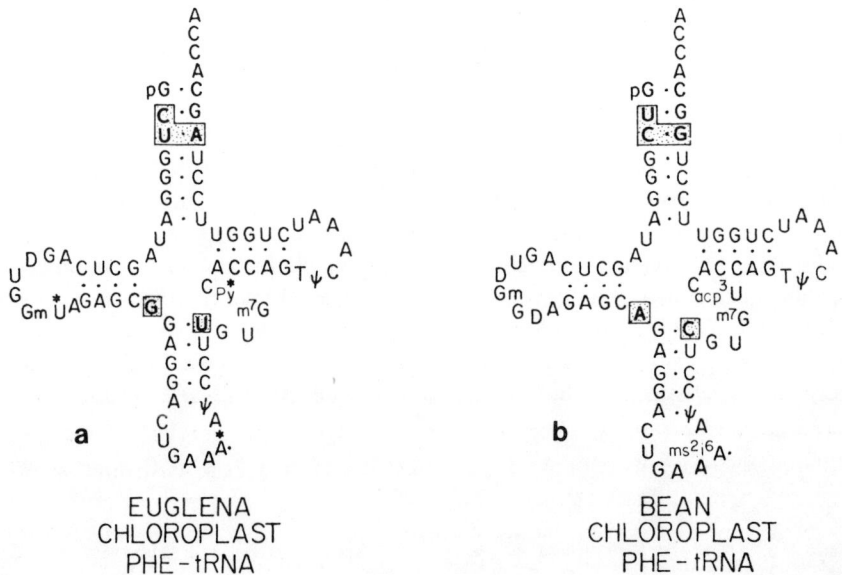

a EUGLENA CHLOROPLAST PHE-tRNA

b BEAN CHLOROPLAST PHE-tRNA

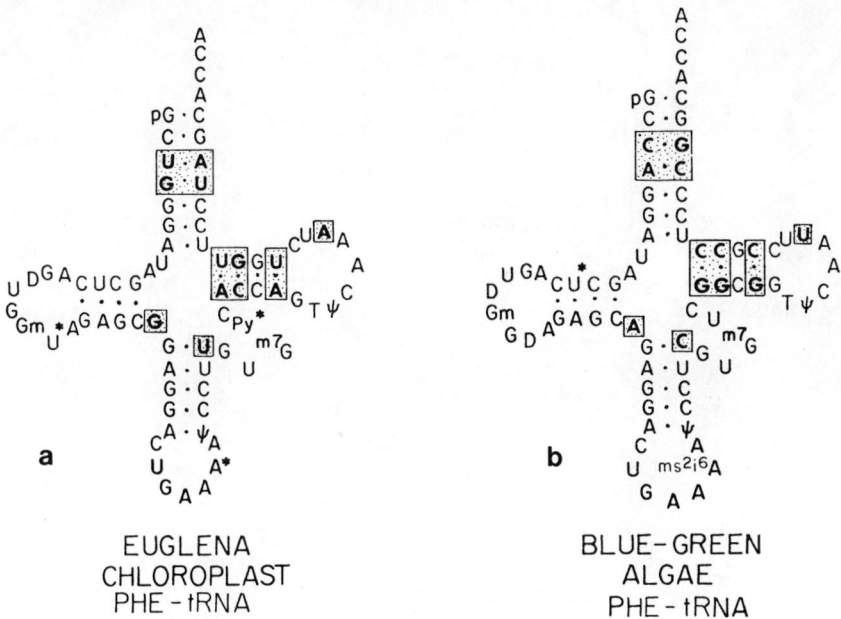

EUGLENA
CHLOROPLAST
PHE-tRNA

BLUE-GREEN
ALGAE
PHE-tRNA

Figure 2. Sequence homology of *Euglena* chloroplast phenylalanine tRNA and blue-green algal (cyanobacterial) phenylalanine tRNA. The shaded areas indicate differences in bases between the two tRNAs.

lutionary experimentation. If, however, one wishes to describe a tRNA simply on the basis of its resemblance to the tRNAs of bacteria or eukaryotes, the analysis becomes a much more accurate and easy procedure. We, among many others, have exploited the marked differences between the two types of tRNAs as a basis for determining the origin and evolution of transfer RNAs of chloroplasts.

In December 1976, in close collaboration with Simon Chang of LSU and Uttam RajBhandary of MIT, our laboratory published the first complete nucleotide sequence of a chloroplast tRNA—phenylalanine tRNA (tRNA^{phe}) from the chloroplast of the alga *Euglena gracilis* (Chang et al., 1976). Within a year, this chloroplastic tRNA^{phe} sequence was followed by that from the chloroplast of the bean *Phaseolus* (Guillemaut and Keith, 1977).

Like all known tRNAs^{phe}, the chloroplast tRNAs^{phe} are 76 nucleotides long. As indicated by the darkened areas in Figure 1, the bean and *Euglena* molecules differ in only five parent nucleotides, three in the acceptor stem at positions 2, 3, and 70, and also at positions 26 and 47. Other differences involve post-transcriptional modifications. This similarity represents a 93.4% sequence homology between the molecules.

Based upon the sequence homologies between chloroplast ribosomal RNA and both ribosomal RNA and total DNA of blue-green algae (cyanobacteria), Bonen

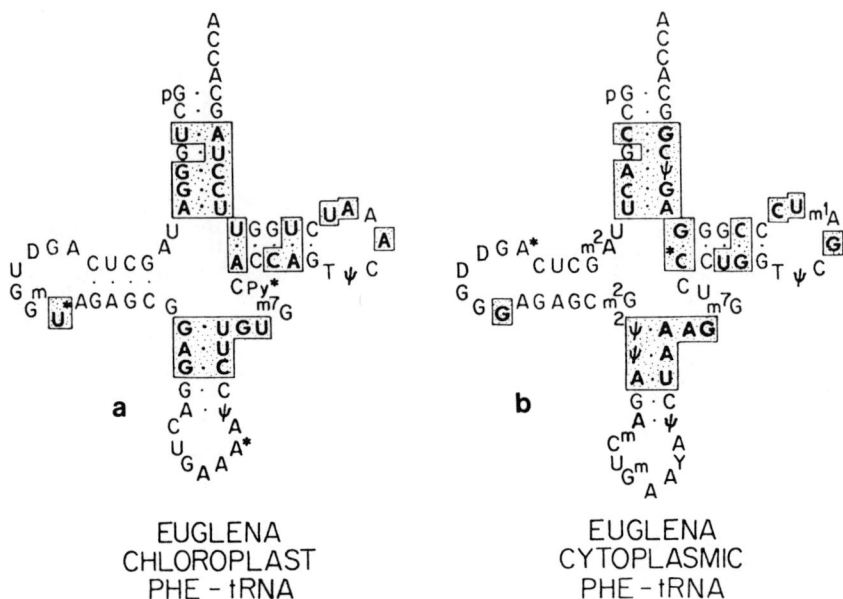

Figure 3. Sequence homology between *Euglena* chloroplast phenylalanine tRNA and *Euglena* cytoplasmic phenylalanine tRNA. The shaded areas indicate differences in bases between the two tRNAs.

and Doolittle (1976) and Phillips and Carr (1975) have proposed that blue-green algae are likely candidates for ancestors of chloroplasts. By determining the sequence of a tRNA[phe] from the blue-green alga *Agmenellum quadruplicatum* (Chang et al., 1980), and comparing it to that of the *Euglena* chloroplast tRNA[phe] (Figure 2), we confirm and support this hypothesis. The blue-green algal tRNA[phe] not only shares an 83% sequence homology with the *Euglena* chloroplast molecule, but also an 87% homology with the bean leaf chloroplast tRNA[phe]. Based upon this molecular evidence, the relatedness of these chloroplast tRNAs to those from prokaryotes is very close.

However, these homologies are in sharp contrast with comparisons between the tRNA[phe] from the *Euglena* cytoplasm (Hecker et al., 1976) and chloroplasts. Figure 3 demonstrates the huge divergence (27 of 76 bases). This difference is typical for comparisons between prokaryotic and eukaryotic tRNAs. The most remarkable structural feature of the cytoplasmic tRNA[phe] is its close similarity to the mammalian tRNA[phe] which has been sequenced by Guillemaut and Keith (1977). Whereas the sequence of bean leaf cytoplasmic tRNA[phe] is identical to that from wheat germ, the sequence of the *Euglena* cytoplasmic tRNA[phe] is almost (94.7%) completely homologous with the mammalian species, differing at only two base pairs (Figure 4).

EUGLENA
CYTOPLASMIC
PHE-tRNA

MAMMALIAN
PHE-tRNA

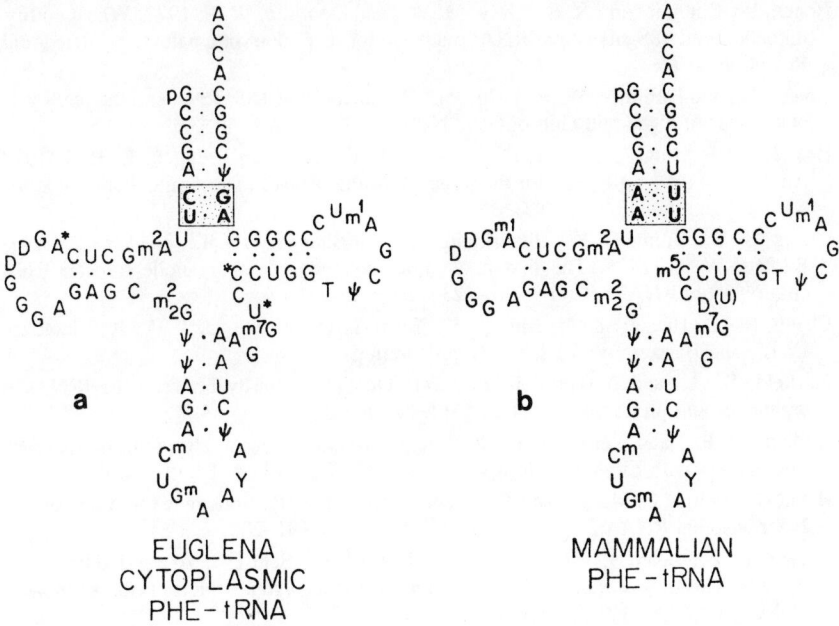

Figure 4. Sequence homology between *Euglena* cytoplasmic phenylalanine tRNA and mammalian phenylalanine tRNA. The shaded areas indicate differences in bases between the two tRNAs.

Thus, at the molecular level the resemblance between chloroplasts and prokaryotes is striking. The large and predictable differences between the sequences of tRNAs from prokaryotes and eukaryotes have been employed by us as a background against which to measure the homology of the chloroplast and prokaryotic tRNAs. Based upon these studies, we can only conclude that chloroplasts originated from a prokaryotic ancestor.

References

Ambler, R. P., and Bartsch, R. G. 1975. Amino acid sequence similarity between cytochrome *f* from a blue-green bacterium and algal chloroplasts. Nature 253:285–288.

Barnett, W. E., Schwartzbach, S. D., and Hecker, L. I. 1976. The tRNAs and aminoacyl-tRNA synthetases of *Euglena* chloroplasts. *In Genetics and Biogenesis of Chloroplasts.* Th. Bücher, W. Neupert, W. Sebald, and S. Werner, (eds.). Amsterdam: Elsevier, pp. 661–666.

Barnett, W. E., Schwartzbach, S. D., and Hecker, L. I. 1978. Prog. Nuc. Acid Res. Mol. Biol. 21:143–179.

Bogorad, L. 1975. Evolution of organelles and eukaryotic genomes. Separation of genes for chloroplast ribosomes in two genomes suggests principles of organelle biology. Science 188:891–898.

Bonen, L., Cunningham, R. S., Gray, M. W., and Doolittle, W. F. 1977. Wheat embryo mitochondrial 18S ribosomal RNA: evidence for its prokaryotic nature. Nucleic Acids Res. 4:663–671.

Bonen, L., and Doolittle, W. F. 1976. Partial sequences of 16S rRNA and the phylogeny of blue-green algae and chloroplasts. Nature 261:669–673.

Bos, J. L., Heyting, C., Borst, P., Arnberg, S. C., and Van Bruggen, E. R. J. 1978. An insert in the single gene for the large ribosomal RNA in yeast mitochondrial DNA. Nature 275:336–338.

Chang, S. H., Brum, C. K., Silberklang, M., RajBhandary, U. L., Hecker, L. I., and Barnett, W. E. 1976. The first nucleotide sequence of an organelle transfer RNA: chloroplastic tRNA[phe]. Cell 9:717–723.

Chang, S. H., Hecker, L. I., Lin, R. K., Furr, T. D., Heckman, J. E., RajBhandary, U. L., and Barnett, W. E. 1980. In preparation.

Fairfield, S. A., and Barnett, W. E. 1971. On the similarity between the tRNAs of organelles and prokaryotes. PNAS 68:2972–2976.

Guillemaut, P. , and Keith, G. 1977. Primary structure of bean chloroplastic tRNA[phe]. Comparison with Euglena chloroplastic tRNA[phe]. FEBS Lett. 84:351–356.

Hahn, U., Azarus, C. M., Lunsdorf, H., and Untzel, H. 1979. Split gene for mitochondrial 24S ribosomal RNA of Neurospora crassa. Cell 17:191–200.

Hecker, L. Z., Egan, J., Reynolds, R. J., Nix, C. E., Schiff, J. A., and Barnett, W. E. 1974. The sites of transcription and translation for Euglena chloroplastic aminoacyl-tRNA synthetases. PNAS 71:1910–1914.

Hecker, L. I., Chang, S. H., Schwartzbach, S. D., RajBhandary, U. L., and Barnett, W. E. 1976. The comparative sequence and evolutionary origin of chloroplast phenylalanine-tRNA from Euglena. Fed. Proc. 35:1466.

Heckman, J. E., and RajBhandary, U. L. 1979. Organization of tRNA and rRNA genes in N. crassa mitochondria: intervening sequence in the large rRNA gene and strand distribution of the RNA genes. Cell 17:583–595.

Holmquist, R., Jukes, T. H., and Pangburn, S. 1973. Evolution of transfer RNA. J. Mol. Biol. 78:91–116.

Jakovcic, S., Casey, J., and Rabinowitz, M. 1975. Sequence homology between mitochondrial DNAs of different eukaryotes. Biochemistry 14:2043–2050.

Jukes, T. H., and Holmquist, R. 1972. Evolution of transfer RNA molecules as a repetitive process. Biochemical & Biophysical Research Communications 49:212–216.

Margulies, L. 1970. The Origin of Eukaryotic Cells. New Haven: Yale University Press.

Phillips, D. O., and Carr, N. G. 1975. Hybridization of prokaryotic and eukaryotic 5S rRNA to Euglena gracilis chloroplast DNA. FEBS Lett. 60:94–97.

Reger, B. J., Fairfield, S. A., Epler, J. L., and Barnett, W. E. 1970. Identification and origin of some chloroplast aminoacyl-tRNA synthetases and tRNAs. PNAS 67:1207–1213.

Rochaix, J. D., and Malnoe, P. 1978. Anatomy of the chloroplast ribosomal DNA of Chlamydomonas reinhardii. Cell 15:661–670.

Schwartz, R. M., and Dayhoff, M. O. 1978. Origins of prokaryotes, eukaryotes, mitochondria, and chloroplasts. A perspective is derived from protein and nucleic acid sequence data. Science 199:395–403.

Schwarz, Z., and Kossel, H. 1979. Sequencing of the 3'-terminal region of a 16S rRNA gene from Zea mays chloroplast reveals homology with E. coli 16S rRNA. Nature 279:520–522.

Schwartzbach, S. D., Hecker, L. I., and Barnett, W. E. 1976. Transcriptional control of *Euglena* chloroplast tRNA's. PNAS 73:1974–1988.
Storti, R. V., and Sinclair, J. H. 1974. Sequence homology between mitochondrial DNA and nuclear DNA in the yeast, *Saccharomyces cerevisiae*. Biochemistry 13:4447–4455.
Zablen, L. B., Kissil, M. S., Woese, C. R., and Buetow, D. E. 1975. Phylogenetic origin of the chloroplast and prokaryotic nature of its ribosomal RNA. PNAS 72:2418–2422.

Discussion of Presentation by Drs. Delehanty, Farmerie, Chang, and Barnett

HALLICK: The estimates for the numbers of tRNA genes on chloroplast DNA are similar, but not exactly the same. For example, results from saturation hybridization experiments are often consistent with approximately 25 tRNA genes, as you showed in your chapter. We can resolve approximately 40 tRNAs from *Euglena* chloroplasts by two-dimensional gel analysis. It may be that the estimates of tRNA gene number based on saturation hybridization experiments are lower limits, and that multiple genes for many tRNAs will be found.

SCHIFF: Is there any evidence at all that tRNAs move from one cellular compartment to another?

tRNAs may be similar because they are under strong selective pressure; any changes would be strongly selected since they directly affect the survival of the organism.

BARNETT: In my opinion there is no evidence that tRNAs move from one cellular compartment to another.

Additional Comments: Evolution of the Regulation of Plastid Development

Jerome A. Schiff

The regulation of plastid development has been discussed in relation to the ecological influences that may have selected the control systems during evolution (Schiff, 1980, 1981). Viewed as prokaryotic residents in eukaryotic cells, plastids might be expected to have control systems very similar to those found in comtemporary prokaryotes.

Most chloroplasts are probably constitutive: they are present whether or not light is available, as in the green algae and in several species of diatoms. In certain organisms, however, the formation of chloroplasts is inducible by light, the substrate for photosynthesis. In darkness, most angiosperms, a few strains of *Euglena, Ochromonas,* and perhaps a few other organisms form arrested chloroplasts called proplastids or etioplasts. On illumination of these organisms the proplastids are induced to develop into chloroplasts. In the green organisms, blue light and red light are most effective in this induction. Since blue light and red light are the substrates for photosynthesis by green organisms, it would have been highly adaptive during evolution for the control systems of the organisms to respond to these wavelengths of light as inducers. This response would be similar to the well-known induction of enzymes in bacteria by their substrates. In chloroplast development, however, light as a substrate must bring about the induction of many systems inside and outside of the chloroplast, all properly coordinated in time, the process we call chloroplast development.

Carrying this idea of substrate induction further, we might suppose that control systems familiar to us from prokaryotes have persisted in the plastids. For instance, pigmented control molecules or repressors might participate in transcriptive control much as their nonpigmented counterparts do. Instead of recognizing an organic molecule like lactose, these control molecules might absorb light of appropriate wavelengths and thereby bring about induction of transcription (Schiff, 1981) and derepression of protein synthesis. If this were so, one would expect that these pigmented control molecules would evolve to absorb light in the same regions of the spectrum as does photosynthesis, since to be adaptive a chloroplast should form only when light of the appropriate quality for photosynthesis is available. By analogy to modern prokaryotes, we might also expect control to be exerted at other levels including translation, processing, and transport of proteins and at the level of substrate control. A particularly interesting possibility for which experimental evidence is emerging is that photocontrol may be exerted at the membrane level with a membrane-localized photoreceptor controlling the transport of materials between compartments. Whatever the mechanism, blue and red light, the most effective substrates for photosynthesis, are the most effective wavelengths for the induction of chloroplast development in green organisms.

But what has selected the photosynthetic pigment systems we see today in contemporary organisms? Building on earlier ideas of Granick we might suppose that each of the intermediates in chlorophyll biosynthesis might have, at some time, served as the primary sensitizer of photosynthesis. Thus the biosynthetic pathway may once have terminated at magnesium protoporphyrin IX with this molecule serving as the primary photosynthetic pigment. Later this molecule would have become further modified and the biosynthetic chain extended to form later intermediates such as photochlorophyll (ide) and, finally, chlorophyll. These intermediates, and ultimately, chlorophyll would have served as the primary sensitizers of photosynthesis probably because they were chemically appropriate to lose and gain electrons to the reactions of photosynthesis given the quantum energies available. As the biosynthetic pathway for these compounds lengthened the long wavelength absorptions of these compounds moved further toward the red region of the spectrum and the strength of absorption increased. The selective pressure for the shift in absorption of these compounds toward longer wavelengths and the concomitant elaboration of the biosynthetic pathway may well have come from the evolution of accessory pigments in response to the ecological needs of various groups of organisms. As the accessory pigments filled in the middle of the spectrum, the absorptions of the primary sensitizers of photosynthesis would be pushed to longer wavelengths to serve as the sink for energy transfer from the accessory pigments. In this way systems of accessory pigments could become established, but the absorption of the primary sensitizer of photosynthetic electron transport would always have to lie at the longest wavelength of all the pigments present in a given organism.

A certain plasticity has evolved, however, in the form and content of photosynthetic accessory pigments. Many organisms can regulate the amounts of their photosynthetic pigments in response to the amount of available light, providing more pigment to enhance the probability of capture of fewer available photons. Among cyanobacteria that have phycobiliproteins such as phycocyanin and phycoerythrin as photosynthetic accessory pigments, chromatic adaptation has evolved as a control mechanism to adjust the proportions of these pigments to suit the available light quality. The control systems, once again, have evolved to use the same wavelengths of light as the pigment systems they control; in this case, green light and orange-red light. Perhaps the phycobilisome that contains these phycobiliproteins evolved to permit extensive changes in accessory pigment composition without disturbing the internal architecture of the thylakoid to which it is attached.

These questions and several others relating to the evolution of photocontrol systems have been discussed more fully elsewhere (Schiff, 1980, 1981). I note only that, from the standpoint of photocontrol of plastid development, the endosymbiotic hypothesis is attractive since many of the properties characteristic of photosynthetic prokaryotes seem to appear in the evolution of chloroplasts. But, as has been emphasized, sufficient evidence is not in hand to exclude some version of the episomal or cluster clone hypotheses. Parsimony makes the endosymbiotic hypothesis attractive, but nature may be more profligate than we presently suppose.

References

Schiff, J.A. 1980. Development, inheritance and evolution of plastids and mitochondria. *In* The Biochemistry of Plants, Vol. 1, N. E. Tobert, ed. New York: Academic Press, pp. 209–272.

Schiff, J. A. 1981. Origin and evolution of the plastid and its function, Ann. N. Y. Acad. Sci. 361:166–192.

Rapporteur's Summary: Molecular Biology and Control of Plastid Development

Harvard Lyman

Discussions of the evolutionary origins of chloroplasts invariably return to a comparison of the molecular biology of chloroplasts and how it compares to that of cyanobacteria, as well as other prokaryotes. Similarities and discrepancies of genome organization, transcription, and translation between prokaryotes and chloroplasts are examined and then used to bolster arguments for an endosymbiotic origin, a cluster-clone origin, or other variations, depending upon the evidence, the strength of one's convictions, the flights of one's imagination, or the lateness of the hour and proximity to dinner or liquid refreshments.

What can one learn from an examination of the molecular biology of chloroplasts?

Bogorad reviewed in some detail the work in his laboratory on the organization of chloroplast DNA, especially that of *Zea mays* (for more details see Bogorad et al., 1979). He discussed the organization of the ribosomal RNA genes, the gene for the large subunit of RuBPcase, and the light-induced 32,000-dalton thylakoid polypeptide. While the rRNA operon contains 5S, 16S, and 23S genes arranged in a manner similar to *E. coli,* Bogorad emphasized that when one compares rRNA gene organization from several plants and algae, differences are seen in the number of repeated sequences, the presence of inverted sequences, the size of spacer regions, and the existence of introns. Although the organization of the rRNA genes shows variability, it should be pointed out that when rRNA from pea, spinach, bean, corn, and lettuce are compared by heterologous hybridization and competition and thermal stability studies, a high degree of sim-

Address reprint requests to Dr. Harvard Lyman, Associate Professor of Biology, Department of Biology, State University of New York, Stony Brook, New York 11794.

ilarity is observed suggesting a conservation of these genes during evolution (Tewari and Meeker, 1979).

With respect to the organization of tRNA genes, Bogorad pointed out that the clustering of tRNAs is different in the plastids of different plants. Setsuko Jolly, working in Bogorad's laboratory, has demonstrated that chloroplast RNA polymerase requires a unique stimulator (a 26,000-dalton polypeptide) to transcribe plastid DNA. Sigma factor from *E. coli* RNA polymerase will not substitute. This also suggests a difference between prokaryotes and chloroplasts. Barnett, however, compared the tRNA^{Phe} from *Euglena* chloroplasts, bean chloroplasts, and a cyanobacterium on the basis of sequence homology, and showed that bean and *Euglena* chloroplast tRNA^{Phe} were essentially identical and quite similar to the cyanobacterial tRNA^{Phe}. Comparing the chloroplast tRNA^{Phe} of *Euglena* with the analogous molecule from the cytoplasm of this organism revealed substantial differences. Based on the similarities of tRNAs from eukaryotic cytoplasms of different species to each other and the similarity of organelle and prokaryotic tRNAs, Barnett concluded that an endosymbiotic origin of organelles is likely.

Hallick (Chapter 00 of the present volume) pointed out that the operons of *Euglena* chloroplast rRNA and that of *E. coli* are strikingly similar in gene arrangement, gene size spacer size, and tRNAs coded in the spacer region. This also suggests a common ancestor.

But does the common ancestor have to be a prokaryote? Bogorad noted that both mitochondria and chloroplasts possess multimeric components, some of whose polypeptides are coded in the organelle DNA, while some are coded in the nuclear DNA. For chloroplasts these include chloroplast ribosomes, Ru-BPcase, and the light-harvesting chl *a/b* protein. How did genes get into separate compartments from their products? Bogorad suggested a gene transfer mechanism or a gene substitution mechanism (Bogorad, 1975).

In a gene transfer process, an organelle gene duplicates and a copy is incorporated into the nuclear genome. The subsequent loss of the gene from the organelle completes the dispersal. If the organelle and the nucleus shared common DNA sequences or a common replication system, or possessed compatible transcription and translation systems, this mode of gene dispersal would be favored. If the nucleus in this case were prokaryotic the above conditions would be likely. If the ancestor of the eukaryotic cell was a prokaryote whose genes had become associated into clusters, then compatible replication, transcription, and translation would exist.

The gene substitution mechanism proposes that an organelle gene mutates so that its product is useless for the organelle. The organelle is then "rescued" by a protein coded by the nucleus which can substitute for the lost protein. In this situation the substitution is more likely to occur if the nucleus and organelle have common metabolic systems and if the membranes are relatively nondiscriminating as far as the transfer of proteins is concerned.

Gene dispersal proposals are more likely if eukaryotes arose by a cluster-clone mechanism, although similar events could occur between symbiont and host.

One barrier to gene transfer either from an organelle or a symbiont would be the existence of restriction-modification systems set up to prevent the entry and replication of foreign DNA. Is there any evidence for this system in eukaryotic nuclei? Eukaryotic nuclear DNAs usually have 5-methyl cytosine. This might imply a restriction-modification system. However, it appears that chloroplasts do have restriction-modification systems, but not for protection against external DNAs. Sager (1979) has presented evidence for a restriction-modification system in the chloroplasts of *Chlamydomonas* which regulates maternal inheritance. Lyman and Srinivas (1978) showed that a light-induced DNAse in *Euglena* may also be part of such a system for the regulation of the amount of chloroplast DNA. It is possible that, early in the evolution of eukaryotes, these systems did serve to prevent invasion by exogenous DNAs and might pose a barrier to gene dispersal from a symbiont. They would not be likely, however, to restrict DNA that was already internal, that is, DNA that originated through a cluster-clone mechanism. It must be remembered, however, that the Ti plasmid of *Agrobacterium tumifaciens* is incorporated into the nuclei of gymnosperms and dicotyledons (Schell et al., 1979). Apparently, no restriction mechanism prevents the entry of this symbiont.

The fact that genes do not seem to undergo dispersal in modern eukaryotes does not mean that such events did not occur in primitive eukaryotes. Subsequent evolution and selection have set up barriers to gene dispersal. Bogorad pointed out several barriers that would prevent genes or gene products today from easily moving from one cellular compartment to another:

1. Barriers were set in place at a certain time in the distant past and change is unlikely.

2. Some products coded and made in one compartment cannot be transported. If their genes were moved elsewhere no substitute would be available.

3. A variety of mechanisms have evolved for locking nuclear-coded products into an organelle. These include transport devices, protein-modifying enzymes, complexing of a nuclear-coded polypeptide with an organelle-coded polypeptide (i.e., the subunits of RuBPcase), and trapping of a nuclear-coded product by the organelle membrane (i.e., assembly of photosynthetic membranes).

4. Selection for the separation of self-assembling components until their assembly is needed.

Little attention is paid to the possible form of the primitive eukaryotic nucleus. If it was a loose association of plasmidlike chromosomes (a clone of some original cluster?), then gene transfer seems more likely. Transfer from another loose cluster or from a prokaryote symbiont might be quite possible. Chloroplasts are light-transducing organelles, and it is not surprising to find that they have evolved mechanisms to also use light to regulate their synthesis. Hallick described the changes in transcription occurring during light-mediated chloroplast synthesis in *Euglena*, while Schiff compared the photocontrol systems in regulating the synthesis of bean, *Euglena, Ochromonas Chlamydomonas,* and some other chloroplasts. Pigments evolved for light-trapping functions have subsequently been adapted to regulate transcription on the organelle genome as well. A problem relating to the origin of chloroplasts concerns the location of coding for the various photocontrol systems: phytochrome, cryptochrome, O'Kelleychrome (O'Kelley and Hardman, 1977; Bjorn, 1979), and protochlorophyll (ide). Some or all of these systems are nuclear-coded. Phytochrome, cryptochrome, and at least the flavoprotein portion of O'Kelleychrome are probably located in the cytoplasm and are presumably nuclear coded. Were they originally a component of the primitive eukaryote before the invasion of a photosynthetic prokaryote (assuming an endosymbiotic origin of plastids)? The flavoproteins are good candidates for being original eukaryotic proteins because of their role in respiration, but were they in mitochondria or in some kind of desmosome? Hemes (as a precursor to phytochrome) could have been in a primitive eukaryote also, but a clustering of genes in a photosynthetic prokaryote would have produced a similar result. The fact that whole algal cells, as well as chloroplasts, can be found as symbionts today (see chapters by Trench and by Muscatine, in the present volume) is not necessarily an indication that *Prochloron*-like organisms were the first chloroplasts. Gene dispersal in a *Prochloron*-type organism could also have given rise to eukaryotic plants.

References

Bjorn, L. O. 1979. Photoreversibly photochromic pigments in organisms: Properties and role in biological light perception. Quart. Rev. Biophys. 12:1–23.

Bogorad, L. 1975. Evolution of organelles and eukaryotic genomes. Science 188:891–897.

Bogorad, L., Link, G., McIntosh, L., and Jolly, S. 1979. Genes on the maize chloroplast chromosome. *In* Extrachromosomal DNA. D. Cummings, P. Borst, I. Dawid, S. Weissman, and C. Fox (eds.). New York: Academic Press.

Lyman, H., and Srinivas, U. 1978. Regulation of chloroplast DNA synthesis: Possible role of chloroplast nucleases in *Euglena*. *In* Chloroplast Development. G. Akoyunoglou and J. Argyroudi-Akoyunoglou (eds.). Amsterdam: Elsevier-North Holland Biomedical Press, pp. 593–607.

O'Kelley, J. C., and Hardman, J. K. 1977. A blue light reaction involving flavin nucleotides and plastocyanin from *Protosiphon botryoides*. Photochem. Photobiol. 25:559–564.

Sager, R. 1979. Methylation and restriction of chloroplast DNA: The molecular basis of maternal inheritance in *Chlamydomonas*. *In* Extrachromosomal DNA. D. Cummings, P. Borst, I. Dawid, S. Weissman, and C. Fox (eds.). New York: Academic Press.

Schell, J., Van Montagu, M., De Picker, A., De Waele, D., Engler, G., Gentello, C., Hernalsteens, J., Holsters, M., Messens, E., Silva, B., Van den Elsacker, S., Van Larebeke, N., and Zaenen, I. 1979. Crown gall: Bacterial plasmids as oncogenic elements for eukaryotic cells. *In* The Molecular Biology of Plants. I. Rubenstein, R. Philips, C. Green, and B. Gengenbach (eds.). New York: Academic Press.

Tewari, K., and Meeker, R. 1979. Chloroplast DNA: Structure, Information Content, and Replication. *In* The Molecular Biology of Plants. I. Rubenstein, R. Philips, C. Green, and B. Gengenbach (eds.). New York: Academic Press.

Addendum

A mutable and treacherous tribe
—Anon. (Doubtless, a microbial taxonomist in a moment of exasperation.)

On the Origin of the Bergey Award
and the Choice of the First Recipient:
Roger Y. Stanier

Remarks by Dr. Arnold W. Ravin

The *Bergey's Manual of Determinative Bacteriology* is an attempt to provide as up-to-date information as possible about the various species of bacteria recognized by microbiologists. The efficient presentation of this information is intended primarily to aid students and researchers seeking to identify the organisms with which they are working. Originally published in 1923 under the auspices of an editorial board appointed by the Society of American Bacteriologists and chaired by David H. Bergey, the *Manual* has undergone seven new editions since that time, the current eighth edition having been published in 1974. During all of this time the publisher has been the Williams & Wilkins Company of Baltimore, Maryland.

After the third edition appeared in 1930 a new organization was devised to assure successive editions of the *Manual*. An educational trust, known as the Bergey's Manual Trust, was instituted to receive the right, title, and interest in the *Manual* and to use its resources, in particular royalties accruing from sales, to ensure preparation of new and improved editions. This trust is under the control of a Board of Trustees, which elects its membership.

While the *Bergey's Manual* has been used in the past largely as a guide for identification purposes, the Board has been aware of the usefulness of the *Manual* to students of taxonomy and systematics of prokaryotes. Consequently, the Board has sought to encourage first-rate modern systematic studies of prokaryotes, especially of such groups that clearly needed extensive research, and to foster publication of the results of such research.

Address reprint requests to Dr. Arnold Ravin, Department of Biology, The University of Chicago, 1103 East 57th Street, Chicago, Illinois 60637.

In September of 1978 the Board of Trustees of the Bergey's Manual Trust established a Bergey Award, to be offered every year or two, in recognition of outstanding work in the field of bacterial taxonomy. The prize, in the form of a scroll and a check in the amount of $2000, was to be contributed jointly by the Trust and the Williams & Wilkins Company.

We are pleased to present the first Bergey Award to Roger Y. Stanier, who has chosen this occasion for the receipt of his prize.

Remarks of Dr. Robert G. E. Murray
Chairman of the Board of the Bergey's Manual Trust

This happy scientific occasion enables us to add to the many well-deserved honors accorded to Roger Stanier, a much respected colleague, friend, and stalwart contender in demanding scientific respect for our good friends, the bacteria. Most of all, he has been a major force, along with other genial heretics, in convincing biologists, and more importantly their students, that there is a real place for bacteria in the living world.

It was not a quick journey—it was more like a determined pilgrimage through thickets of uncertain knowledge, filling gaps along the way, committing and abolishing heresies, with clear vision from time to time of what is, what might be, and what could be done. The compelling thread of scholarly intent from the beginning of Dr. Stanier's Stanford days has been bacterial taxonomy. He expressed the sure feeling that it was not technical competence that creates order out of chaos, but rather the perceptive use of everything that we know.

Probably he owes a lot to the redoubtable Kees van Niel; thus he is a direct descendant of the Delft school and an alumnus of the van Niel Course. From that early (1941) influential systematic essay with van Niel to the definitive essay (1962), which hammered home the message that you can't describe and classify bacteria unless you can describe them "as cells," you can appreciate Roger's mind at work and his hand on the pen. A facility at expressing himself has done him no harm and has served all of us very well. Indeed, his conceptual talents, literacy, and energy has led to *The Microbial World*—a remarkable text with much in it to earn the respect of all of us, students of all ages.

Every series of papers—whether concerned with photosynthesis, with pigments, with metabolic pathways and the phenomena of simultaneous adaptation, with differentiation, or with a number of microbial groups—shows a concern for comparative biology and the order of life. He practices an old-fashioned discipline in a new-fashioned way to our great benefit.

He has chosen his students and colleagues well and a number of them are here today to share their understanding with him. This form of hindsight (like choosing the right grandparents) is not to be underrated. All these encounters, with an unselfish sharing and elaboration of concepts and information to beget ideas, has led to an arborization and extension of the central theme of his life's work in a host of young and not-so-young colleagues.

He was, for years, a Trustee of *Bergey's Manual* and what was conceptually valuable in the structure of that eighth edition owes more to Roger than to most of us. This was not done without effort and some pain, often against formidable and conservative odds. With publication in sight for 1974 he left us to concentrate his efforts on the "blue-green algae."

The recent years, culminating in a remarkable series of papers on what he unblushingly calls the cyanobacteria, show the real stuff that he is made of and the depth of his scholarly intent. In essence, this was old ground that needed to be cultivated, seeded, grown, and harvested with a more-than-ordinary perception of what microbial life is all about.

All these things, and more, convince us of the rightness of our decision. Modern taxonomy, and a new respect for the expertise and intellectual vigor required for its study, has been evolving under his watchful eye. (Students have even been observed to enjoy it!) The broad descriptions of cells that Roger forced us to begin, and which generated much definitive work, are now joined by molecular data, catalogues of sequences, and much more to come.

The Bergey's Manual Trust, which has a big stake in the consequences of taxonomic and systematic research, and the Williams & Wilkins Company, the scientific publishers who have put the *Manual* into print from its beginning some 56 years ago, take great pleasure in making the first Bergey Award for remarkable contributions to taxonomy to Roger Stanier.

We hope that it will not be counted among the least of his many honors.

EPILOGUE

Lest we take ourselves and our conclusions too seriously, we end on a note of delicate irony. Whenever scholars seek to reconstruct the past, whether this is the more recent past of man's cultural history measured in centuries or the more distant geological and evolutionary past reckoned in eons, they run the risk of being diminished by the great events they seek to record, understand and preserve:

When eras die, their legacies
   Are left to strange police.
Professors in New England guard
   The glory that was Greece.

Clarence Day

Index